Next-Generation Internet

With ever-increasing demands on capacity, quality of service, speed, and reliability, current Internet systems are under strain and under review. Combining contributions from experts in the field, this book captures the most recent and innovative designs, architectures, protocols, and mechanisms that will enable researchers to successfully build the next-generation Internet. A broad perspective is provided, with topics including innovations at the physical/transmission layer in wired and wireless media, as well as the support for new switching and routing paradigms at the device and subsystem layer. The proposed alternatives to TCP and UDP at the data transport layer for emerging environments are also covered, as are the novel models and theoretical foundations proposed for understanding network complexity. Finally, new approaches for pricing and network economics are discussed, making this ideal for students, researchers, and practitioners who need to know about designing, constructing, and operating the next-generation Internet.

Byrav Ramamurthy is an Associate Professor in the Department of Computer Science and Engineering at the University of Nebraska-Lincoln (UNL). He is a recipient of the UNL CSE Department Student Choice Outstanding Teaching Award, the author of *Design of Optical WDM Networks* (2000), and a co-author of *Secure Group Communications over Data Networks* (2004). His research areas include optical networks, wireless/sensor networks, network security, distributed computing, and telecommunications.

George N. Rouskas is a Professor of Computer Science at North Carolina State University. He has received several research and teaching awards, including an NSF CAREER Award, the NCSU Alumni Outstanding Research Award, and he has been inducted into the NCSU Academy of Outstanding Teachers. He is the author of the book *Internet Tiered Services* (2009) and his research interests are in network architectures and protocols, optical networks, and performance evaluation.

Krishna Moorthy Sivalingam is a Professor in the Department of Computer Science and Engineering at the Indian Institute of Technology (IIT), Madras. Prior to this, he was a Professor at the University of Maryland, Baltimore County, and also conducted research at Lucent Technologies and AT&T Bell Labs. He has previously edited three books and holds three patents in wireless networking. His research interests include wireless networks, wireless sensor networks, optical networks, and performance evaluation.

Next-Generation Internet
Architectures and Protocols

Edited by

BYRAV RAMAMURTHY
University of Nebraska-Lincoln

GEORGE ROUSKAS
North Carolina State University

KRISHNA M. SIVALINGAM
Indian Institute of Technology Madras

CAMBRIDGE
UNIVERSITY PRESS

CAMBRIDGE UNIVERSITY PRESS
Cambridge, New York, Melbourne, Madrid, Cape Town, Singapore,
São Paulo, Delhi, Dubai, Tokyo, Mexico City

Cambridge University Press
The Edinburgh Building, Cambridge CB2 8RU, UK

Published in the United States of America by Cambridge University Press, New York

www.cambridge.org
Information on this title: www.cambridge.org/9780521113687

First published 2011

Printed in the United Kingdom at the University Press, Cambridge

A catalog record for this publication is available from the British Library

Library of Congress Cataloging in Publication data
  Next-generation internet : architectures and protocols / edited by Byrav Ramamurthy,
  George N. Rouskas, Krishna Moorthy Sivalingam.
     p.   cm.
  Includes bibliographical references and index.
  ISBN 978-0-521-11368-7 (hardback)
  1. Internet–Technological innovations.   2. Internetworking–Technological innovations.
  I. Ramamurthy, Byrav.   II. Rouskas, George N.   III. Sivalingam, Krishna M.
  TK5105.875.I57N525 2010
  004.67'8–dc22   2010038970

ISBN 978-0-521-11368-7 Hardback

Contents

**3   The optical control plane and a novel unified control plane
architecture for IP/WDM networks**   42

Georgios Ellinas, Antonis Hadjiantonis, Ahmad Khalil, Neophytos Antoniades,
and Mohamed A. Ali

4   Cognitive routing protocols and architecture   72

Suyang Ju and Joseph B. Evans

7   Contract-switching for managing inter-domain dynamics   136
Murat Yuksel, Aparna Gupta, Koushik Kar, and Shiv Kalyanaraman

8   PHAROS: an architecture for next-generation core optical networks   154
Ilia Baldine, Alden W. Jackson, John Jacob, Will E. Leland, John H. Lowry, Walker
C. Milliken, Partha P. Pal, Subramanian Ramanathan, Kristin Rauschenbach, Cesar A.
Santivanez, and Daniel M. Wood

Contributors

Mohamed A. Ali, City University of New York, USA
Neophytos Antoniades, City University of New York, USA
Anish Arora, The Ohio State University, USA
Onur Ascigil, University of Kentucky, USA
Marcelo Bagnulo, Universidad Carlos III de Madrid, Spain
Ilia Baldine, Renaissance Computing Institute, USA
Kenneth L. Calvert, University of Kentucky, USA
Mung Chiang, Princeton University, USA
Davide Cuda, Politecnico di Torino, Turin, Italy
Rudra Dutta, North Carolina State University, USA
Georgios Ellinas, University of Cyprus, Cyprus
Joseph B. Evans, University of Kansas, USA
Alberto García-Martínez, Universidad Carlos III de Madrid, Spain
Roberto Gaudino, Politecnico di Torino, Turin, Italy
Guido A. Gavilanes Castillo, Politecnico di Torino, Turin, Italy
James Griffioen, University of Kentucky, USA
Aparna Gupta, Rensselaer Polytechnic Institute, USA
Andrei Gurtov, University of Oulu, Finland
Antonis Hadjiantonis, University of Nicosia, Cyprus
Thomas R. Henderson, Boeing, USA
Alden W. Jackson, BBN Technologies, USA
John Jacob, BAE Systems, USA
Suyang Ju, University of Kansas, USA
Xi Ju, The Ohio State University, USA
Shiv Kalyanaraman, IBM Research, India
Koushik Kar, Rensselaer Polytechnic Institute, USA
Frank Kelly, University of Cambridge, UK
Ahmad Khalil, City University of New York, USA
William Leal, The Ohio State University, USA
Will E. Leland, BBN Technologies, USA
Baochun Li, University of Toronto, Canada
Zongpeng Li, University of Calgary, Canada
John H. Lowry, BBN Technologies, USA
Walter C. Milliken, BBN Technologies, USA

Biswanath Mukherjee, University of California Davis, USA
John Musacchio, University of California Santa Cruz, USA
Fabio Neri, Politecnico di Torino, Turin, Italy
Pekka Nikander, Ericsson Research, Finland
P. Pal, BBN Technologies, USA
Gaurav Raina, Indian Institute of Technology Madras, India
Byrav Ramamurthy, University of Nebraska-Lincoln, USA
Subramanian Ramanathan, BBN Technologies, USA
Rajiv Ramnath, The Ohio State University, USA
K. Rauschenbach, BBN Technologies, USA
Anusha Ravula, University of Nebraska-Lincoln, USA
Abu (Sayeem) Reaz, University of California Davis, USA
Cesar A. Santivanez, BBN Technologies, USA
Galina Schwartz, University of California Berkeley, USA
Lei Shi, University of California Davis, USA
Mukundan Sridharan, The Ohio State University, USA
Francisco Valera, Universidad Carlos III de Madrid, Spain
Iljitsch van Beijnum, IMDEA Networks, Spain
Jean Walrand, University of California Berkeley, USA
Damon Wischik, University College London, UK
Daniel Wood, Verizon Federal Network Systems, USA
Tilman Wolf, University of Massachusetts Amherst, USA
Hong Xu, University of Toronto, Canada
Yung Yi, Korea Advanced Institute of Science and Technology (KAIST),
South Korea
Song Yuan, University of Kentucky, USA
Murat Yuksel, University of Nevada-Reno, USA
Wenjie Zeng, The Ohio State University, USA
Hongwei Zhang, The Ohio State University, USA

Preface

The field of computer networking has evolved significantly over the past four decades since the development of ARPANET, the first large-scale computer network. The Internet has become a part and parcel of everyday life virtually worldwide, and its influence on various fields is well recognized. The TCP/IP protocol suite and packet switching constitute the core dominating Internet technologies today. However, this paradigm is facing challenges as we move to next-generation networking applications including multimedia transmissions (IPTV systems), social networking, peer-to-peer networking and so on. The serious limitations of the current Internet include its inability to provide Quality of Service, reliable communication over periodically disconnected networks, and high bandwidth for high-speed mobile devices.

Hence, there is an urgent question as to whether the Internet's entire architecture should be redesigned, from the bottom up, based on what we have learned about computer networking in the past four decades. This is often referred to as the "clean slate" approach to Internet design. In 2005, the US National Science Foundation (www.nsf.gov) started a research program called Future Internet Network Design (FIND) to focus the research community's attention on such activities. Similar funding activities are taking place in Europe (FIRE: Future Internet Research and Experimentation), Asia, and other regions across the globe. This book is an attempt to capture some of the pioneering efforts in designing the next-generation Internet. The book is intended to serve as a starting point for researchers, engineers, students, and practitioners who wish to understand and contribute to the innovative architectures and protocols for the next-generation Internet.

Book organization

The book is divided into four parts that examine several aspects of next generation networks in depth.

Part I, titled "Enabling technologies," consists of five chapters that describe the technological innovations which are enabling the design and development of next-generation networks.

Chapter 1, "Optical switching fabrics for terabit packet switches," describes photonic technologies to realize subsystems inside high-speed packet switches and routers. The proposed architectures using optical interconnections remain fully compatible with current network infrastructures. For these architectures, the authors conduct scalability analysis and cost analysis of implementations based on currently available components.

Chapter 2, "Broadband access networks: current and future directions," describes Long-Reach Passive Optical Network (LR-PON) technology which brings the high capacity of optical fiber closer to the user. The authors propose and investigate the Wireless-Optical Broadband Access Network (WOBAN) which integrates the optical and wireless access technologies.

Chapter 3, "The optical control plane and a novel unified control plane architecture for IP/WDM networks," provides an overview of current protocols utilized for the control plane in optical networks. The authors also propose and investigate a new unified control plane architecture for IP-over-WDM networks that manages both routers and optical switches.

Chapter 4, "Cognitive routing protocols and architecture," describes the operation of wireless networks in which cognitive techniques are becoming increasingly common. The authors present cognitive routing protocols and their corresponding protocol architectures. In particular, the authors propose and investigate the mobility-aware routing protocol (MARP) for cognitive wireless networks.

Chapter 5, "Grid networking," describes Grid networks which are enabling the large-scale sharing of computing, storage, communication and other Grid resources across the world. Grid networks based on optical circuit switching (OCS) and optical burst switching (OBS) technologies are discussed. The authors describe approaches for resource scheduling in Grid networks.

Part II, titled "Network architectures," consists of five chapters that propose and investigate new architectural features for next-generation networks.

Chapter 6, "Host identity protocol (HIP): an overview," describes a set of protocols that enhance the original Internet architecture by injecting a name space between the IP layer and the transport protocols. This name space consists of cryptographic identifiers that are used to identify application endpoints, thus decoupling names from locators (IP addresses).

Chapter 7, "Contract switching for managing inter-domain dynamics," introduces contract switching as a new paradigm for allowing economic considerations and flexibilities that are not possible with the current Internet architecture. Specifically, contract switching allows users to indicate their value choices at sufficient granularity, and providers to manage the risks involved in investments for implementing and deploying new QoS technologies.

Chapter 8, "PHAROS: an architecture for next-generation core optical networks," presents the Petabit/s Highly-Agile Robust Optical System (PHAROS), an architectural framework for future core optical networks. PHAROS, which is designed as part of the DARPA core optical networks (CORONET) program, envisions a highly dynamic network with support for both wavelength and IP services, very fast service setup and teardown, resiliency to multiple concurrent network failures, and efficient use of capacity reserved for protected services.

Chapter 9, "Customizable in-network services," proposes the deployment of custom processing functionality within the network as a means for enhancing the ability of the Internet architecture to adapt to novel protocols and communication paradigms. The chapter describes a network service architecture that provides suitable abstractions for specifying data path functions from an end-user perspective, and discusses technical challenges related to routing and service composition along the path.

Chapter 10, "Architectural support for continuing Internet evolution and innovation," argues that, while the current Internet architecture houses an effective design, it is not in itself effective in enabling evolution. To achieve the latter goal, it introduces the SILO architecture, a meta-design framework within which the system design can change and evolve. SILO generalizes the protocol layering concept by providing each flow with a customizable arrangement of fine-grain, reusable services, provides support for cross-layer interactions through explicit control interfaces, and decouples policy from mechanism to allow each to evolve independently.

Part III, titled "Protocols and practice," deals with different aspects of routing layer protocols and sensor network infrastructures.

Chapter 11, titled "Separating Routing Policy from Mechanism in the Network Layer", describes a network layer design that uses a flat endpoint identifier space and also separates routing functionality from forwarding, addressing, and other network layer functions. This design is being studied as part of the Postmodern Internet Architecture (PoMo) project, a collaborative research project between the University of Kentucky, the University of Maryland, and the University of Kansas. The chapter also presents results from experimental evaluations, using a tunneling service that runs on top of the current Internet protocols.

Chapter 12, titled "Multi-path BGP: motivations and solutions," discusses the motivation for using multi-path routing in the next generation Internet, in the context of the widely-used Border Gateway Protocol (BGP) routing protocol. The chapter then presents a set of proposed mechanisms that can interoperate with the existing BGP infrastructure. Solutions for both intra-domain and inter-domain multi-path routing are presented. These mechanisms are being implemented as part of the ongoing TRILOGY testbed, a research and development project funded by the European Commission.

Chapter 13, titled "Explicit congestion control: charging, fairness, and admission management," presents theoretical results on explicit congestion control mechanisms. These are a promising alternative to the currently used implicit congestion control mechanism of TCP. Examples of protocols using explicit congestion control are eXplicit Control Protocol (XCP) and the Rate Control Protocol (RCP). This chapter presents a proportionally fair rate control protocol and an admission management algorithm that deals with the tradeoff between maximizing resource utilization and admission of burst arrivals.

Chapter 14, titled "KanseiGenie: software infrastructure for resource management and programmability of wireless sensor network fabrics," presents a software framework that allows a community of users to develop applications based on a network of deployed wireless sensor nodes. The framework allows the sensor nodes to be shared by multiple applications, using slicing and virtualization of the nodes' resources. This project has been implemented as part of the NSF GENI initiative and promises to change the way in which sensor networks will operate in the future.

Finally, **Part IV**, titled "Theory and models", deals with theoretical foundations and models, as applicable to next generation Internet protocol design.

Chapter 15, "Theories for buffering and scheduling in Internet switches," presents interesting theoretical results on the use of small buffer sizes in the design of next generation routers/switches. It presents results on the interactions between a router's buffer size and TCP's congestion control mechanisms, and results on queueing theory based analysis on the fluctuation of traffic arrivals at a router. Based on these results, an active queue management mechanism is presented.

Chapter 16, "Stochastic network utility maximization and wireless scheduling," discusses network utility maximization (NUM) as a refinement of the layering as optimization decomposition principle that is applicable to dynamic network environments. The chapter provides a taxonomy of this research area, surveys the key results obtained over the last few years, and discusses open issues. It also highlights recent progress in the area of wireless scheduling, one of the most challenging modules in deriving protocol stacks for wireless networks.

Chapter 17, "Network coding in bi-directed and peer-to-peer networks," examines the application of network coding principles to bi-directed and peer-to-peer (P2P) networks. With the increasing use of P2P networks, it is essential to study how well network coding can be useful in such networks. The chapter discusses fundamental limitations of network coding for such networks and derives performance bounds. For P2P networks, the chapter presents practical network coding mechanisms for peer-assisted media streaming and peer-assisted content distribution.

Chapter 18, "Network economics: neutrality, competition, and service differentiation," argues that the current Internet is not living up to its full potential because it delivers insufficient or inconsistent service quality for many applications of growing importance. The author explores how pricing can help expose hidden externalities and better align individual and system-wide objectives by structuring payments between content providers, ISPs, and users, to create the right incentives. The role of service differentiation in remedying problems that arise when users have heterogeneous requirements or utility functions is also examined.

Acknowledgements

We are grateful to the contributing authors for their efforts and diligence. This book would not have been possible without their help and cooperation – our sincere thanks to all for their contributions.

We also acknowledge the National Science Foundation which has provided partial support for our work on this project under grant CNS-0626741.

We thank Cambridge University Press for the opportunity to publish this book. We are especially grateful to Philip Meyler, Publishing Director for Engineering, Mathematical and Physical Sciences, Sabine Koch, Sarah Matthews, Sarah Finlay, Jonathan Ratcliffe, Joanna Endell-Cooper and the Cambridge TEXline staff for their guidance throughout this task. We also like to thank Ms. Yuyan Xue (UNL) and Mrs. Joyeeta Mukherjee (UNL) who helped us in putting together the book materials. Thanks are also due to Google, Inc. for the *Google Docs* and *Google Sites* services, which facilitated cooperation and communication among the book editors and authors. We also acknowledge our respective institutions for providing us with the necessary computing infrastructure.

Part I

Enabling technologies

1 Optical switching fabrics for terabit packet switches

Davide Cuda, Roberto Gaudino, Guido A. Gavilanes Castillo, and Fabio Neri

Politecnico di Torino, Turin, Italy

A key element of past, current, and future telecommunication infrastructures is the switching node. In recent years, packet switching has taken a dominant role over circuit switching, so that current switching nodes are often packet switches and routers. While a deeper penetration of optical technologies in the switching realm will most likely reintroduce forms of circuit switching, which are more suited to realizations in the optical domain, and optical cross-connects [1, Section 7.4] may end up playing an important role in networking in the long term, we focus in this chapter on high-performance packet switches.

Despite several ups and downs in the telecom market, the amount of information to be transported by networks has been constantly increasing with time. Both the success of new applications and of the peer-to-peer paradigm, and the availability of large access bandwidths (few Mb/s on xDSLs and broadband wireless, but often up to 10's or 100's of Mb/s per residential connection, as currently offered in Passive Optical Networks – PONs), are causing a constant increase of the traffic offered to the Internet and to networking infrastructures in general. The traffic increase rate is fast, and several studies show that it is even faster than the growth rate of electronic technologies (typically embodied by Moore's law, predicting a two-fold performance and capacity increase every 18 months).

Optical fibers are the dominant technology on links between switching nodes, and, while the theoretical capacity of several tens of Tb/s on each fiber is practically never reached, still very high information densities are commercially available: 10–40 Gb/s, and soon 100 Gb/s, per wavelength channel are commonplace in WDM (Wavelength Division Multiplexing) transmission systems capable of carrying up to tens of channels on a single fiber, leading to Tb/s information rates on a single optical fiber.

Very good and mature commercial offerings are on the market today for packet switches and routers, with total switching capacities up to few Tb/s. These devices are today fully realized in the electronic domain: information received from optical fibers is converted in linecards to the electronic domain, in

Next-Generation Internet Architectures and Protocols, ed. Byrav Ramamurthy, George Rouskas, and Krishna M. Sivalingam. Published by Cambridge University Press. © Cambridge University Press 2011.

which packets are processed, stored to solve contentions, and switched through a switching fabric to the proper output port in a linecard, where they are converted back to the optical domain for transmission.

The fast traffic growth however raises concerns on the capability of electronic realizations of packet switches and routers to keep up with the amount of information to be processed and switched. Indeed, the continuous evolution of high-capacity packet switches and routers is today bringing recent realizations close to the fundamental physical limits of electronic devices, mostly in terms of maximum clock rate, maximum number of gates inside a single silicon core, power density, and power dissipation (typically current large routers need tens of kW of power supply; a frequently cited example is the CRS-1 System – see [2, page 10]). Each new generation of switching devices shows increasing component complexity and needs to dissipate more power than the previous one. The current architectural trend is to separate the switching fabric from the linecards and often to employ optical point-to-point interconnections among them (see [3, page 6] again for the CRS-1). This solution results in a large footprint (current large switches are often multi-rack), poses serious reliability issues because of the large number of active devices, and is extremely power-hungry.

A lively debate on how to overcome these limits is ongoing in the research community. Optical technologies today are in practice limited to the implementation of transmission functions, and very few applications of photonics in switching can be found in commercial offerings. Several researchers however claim that optical technologies can bring significant advantages also in the realization of switching functions: better scalability towards higher capacities, increased reliability, higher information densities on internal switch interconnections and backplanes, reduced footprint and better scaling of power consumption [7, 8].

In this chapter[1] we consider photonic technologies to realize subsystems inside packet switches and routers, as recently has been done by several academic and industrial research groups. In particular, we refer to a medium-term scenario in which packet switching according to the Internet networking paradigm dominates. Hence we assume that packets are received at input ports according to current formats and protocols (such as IP, Ethernet, packet over Sonet). We further assume that packets are converted in the electronic domain at linecards for processing and contention resolution. We propose to use optical interconnections among linecards, hence to implement an optical switching fabric internally to the switch. At output linecards, packets are converted back to legacy formats and protocols, so that the considered architectures remain fully compatible with current network infrastructures. For these architectures we will evaluate the maximum achievable switching capacities, and we will estimate the costs (and their scaling laws) of implementations based on currently available discrete components.

[1] Preliminary parts of this chapter appeared in [4, 5, 6]. This work was partially supported by by the Network of Excellence BONE ("Building the Future Optical Network in Europe"), funded by the European Commission through the Seventh Framework Programme.

1.1 Optical switching fabrics

To study the suitability of optical technologies for the realization of switching fabrics inside packet switching devices, we focus on three optical interconnection architectures belonging to the well-known family usually referred to as "tunable transmitter, fixed receiver" (TTx-FRx), which was widely investigated in the past (the three specific architectures were studied [9, 10] within the e-Photon/ONe European project). In particular, we consider optical interconnection architectures based on the use of broadcast-and-select or wavelength-routing techniques to implement packet switching through a fully optical system in which both wavelength division and space multiplexing are used. Indeed, WDM and space multiplexing have proven to make best use of the distinctive features of the optical domain.

Our architectures are conceived in such a way that switching decisions can be completely handled by linecards, enabling the use of distributed scheduling algorithms. We will, however, not deal with control and resource allocation algorithms, thus focusing on the switching fabric design, for which we will consider in some detail the physical-layer feasibility, the scalability issues, and the costs related to realizations with currently available components.

Many optical switching experiments published in the last 10–15 years have used optical processing techniques such as wavelength conversion or 3R regeneration [1, Chapter 3], and even optical label recognition and swapping, or all-optical switch control, and have been often successfully demonstrated in a laboratory environment. However, they are far from being commercially feasible, since they require optical components which are still either in their infancy or simply too expensive. We take a more conservative approach, restricting our attention to architectures that are feasible today, as they require only optical components commercially available at the time of this publication. Fast-tunable lasers with nanosecond switching times are probably the only significant exception, since they have not yet a real commercial maturity, even though their feasibility has already been demonstrated in many experimental projects [7] and the first products are appearing on the market.

The reference architecture of the optical switching fabric considered in this chapter is shown in Figure 1.1: a set of N input linecards send packets to an optical interconnection structure that provides connectivity towards the N output linecards using both WDM and optical space multiplexing techniques. The optical switching fabric is organized in S switching planes, to which a subset of output linecards are connected. As we mainly refer to packet switching, fast optical switches (or in general plane distribution subsystems) allow input linecards to select the plane leading to the desired output linecard on a packet-per-packet basis. Obviously, slower switching speeds would suffice in case of circuit switching. Within each plane, wavelength routing techniques select the proper output. Thus packet switching is controlled at each input linecard by means of a fast tunable laser (i.e., in the wavelength domain) and, for $S > 1$, of a fast optical switch (i.e., in the space domain). Each linecard is equipped with one tunable transmitter (TTx) and one fixed-wavelength burst-mode receiver (BMR) operating at the data rate of a single WDM channel. Burst-mode operation is required on a packet-by-packet basis. Note that both TTx's and BMR's have recently appeared in the market to meet the demand for flexible WDM systems (for TTx's) and for upstream receivers in PONs (for BMR's).

For simplicity,[2] we assume that all the considered architectures have a synchronous and time-slotted behavior, as reported in [4] and [5]: all linecards are synchronized to a common clock signal which can be distributed either optically or electrically.

Packet transmissions are scheduled so that at most one packet is sent to each receiver on a time slot (i.e., contentions are solved at transmitters). Packet scheduling can be implemented in a centralized fashion as in most current packet switches. In this case, an electronic scheduler is required so that, after receiving status information from linecards, it decides a new permutation, i.e., an input/output port connection pattern, for each time slot. Centralized schemes can potentially offer excellent performance in terms of throughput, but the electronic complexity of the scheduler implementation can upper bound the achievable performance [8]. Furthermore, centralized arbitration schemes require signaling bandwidth to collect status information and to distribute scheduling decisions; these introduce latencies due to the time needed to propagate such information and to execute the scheduling algorithm. In this context, the implementation of a distributed scheduling scheme becomes a crucial issue to assess the actual value of proposed optical interconnection architectures. Distributed schemes that exhibit fair access among linecards using only locally available information (see, e.g., [11, 12]) have been proposed for architectures similar to those considered in this chapter; they avoid bandwidth waste due to the signaling process and limit the scheduling algorithm complexity, thus improving the overall fabric scalability.

The lasers tuning range, i.e, the number of wavelengths a transmitter is required to tune to, can be a practical limiting factor. Even for laboratory prototypes, the maximum tuning range for tunable lasers is in the order of a few tens of wavelengths [13]. As a result, the wavelength dimension alone could not ensure input/output connectivity when the number N of linecards is large. Multiple switching planes, i.e., the space diversity dimension, were indeed introduced to overcome this limitation. By doing so, since the same wavelengths can be reused on each switching plane, if S is the number of switching planes, a wavelength tunability equal to N/S (instead of N) is required.

In the three considered architectures, shown in Figs. 1.2–1.4, transmitters reach the S different planes by means of proper optical *distribution* stages that

---

[2] This assumption is not strictly necessary, but we introduce it to describe in a simpler way the operation of the switching fabric.

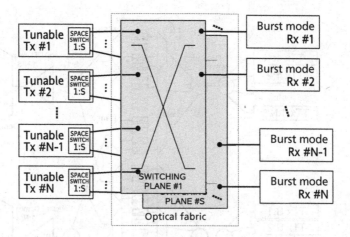

Figure 1.1 Multi-plane optical fabric architecture.

actually differentiate each of the three architectures. At the output of the switching plane, an $S : 1$ optical coupler collects the packets, that are amplified with an Erbium Doped Fiber Amplifier (EDFA), and then distributed by a WDM demultiplexer to the N/S receivers. These fabric architectures are designed so that the number of couplers and other devices that packets have to cross is the same for all input/output paths. The EDFA WDM amplification stages can thus add equal gain to all wavelength channels since all packets arrive with the same power level.

1.1.1 Wavelength-selective (WS) architecture

This architecture, presented in Figure 1.2, was originally proposed for optical packet switched WDM networks inside the e-Photon/ONe project [9]. This optical switching fabric connects N input/output linecards by means of broadcast-and-select stages; groups of N/S TTx's are multiplexed in a WDM signal by means of an $N/S : 1$ coupler; then this signal is split into S copies by means of a $1 : S$ splitter. Each copy is interfaced to a different switching plane by means of wavelength selectors, composed by a demux/mux pair separated by an array of N/S Semiconductor Optical Amplifier (SOA) gates, which are responsible for both plane and wavelength selection.

Wavelength-selective architecture is a blocking architecture. Since within a group of input linecards all the transmitters must use a different wavelength, it is not possible for transmitters located in the same input group to transmit to output ports located on different switching planes that receive on the same wavelength. Proper scheduling strategies can partly cope with this issue, but we do not discuss them here.

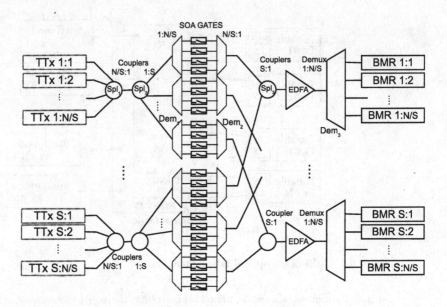

Figure 1.2 Wavelength-selective (WS) architecture.

1.1.2   Wavelength-routing (WR) architecture

In this case (see Figure 1.3) the wavelength-routing property of Arrayed Waveguide Gratings (AWGs), the main component of the distribution stage, is exploited to perform both plane and destination selection: no space switch is necessary. Each collecting stage gathers all the packets for a specific plane.

The transfer function of AWGs [14] exhibits a cyclic routing property: several homologous wavelengths belonging to periodically repeated Free Spectral Ranges (FSR) are identically routed to the same AWG output. The WR fabric can either exploit this property or not, with conflicting impacts on tuning ranges and crosstalk, respectively.

The former situation (i.e., full exploitation of the AWG cyclic property), called WR Zero-Crosstalk (WR-ZC) is depicted in Figure 1.3, which shows an instance of the WR architecture with $N = 9$ and $S = 3$. Note that each transmitter in a group of N/S transmitters uses N wavelengths in a different FSR; in other words we have N/S FSRs comprising N wavelengths each. The tunability range of each transmitter is N, and each receiver is associated with a set of N/S different wavelengths (for receivers this is not a real limitation because the optical bandwidth of photodiodes is actually very large). In this way, by exploiting the cyclic property of AWGs, we can prevent linecards from reusing the same wavelength at different input ports of the AWG. The advantage is that no coherent crosstalk (see [4]) is introduced in the optical fabric, and only out-of-band crosstalk is present, which introduces negligible penalty (see Section 1.2.1). Even though this solution can be almost optimal with respect to physical-layer impairments,

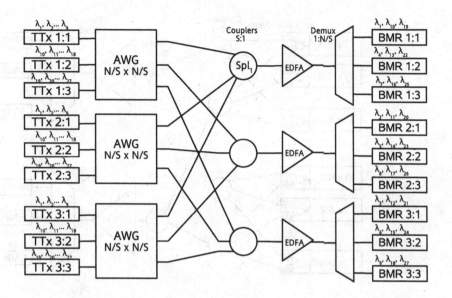

Figure 1.3 Wavelength-routing Zero-Crosstalk (WR-ZC) architecture.

the AWGs must exhibit an almost identical transfer function over N/S FSRs, and the EDFA amplifying bandwidth has to be significantly larger.

If instead the AWG cyclic property is exploited only in part (we call this architecture simply WR), and the system operation is limited to a single FSR (all TTx's are identical with tunability N), some in-band crosstalk can be introduced, which can severely limit scalability [4] when two or more TTx's use the same wavelength at the same time. However, proper packet scheduling algorithms, which avoid using the same wavelengths at too many different fabric inputs at the same time [15], can prevent the switching fabric from operating in high-crosstalk conditions.

1.1.3   Plane-switching (PS) architecture

In the PS fabric, depicted in Figure 1.4, the distribution stage is implemented in the linecards by splitting transmitted signals in S copies by a $1 : S$ splitter, and then sending the signals to SOA gates. A given input willing to transmit a packet to a given output must first select the destination plane by turning off all the SOA gates except the one associated with the destination plane and, then, use wavelength tunability to reach the desired output port in the destination plane. Coupling stages are organized in two vertical sections: the first is a distribution stage used by inputs to reach all planes, while the second is a collecting stage to gather packets for each destination plane. Each plane combines at most N/S packets coming from the N input linecards.

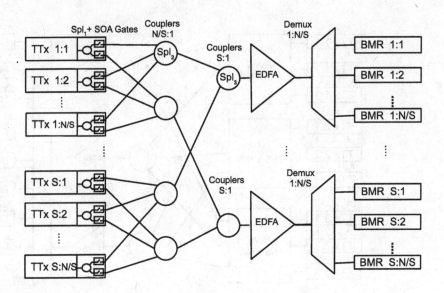

Figure 1.4 Plane-switching (PS) architecture.

1.2   Modeling optical devices

None of the proposed optical fabrics includes any signal regeneration besides pure linear optical amplification. Using the common terminology introduced in [1], we have at most 1R regeneration of the signals inside the optical fabric, while we exclude 2R and 3R regeneration. As a result, physical layer impairments may accumulate when increasing the port count N or the number of planes S, so that the characterization of the used optical devices becomes crucial to effectively assess each architecture's ultimate scalability. In performing the analysis described in this chapter, we observed that a first-order scalability assessment based on theoretical insertion loss values gives unrealistic results. As a clear example, the AWG in the WR architecture has an insertion loss that in a first approximation does not depend on the number of input/output ports, thus leading to a theoretical "infinite scalability." Clearly, we needed a more accurate second-order assessment capable of capturing other important effects that characterize commercial devices, such as polarization dependence, excess losses, channel uniformity, and crosstalk. Despite their different nature, all these effects can be expressed as an input/output equivalent power penalty which accounts for both actual physical power loss and the equivalent power penalty introduced by other second-order transmission impairments, as described below. We only focused our study on optical components, as fiber-related effects (e.g., dispersion, attenuation, non-linearities, cross-phase modulation, etc.) are likely to be negligible in the proposed architectures, mainly due to the short distances involved.

1.2.1   Physical model

The following physical-layer effects are taken into account in our analysis. See [4] for details.

Insertion Loss (IL): We indicate as insertion loss the total worst-case power loss, which includes all effects related to internal scattering due to the splitting process and also non-ideal splitting conditions, such as material defects, or manufacturing inaccuracies. In the case of n-port splitters, the splitting process gives a minimum theoretical loss increasing with $10 \log n$ dB, but extra loss contributions due to non-ideal effects, often referred to as Excess Losses (EL), must also be considered.

Uniformity (U): Due to the large wavelength range typically covered by multi-port devices, different transmission coefficients exist for different wavelengths. Over the full WDM comb, the propagation conditions vary slightly from center channels to border ones. Similar uneven behaviors appear in different spatial sections of some components. These differences are taken into account by the U penalty component, which is often referred to as the maximum IL variation over the full wavelength range in all paths among inputs and outputs.

Polarization Dependent Loss (PDL): The attenuation of the light crossing a device depends on its polarization state due to construction geometries, or to material irregularities. Losses due to polarization effects are counted as a penalty in the worst propagation case.

Crosstalk (X): A signal out of a WDM demultiplexing port always contains an amount of power, other than the useful one, belonging to other channels passing through the device. This effect is generally referred to as crosstalk. For a given useful signal at wavelength λ, the crosstalk is usually classified [1] as either out-of-band, when the spurious interfering channels appear at wavelengths spectrally separated from λ, or as in-band crosstalk, when they are equal to λ. For the same amount of crosstalk power, this latter situation is much more critical in terms of overall performance [16]. Both types of crosstalk translate into a power penalty at receivers dependent on the amount of interfering power.

For out-of-band crosstalk, also called incoherent crosstalk, the contribution from adjacent wavelength channels X_A is usually higher than the contribution from non-adjacent channels X_{NA}. Following the formalism presented in [17], the overall crosstalk relative power level, expressed in dimensionless linear units, can be approximated as follows

$$X(w) = 2X_A + (w - 3)X_{NA}, \qquad (1.1)$$

where $X(w)$ is the total amount of crosstalk power present on a given port, normalized to the useful signal power out of that port; w is the number of wavelength channels, which is typically equal to the number n of ports of the device. Out-of-band crosstalk is present on any WDM filtering device, such as WDM demultiplexers, $1 : n$ AWGs, and optical filters, due to the fact that their ability to transmit/reject out-of-band signals does not behave as an ideal step transfer function. As such, incoherent crosstalk is present in all our proposed architectures. Typical values for multiplexers [7] are -25 dB and -50 dB for adjacent X_A and non-adjacent X_{NA} channels, respectively. The equation above can be transformed into an equivalent power penalty (in dB), labeled OX. Following the approximations presented in [17], OX is equal to:

$$OX(w)|_{dB} = 10\log_{10}(1 + X(w)). \tag{1.2}$$

In-band crosstalk, or coherent crosstalk, is caused by interference from other channels working on the same wavelength as the channel under consideration. In the WR case, the same wavelength can be generated simultaneously at the input of many AWG ports. Due to the AWG actual transfer functions, some amount of in-band power leaks to other device ports; this behavior is described in data sheets as adjacent/non-adjacent port crosstalk (X_A and X_{NA}, respectively, defined for physical ports instead of wavelength channels as for incoherent crosstalk). For the other architectures, space switching is not ideal so that a small portion of the useful linecard input power leaks into other switching planes, as will be explained in Section 1.2.3. The impact of this crosstalk is typically high given its in-band characteristics, and the equivalent IX power penalty (in dB) for optimized decision-threshold in the receiver can be estimated [16] by

$$IX(n)|_{dB} = -10\log_{10}(1 - X(n)Q^2), \tag{1.3}$$

where Q is the target eye-opening quality factor in linear units, determining the target Bit Error Rate (BER) (typically Q lies in the range from 6 to 7 in linear scale for BER between 10^{-9} and 10^{-12}). $X(n)$ represents here the normalized crosstalk power from the other n in-band sources (for example multi-plane non-ideal switches or crosstalk between spatially adjacent ports in AWG devices) relative to the useful signal in consideration (see Section 1.2.3).

1.2.2   Device characterization

The previously described power penalties enable the characterization of passive optical devices of a given number n of ports in terms of the overall power penalty, which takes into account all actual passive losses introduced by the components plus equivalent losses introduced by crosstalk impairments. We will denote with $L_{Dem}(n)$, $L_{Spl}(n)$, and $L_{AWG}(n)$ the "equivalent" losses introduced by muxes/demuxes, couplers/splitters, and AWGs of n ports, respectively. To estimate these power penalties, a detailed study has been carried out in order to find reasonable values for realistic commercial devices, by analyzing a large number of commercial device datasheets [18, 19, 20]. As a result, we collected typical realistic values of each parameter for the different devices. Linear and logarithmic regression methods have been used to derive analytical formulas that fit well on datasheet values and can estimate unknown ones. For the same device type,

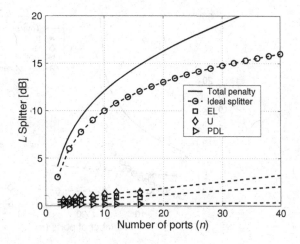

Figure 1.5 Power penalties for $1:n$ couplers/splitters.

the values reported in datasheets from different vendors were usually very similar, and are often dictated by the the specification of some relevant international standard. For instance, most commercial $1:n$ splitters have values that are set by current PON standards. Thus, the values that we considered can be assumed as fairly general and consistent among different optical component vendors.

As an example, the estimated losses for coupler/splitter devices are shown in Figure 1.5. These plots report the contribution of each of the individual effects described in Section 1.2.1, and the resulting total equivalent power penalty. In both cases, the ideal $\log n$-like loss dominates over the contributions of other parameters like U, PDL, and EL. However, as the number of ports increases, so does the relative contribution of these second-order parameters. For instance, 20 ports contribute 3–4 dB of additional penalty with respect to the ideal case.

The characterization of AWGs is shown in Figure 1.6. While the power penalty of these devices should ideally be independent of the number of ports due to the wavelength routing property, we observe that the IL values inferred from datasheets show a dependency on the number of ports that contributes logarithmically to the power penalty. More notably, we found that out-of-band crosstalk effects are negligible, while in-band crosstalk has a significant impact in the case of AWGs of size $n:n$. In this case, crosstalk increases exponentially the power penalty, limiting the realistically useful size of the AWG device to about 10–15 ports (which cause 13–18 dB of equivalent losses). This rather strong limit is confirmed by several experimental works, such as [7, 16], and is in contrast with many studies on switching architectures in which AWGs with large port counts are indicated as very promising components for the implementation of large optical switches.

Tunable transmitters are a key component in these architectures. They are modeled as sources characterized by a given Optical Signal-to-Noise Ratio

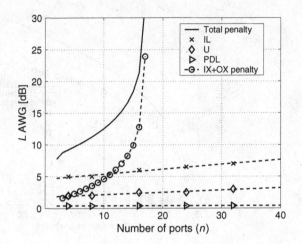

Figure 1.6 Power penalties for $n : n$ AWGs.

$OSNR_{TX}$, corresponding to the ratio between the useful laser power and the noise floor due to spontaneous emission inside the laser. Though in standard WDM transmission $OSNR_{TX}$ gives negligible effects, we will show later that it can be relevant for very large optical fabrics, due to the presence of N lasers, each one contributing a low (but non-negligible) noise floor.

For the receivers, in order to address scalability at different bit rates, we followed the analysis presented in [21] which, inferring from many different commercial datasheets, proposes a sensitivity slope vs. bit-rate R_b of 13.5 dB for a ten-fold bit-rate increase. Following this model, and assuming a given sensitivity at, say, 10 Gb/s, the receiver sensitivity at other bitrates in dBm is estimated as follows:

$$P_S(R_b)|_{dBm} = P_S(10\,\text{Gb/s})|_{dBm} + 13.5 \log_{10} \frac{R_b}{10\,\text{Gb/s}}. \tag{1.4}$$

All architectures exhibit a common amplification and demultiplexing output stage. Regarding EDFA amplifiers, we have assumed that the EDFAs operate in the saturated regime; that is, EDFAs show a constant output power, which is split over the N/S channels that cross it simultaneously. The nominal EDFA output power is assumed to be $P_{tot,out}^{EDFA}$, which can be set by means of gain locking techniques. Let A_{EDFA} be the EDFA power gain. The A_{EDFA} used in noise calculations is obtained considering the ratio between the total EDFA output power, $P_{tot,out}^{EDFA}$, and the total EDFA input power, $P_{tot,in}^{EDFA}$. Furthermore, we characterized EDFAs by a noise figure F_{EDFA}.

Finally, regarding the SOA-based space switches, we based our analysis on the characteristics of one of the few commercially available specific SOA-based switches [22]. In the "on" state, the SOA is assumed to have a noisy behavior characterized by a noise figure F_{SOA}. In the "off" state, a realistic switching Extinction Ratio (ER) is considered. This ratio turns out to be relevant for large multi-plane solutions, since it generates in-band and out-of-band crosstalk. Besides, we assumed a gain transparency condition for the full switch, where the SOA gain compensates the passive losses of the $1:S$ splitter required to implement space switching inside the linecard.

1.2.3 Multi-plane-specific issues

Some considerations must be introduced when considering multi-plane architectures.

Switching extinction ratio: Due to the finite extinction ratio of SOA-based switching devices, crosstalk arises across planes just before the EDFA. In the worst case, the number of crosstalk contributions is one in-band component for each plane, and the resulting crosstalk impact is given by the coherent expression IX as shown before in Eq. (1.3), that here depends on S and on the nominal switch extinction ratio ER (in linear units). As a result, the crosstalk penalty in multi-plane configurations due to this effect can be estimated as:

$$IX(S)|_{dB} = -10\log_{10}(1 - (S-1)ER\, Q^2). \qquad (1.5)$$

Cross noise floor accumulation: In general, lasers and SOAs as optical sources generate Amplified Spontaneous Emission (ASE) noise when operating in the "on" state, which in turn is sent to the selected planes. Although individually the resulting noise floor levels are quite low, all their spectra add up, and for a high number of planes the noise accumulates, resulting in an intrinsic limitation to scalability. We took this effect into account in our model: we considered the maximum number of noise floor sources per plane (corresponding to all linecards transmitting), and evaluated the noise accumulation accordingly on each switching plane.

Optimum number of switching planes: The choice of the optimal number of switching planes S is a critical design choice and depends on the following considerations. First, a large S brings the advantage of reducing TTx tunability, but it increases the amount of coherent crosstalk. Second, smaller values of S reduce the number of optical components, hence the overall complexity. Third, both very large and very small values of S introduce larger excess losses, thereby reducing the power budget.

A numerical analysis suggests that there is an optimum value of S, which also depends on the linecard bit-rate R_b, and is roughly close to $S \approx \sqrt{N}$; this value of S shows the maximum OSNR at receivers, and thus allows better scalability in terms of aggregate bandwidth.

1.3   Scalability analysis

By considering the physical impairments and the other effects mentioned in previous sections we conducted a scalability and feasibility study of the optical switching fabrics described in Section 1.1, assuming typical values for the parameters characterizing the behavior of optical devices.

In our architectures, physical scalability is limited by both the useful signal power at the receiver's photodiode, and by the ASE noise present at the EDFA output. Firstly, the received signal power must be larger than the receiver's sensitivity P_S; thus, since the considered architectures have the same output EDFA amplification stage, at each output the following must hold:

$$P_{ch,out}^{EDFA}|_{\text{dBm}} - L_{Dem}(N/S)|_{\text{dB}} - \mu \geq P_S|_{\text{dBm}}, \tag{1.6}$$

where $P_{ch,out}^{EDFA}|_{\text{dBm}} = P_{tot,out}^{EDFA}|_{\text{dBm}} - 10\log(N/S)|_{\text{dB}}$ is the power per channel after amplification, and μ is a 3 dB margin to consider component aging, penalties of burst-mode receivers with respect to standard continuous receivers, laser misalignments, and other effects that might degrade the received signal. We assumed for the EDFA total output power $P_{tot,out}^{EDFA}$ a typical value of 17 dBm, $P_S = -26$ dBm at 10 Gb/s, and we used (1.4) to infer the receiver sensitivity at other bit-rates.

Regarding WDM tunable transmitters, an average transmitted power P_{TX} of 3 dBm is assumed. Generally, a typical tunable laser peak output power is of the order of +10 dBm, but we need to consider 6–7 dB of equivalent loss introduced by an external modulator (3 dB due to on/off keying and 3–4 dB due to additional insertion losses).

Regarding the SOA-based gates (used in PS and WS architectures), in the "off" state, we assumed an ER for SOAs of 35 dB in order to calculate the crosstalk penalty from Eq. (1.5).

Even though the sensitivity constraint is the same for all the considered architectures, the noise-related constraint is not. We require a commonly accepted minimum optical signal-to-noise ratio T_{OSNR} of 17 dB (defined over a bandwidth equal to the bit-rate) to guarantee the target T_{BER} of 10^{-12}.

WDM tunable transmitters were modeled then as optical sources characterized by a given optical signal-to-noise ratio $OSNR_{TX}$, corresponding to the ratio between the useful laser power and the noise power due to spontaneous emission inside the laser. Using the laser $OSNR_{TX}$, we calculate the equivalent noise power spectral density (N_{0TX}) as $OSNR_{TX} = P_{TX}/N_{0TX}B$. Moreover, we assumed a flat noise behavior ranging from a typical $OSNR_{TX}$ value of 40 dB to an ideal value of 60 dB over a 0.1 nm reference bandwidth B (in Hz).

The noise spectral density $G_N(f)$ (in linear units, f being the optical frequency in Hz) was evaluated for each architecture. Generally, the noise power P_N over a bandwidth equal to the bit-rate R_b can be evaluated as $P_N = \int_{R_b} G_N(f)df$.

Now let $G_{N,out}^{Amp}(f)$ be an amplifier output noise spectral density [17]:

$$G_{N,out}^{Amp}(f) = G_{N,in}^{Amp}(f) \times A_{Amp} + hf(A_{Amp} - 1) \times F_{Amp} \qquad (1.7)$$

where h is the Planck constant, A_{Amp} and F_{Amp} (either $Amp = EDFA$ or $Amp = SOA$) are the amplifier gain and noise figure, respectively, and $G_{N,in}^{Amp}(f)$ is the noise power spectral density entering the amplifier. Typical noise figure values for optical amplifiers are $F_{SOA} = 9$ dB and $F_{EDFA} = 5$ dB.

In the case of final EDFA (common to all architectures), $G_{N,in}^{EDFA}(f)$ is different for each architecture and we call it $G_N^{WS}(f)$, $G_N^{PS}(f)$, or $G_N^{WR}(f)$ for the WS, the PS, or the WR fabric, respectively. They are derived below.

The overall EDFA noise power can be evaluated by integrating $G_{N,out}^{EDFA}(f)$ over the whole bandwidth; so, the OSNR constraint can be finally expressed as:

$$P_{ch,out}^{EDFA}|_{dBm} - P_{N,out}^{EDFA}|_{dBm} - \mu \geq T_{OSNR}|_{dB} \qquad (1.8)$$

with μ having the same meaning as in Eq. (1.6).

Estimation of $G_{N,in}^{EDFA}(f)$ for the three architectures

The **WS** architecture starts to accumulate noise after Spl_1 (see Fig. 1.2), i.e, after the first coupler following the TTx; this is due to the coupling of the N/S signals coming from the N/S TTx belonging to the same input group. Henceforth, the noise power spectral density after Dem_1, $G_{N,out}^{Dem_1}(f)$ can be evaluated for each channel as:

$$G_N^{Dem_1}(f) = \frac{N_{0TX}}{L_{Spl_1}(N/S)} \times \frac{N}{S} \times \frac{1}{L_{Spl_2}(S)L_{Dem_1}(N/S)}. \qquad (1.9)$$

The out-of-band noise contributions introduced by all N/S SOAs are filtered through Dem_2 and then propagated through Spl_3; therefore, noise contributions affecting an arbitrary channel must account for the effect from itself and all other $N/S - 1$ channels that, at most, can be present after Spl_3.

As discussed in Section 1.2.1, in order to model the propagation of noise through Dem_2 the overall noise contribution affecting a channel (in the worst case) is modified by $X_{OSNR} = 1 + 2X_A + (N/S - 3) \times X_{NA}$, where X_A and X_{NA} are channel isolation ratios for adjacent and non-adjacent channels at Dem_2. Thus the noise power spectral density can be evaluated as:

$$G_N^{WS}(f) = \frac{G_N^{Dem_1}(f)G_{SOA} + hf(G_{SOA} - 1)F_{SOA} \times X_{OSNR}}{L_{Dem_2}(N/S)L_{Spl_3}(S)}. \qquad (1.10)$$

In the **WR** architecture, due to the AWG wavelength routing characteristic, all linecard inputs have noise contributions over all wavelengths. These noise contributions are thus propagated through all the AWG output ports as normal signals; therefore, N noise sources are present on each plane and accumulate after Spl_1, yielding:

$$G_{N,out}^{WR}(f) = N_{0TX} \times \frac{N}{L_{AWG}(N/S) L_{Spl_1}(S)}. \qquad (1.11)$$

For the **PS** architecture, we need to consider the accumulation of the noise floors generated by the TTx's. Indeed, a transmitter signal is copied to the S SOA gates by means of the first coupler; thus, the noise power spectral density present at each SOA input is equal to $G_{N,in}^{SOA}(f) = N_{0TX}/L_{Spl_1(S)}$. Hence, the overall noise after $Spl_3(S)$ is given by:

$$G_N^{PS}(f) = \frac{G_{N,in}^{SOA}(f)G_{SOA} + hf(G_{SOA}-1)F_{SOA}}{L_{Spl2}(N/S)L_{Spl3}(S)} \times \frac{N}{S}. \qquad (1.12)$$

The term N/S on the right-hand side is the noise floor accumulation on each switching plane.

1.4   Cost analysis

Beyond scalability of the switching capacity, a fair assessment of the considered architectures cannot neglect the complexity of their implementations. In this section we evaluate the CAPital EXpenditure (CAPEX) of our switching fabrics: a cost model for each device is considered, and it is then used to assess the cost of the entire architecture obtaining a rough "bill of material" estimation for the cost of photonic components. Note that by so doing we consider an implementation based upon discrete components, neglecting the economical effects of integration and large-scale production. Component costs in our model rely on the current market situation (discrete components, no on-chip integration, no economies of scale); thus, they should be used as a rough guide to assess the relative costs of the different architectures. Our goal is to investigate the main factors impacting the cost of an optical fabric and to provide insight into the cost-wise optimal choices to be taken when designing such architectures.

In our analysis we took into account the fact that the operational bandwidth of every device is finite, hence only a limited number of wavelengths can in general be amplified or routed. In particular, commercially available EDFAs and AWGs are usually designed to be used over one of the C (1535–1565 nm), L (1565–1600 nm), XL (above 1600 nm), or S (below 1535 nm) bands (in order of preferred deployment).

Splitters, multiplexers and demultiplexers. We assume that the main factor affecting both splitter and mux/demux costs is their port count n. From market data [23] (taken in summer 2008), by using a Minimum Square Error (MSE) fitting method, we inferred the following cost model:

$$C_{Spl}(n) = \frac{100}{4^{\frac{1}{1.7}}} \times n^{\frac{1}{1.7}} \qquad [USD]. \qquad (1.13)$$

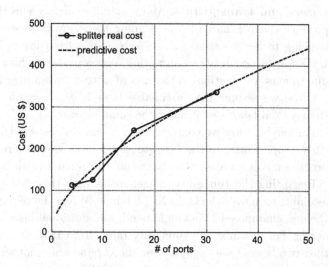

Figure 1.7 Predictive model vs. market data for splitters.

Figure 1.7 shows real and fitted market costs for optical splitters. In order to simplify, but also be compliant with market prices [23], we assumed that mux/demux costs behave as the splitter ones, scaled by a factor α_{Dem}, i.e.,

$$C_{Dem}(n) = \alpha_{Dem} \times C_{Spl}(n), \qquad (1.14)$$

where $\alpha_{Dem} = 10$ in our computations. As an example, we assume that the cost of a 40-port splitter is 400 USD, while for a 40-port WDM multiplexer it is 4000 USD.

Arrayed waveguide gratings. AWG costs depend on both the number of ports and the number of wavelengths it must be able to route, hence on the number of FSRs over which the AWG Transfer Function (TF) should be approximately constant. We heuristically assume that the AWG dependency on the number of ports follows the mux/demux law scaled by a factor $\beta_{AWG} = 2.5$. Let N_{FSR} be the number of FSRs over which the AWG TF has to be flat. In the WR architecture, if the AWG cyclic property is exploited to avoid crosstalk (WR-ZC), $N_{FSR} = (N/S)^2$; otherwise (WR), $N_{FSR} = N/S$. For commercially available AWGs, the TF sharply decreases outside the bandwidth over which the AWG was designed; thus, the AWG TF was assumed to be approximately flat over at most 3 FSRs. If an architecture requires a wider TF, we assume that more AWGs must be used in parallel. The whole AWG cost is thus modeled as:

$$C_{AWG}(n, N_{FSR}) = \left(1 + \left\lfloor \frac{N_{FSR} - 1}{3} \right\rfloor \gamma_{AWG}\right) \beta_{AWG} C_{Dem}(n), \qquad (1.15)$$

where $\gamma_{AWG} = 1.5$; indeed, the cost of adding more AWGs in parallel is larger than simply multiplying the number of AWGs by their cost (some other devices such as splitters or filters might be needed).

Lasers and transmitters. Many manufacturers claim that the technology to produce cheap tunable lasers is going to be available soon and that their cost is going to be less than twice the cost of a fixed laser. Thus, if $C_{\text{f-TX}} \approx 1000$ USD [23] (a pessimistic assumption, not considering future plausible discounts due to mass production) is the cost of a fixed transmitter at a bit-rate of $R_{\text{fix}} = 1$ Gb/s, we assume that a tunable laser is $\delta_{t-TX} = 1.5$ times more expensive than a fixed laser operating at the same bit-rate.

For the bit-rate, we conservatively assume a cost for transmitters that is linearly proportional to the bit-rate and follows the Ethernet market rule which predicts a cost increase by a factor of 3 for a bandwidth boost by a factor of 10.

Regarding the tunability range, we assume that fast tunable lasers will be available soon over the C, L, XL, S bands. When the required tunability exceeds the one guaranteed by a single band, an additional laser is added to the transmitter. For instance, if a tunability range from 1530 nm to 1590 nm is required, then two lasers (one tunable over the C band and another one tunable over the L band) are employed. Finally, the total TTx cost can be evaluated as follows:

$$C_{t-TX}(W, R_b) = \left(1 + \left\lfloor \frac{W-1}{W_b} \right\rfloor \alpha_{TX}\right) \delta_{t-TX} C_{\text{f-TX}} \times \frac{2}{9}\frac{R_b}{R_{\text{fix}}} + \frac{7}{9}, \qquad (1.16)$$

where W is the number of wavelengths the transmitter has to be able to tune to, W_b is the number of wavelengths in each band, and R_b is the transmitter bit-rate. For the multi-band case, we set a multiplicative factor $\alpha_{TX} = 1.3$, since the cost of integrating more than one laser in parallel is larger than simply the product of the number of integrated lasers times the cost of a single laser.

Optical amplifiers are usually implemented using either EDFA or SOA. In our optical fabrics, SOAs are used as on/off gates to perform fast plane selection, and they have to deal with just one wavelength at a time; hence, we consider bandwidth as a cost factor only for EDFAs. The SOA cost is therefore fixed to $C_{SOA} = 1000$ USD (the complexity and technological skills needed to manufacture a SOA should not exceed by much those needed for a laser). Commercially available EDFAs are today designed to operate over one of the four different optical bands; thus, if a large bandwidth is needed, additional EDFAs are used in parallel and the device cost can be expressed as:

$$C_{EDFA}(W) = \left(1 + \left\lfloor \frac{W-1}{W_b} \right\rfloor \alpha_{EDFA}\right) C_{\text{fix}}, \qquad (1.17)$$

where $C_{\text{fix}} = 10\,000$ USD is the cost assumed for a single-band, gain-clamped flat EDFA, W is the number of channels to be amplified, W_b is the number of wavelengths in each optical band, and $\alpha_{EDFA} = 1.5$.

We now compute the cost of each architecture, omitting the receiver part after the final EDFA, which is the same for all the considered architectures and equal to $S \times C_{Dem}(N/S) + N \times C_{BMR}$, where C_{BMR} is the cost of a BMR.

The cost of our architectures with N ports and S planes can be easily found to be given by:

$$C^{WS} = N \times C_{laser}(N/S, R_b) + S \times C_{Spl}(N/S) + 2S \times C_{Spl}(S)$$
$$+ 2S^2 \times C_{Dem}(S) + (NS) \times C_{SOA} + S \times C_{EDFA}(N/S), \quad (1.18)$$

$$C^{WR} = N \times C_{laser}(N, R_b) + S \times C_{AWG}(N/S, N/S)$$
$$+ S \times C_{Spl}(S) + S \times C_{EDFA}(N) \quad (1.19)$$

$$C^{WR-ZC} = N \times C_{laser}(N, R_b) + S \times C_{AWG}(N/S, (N/S)^2)$$
$$+ S \times C_{Spl}(S) + S \times C_{EDFA}(N^2/S) \quad (1.20)$$

$$C^{PS} = N \times C_{laser}(N/S, R_b) + (S + N) \times C_{Spl}(S)$$
$$+ (NS) \times C_{SOA} + S^2 \times C_{Spl}(N/S) + S \times C_{EDFA}(N/S). \quad (1.21)$$

We note that transmitters have a significant impact on the cost of all the considered architectures. Moreover, the cost of PS and WS is heavily affected by SOAs, while the cost of the WR architecture is mainly affected by the AWGs. WR-ZC could be more costly than WR; indeed, it requires a larger tunability range, a larger EDFA bandwidth and a larger number of AWGs (to be able to fully exploit the AWG cyclic property).

1.5   Results

1.5.1   Scalability of the aggregate switching bandwidth

We first discuss the feasibility and scalability of the considered optical fabrics. Each architecture was evaluated in terms of its total bandwidth, which depends on R_b, S, and N. Given a set of line bitrates, we evaluated the maximum achievable bandwidth (corresponding to the maximum number of linecards that can be supported). For all architectures, the main limiting effect was observed to be the OSNR at the receiver, i.e., the maximum achievable bandwidth is mainly limited by the optical noise accumulation, and not by receivers' sensitivity.

Although the number of switching planes S in principle can range from 1 to N, for the architecture configurations satisfying the limits given in Eq. (1.6) and Eq. (1.8) (feasible configurations), there is a value of the triple (N, S, R_b) for which the total capacity is maximum. The dependence of the loss from S and N/S in the denominators of Eqs. (1.10)–(1.12) is the main factor to determine the optimal (i.e., leading to maximum OSNR at receivers) value for the number of planes S. The effect of the number of planes for the PS architecture can be observed in Figure 1.8, in which the total switching capacity is plotted vs. S for linecard bitrates equal to 2.5, 10, 40, and 100 Gb/s. The best configurations reach aggregate capacities up to 5 Tb/s. The saw-tooth decreasing behavior for increasing values of S is due to reductions in the number of linecards N when

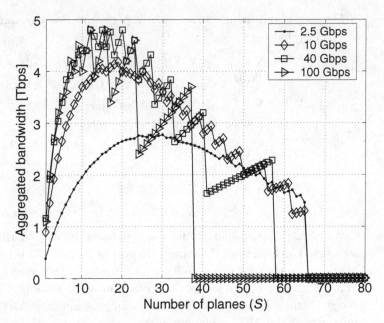

Figure 1.8 Aggregate bandwidth for the PS architecture as a function of the number of planes S for different linecard bitrates.

the combined effects of splitting/coupling losses, crosstalk and OSNR exceed the feasibility constraints.

The nominal transmitter $OSNR_{TX}$, being the earliest noise contribution in the fabric, deeply impacts the optical fabric scalability. Figure 1.9 shows the maximum achievable aggregate bandwidth vs. $OSNR_{TX}$ for different bitrates, with the optimum number of switching planes on each case. Firstly, for low bitrates, the performance of both the WS and the PS architectures is quite independent of the TTx noise, while for higher bitrates, the better the $OSNR_{TX}$, the larger the maximum achievable bandwidth. However, their performance dependency on transmitter noise is very weak; indeed, the number of noise floor sources is equal to N/S. Conversely, WR and WR-ZC performance is deeply affected by the $OSNR_{TX}$. If the transmitter noise is very low, WR and WR-ZC achieve the largest maximum aggregate bandwidth of several tens of Tb/s, but as the $OSNR_{TX}$ become closer to realistic values, performance severely falls. In these architectures the noise floor of all the N sources (instead of N/S as in the case of WS and PS) accumulates as the signal propagates through the optical fabric. Note that up to now, even technologically advanced transmitters show an OSNR of at most 40 dB (this is the value used for Figure 1.8); thus, PS and WS might be the best solutions to implement an optical fabric in the near future.

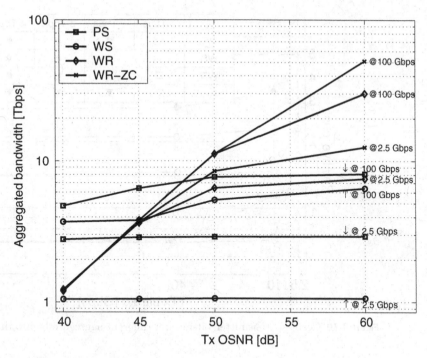

Figure 1.9 Aggregate bandwidth as a function of TTx noise $OSNR_{TX}$ for different architectures and linecard bitrates equal to 2.5 and 100 Gb/s.

1.5.2   CAPEX estimation

Figure 1.10 shows the relative cost of the architectures vs. R_b for different aggregate bandwidths (from 1 Tb/s to 5 Tb/s) for TTx characterized by $OSNR_{TX} = 50$ dB. Different marker shapes identify the different architectures, while different line styles distinguish different aggregate capacities. Given the bit-rate and aggregate bandwidth, cost was calculated by means of the models presented in Section 1.4 for all feasible configurations. Then those showing minimum cost for the same bit-rate appear in the figure. For the PS, WS, and WR architectures, cost significantly decreases as the bit-rate increases. Even though higher bit-rate transmitters (and receivers) are more expensive (remember that we assumed that cost is linearly dependent on the bit-rate), our calculations show that cost savings due to decreasing the number of transmitters, splitters, amplifiers and SOAs dominate over the additional cost of higher bit-rate transmitters (and receivers). The WR fabric costs decrease as lower tunability and lower number of AWGs is needed with higher bit-rate. Consistently with the results shown by Figure 1.9, for transmitters with a realistic OSNR, Figure 1.10 shows that the WR-ZC (especially) and WR can support higher aggregate bandwidths than the PS and WS architectures; indeed, the WR-ZC switching fabric achieves a switching capacity of 8 Tb/s, even though it is generally more expensive.

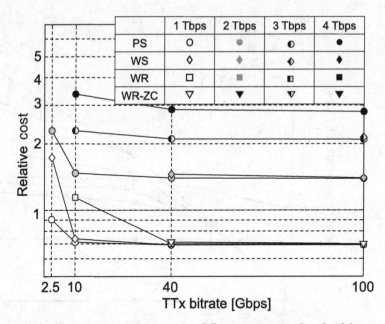

Figure 1.10 Cost vs. transmitter bit-rate for different aggregate bandwidths.

It is interesting to note that costs increase linearly or less than linearly with the aggregate capacity.

We focused the discussion above on relative trends of the curves in Figure 1.10 because the absolute cost values are still quite high: the vertical scale was normalized to one million dollars. We must however remark that our cost analysis considered discrete components as they are commercially available today. It is reasonable to expect that economy of scale effects, technological improvements, and a larger integration of optical components can significantly cut down costs in the near future.

1.6   Conclusions

In this chapter we dealt with physical-layer constraints and realization costs of optical switching fabrics to be used in future high-capacity packet switches. We explored tradeoffs in the use of the wavelength and space dimensions to optimize the design of specific simple switching architectures based on commercially available optical devices. To overcome the current technological limits of laser tunability and amplifier bandwidth, space multiplicity was introduced, partitioning the fabrics in multiple switching planes, thereby enabling wavelength reuse.

Feasibility and scalability studies were conducted to evaluate the maximum aggregate bandwidth of the considered optical fabrics, which were shown to be able to support several Tb/s of aggregate capacity. Moreover, being optical devices (in contrast to electronic ones) far from their intrinsic physical limits, they can still be improved (for instance by reducing the components' noise or by increasing the devices' bandwidth), so that the capacity of the optical fabrics can further increase. Unfortunately, these improvements are today not fostered by a strong market demand.

In addition to the scalability analysis, we performed simple CAPEX estimations for the considered optical fabrics. Even though high costs are still a matter of fact, trends show an almost sub-linear dependence on the aggregate switching capacity, and a limited dependence on the line bit-rate (in particular we found that it is more convenient to employ few high-bitrate linecards than several low-bitrate linecards). Again, several optical devices are still in the first phases of their commercial maturity, and there are still large margins to reduce the costs. We can assume that larger on-chip photonic integration, economy of scale, and volume discount can knock down device costs.

Other advantages of using optical technologies, not discussed in this chapter, are negligible impairments due to the physical distance between boards and racks in the switch, and possibly a reduced power consumption.

In summary, considering the fact that electronic technologies in switching architectures are approaching intrinsic physical limits, and the fact that there is still room to reduce costs of optical fabrics, we think that optical technologies can play an important role in the realization of future generations of high-performance packet switches and routers.

As a final remark, we must note that often there are cultural barriers that prevent an earlier introduction of optical technologies in the switching realm: most device manufacturers have a consolidated expertise in electronic design, and they feel more confident in pushing electronics to extreme limits rather than adopting new technologies (hence requiring new skills) in the design, manufacturing, and customer support chains.

References

[1] R. Ramaswami, and K. N. Sivarajan, *Optical Networks – A Practical Perspective*, second edition, Morgan Kaufman, 2002

[2] Online: `www.cisco.com/en/US/prod/collateral/routers/ps5763/prod_brochure0900aecd800f8118.pdf`

[3] Online: `www.cisco.com/en/US/docs/routers/crs/crs1/mss/16_slot_fc/system_description/reference/guide/msssd1.pdf`

[4] J. M. Finochietto, R. Gaudino, G. A. Gavilanes Castillo, F. Neri, "Can Simple Optical Switching Fabrics Scale to Terabit per Second Switch Capacities?", *IEEE/OSA Journal of Optical Communications and Networking (JOCN)*, vol. 1, no. 3, 2009, B56–B69, DOI: 10.1364/JOCN.1.000B56

[5] J. M. Finochietto, R. Gaudino, G. A. Gavilanes Castillo, F. Neri, "Multiplane Optical Fabrics for Terabit Packet Switches," *ONDM 2008*, Vilanova, Catalonia, Spain, March 2008

[6] D. Cuda, R. Gaudino, G. A. Gavilanes Castillo, *et al.*, "Capacity/Cost Trade-offs in Optical Switching Fabrics for Terabit Packet Switches," *ONDM 2009*, Braunschweig, Germany, February 2009

[7] J. Gripp, M. Duelk, J. E. Simsarian, *et al.*, "Optical Switch Fabrics for Ultra-high-capacity IP Routers," *Journal of Lightwave Technology*, vol. 21, no. 11, 2003, 2839–2850

[8] N. McKeown, "Optics inside Routers," *ECOC 2003*, Rimini, Italy, September 2003

[9] Online: `www.e-photon-one.org/ephotonplus/servlet/photonplus. Generar`

[10] C. Matrakidis, A. Pattavina, S. Sygletos, *et al.*, "New Approaches in Optical Switching in the Network of Excellence e-Photon/ONe," *Optical Network Design and Modeling (ONDM) 2005*, Milan, Italy, February 2005, pp. 133–139

[11] M. Ajmone Marsan, A. Bianco, E. Leonardi, L. Milia, "RPA: A Flexible Scheduling Algorithm for Input Buffered Switches," *IEEE Transactions on Communications*, vol. 47, no. 12, December 1999, 1921–1933

[12] A. Bianco, D. Cuda, J. M. Finochietto, F. Neri, C. Piglione, "Multi-Fasnet Protocol: Short-Term Fairness Control in WDM Slotted MANs," *Proc. IEEE GLOBECOM 2006*, November 2006

[13] A. Bhardwaj, J. Gripp, J. Simsarian, M. Zirngibl, "Demonstration of Stable Wavelength Switching on a Fast Tunable Laser Transmitter," *IEEE Photonics Technology Letters*, vol. 15, no. 7, 2003, 1014–1016

[14] K. A. McGreer, "Arrayed waveguide gratings for wavelength routing," *IEEE Communications Magazine*, vol. 36, no. 12, 1998, 62–68

[15] D. Hay, A. Bianco, F. Neri, "Crosstalk-Preventing Scheduling in AWG-Based Cell Switches," *IEEE GLOBECOM '09*, Optical Networks and Systems Symposium, Honolulu, Hawaii, USA, December 2009

[16] H. Takahashi, K. Oda, H. Toba, "Impact of Crosstalk in an Arrayed-waveguide Multiplexer on $N \times N$ Optical Interconnection," *Journal of Lightwave Technology*, vol. 14, no. 6, 1996, 1097–1105

[17] G. P. Agrawal, *Fiber-Optic Communication Systems*, John Wiley & Sons, 2002

[18] Online: ACCELINK, 100GHz DWDM Module, Product Datasheet, `www.accelink.com`

[19] Online: JDSU, WDM Filter 100 GHz Multi-channel Mux/Demux Module, Product Datasheet, `www.jdsu.com`

[20] Online: ANDevices, $N \times N$ AWG multiplexers and demultiplexers Router Module, Product Datasheet, `www.andevices.com`

[21] E. Sackinger, *Broadband Circuits for Optical Fiber Communication*, John Wiley & Sons, 2005

[22] Online: Alphion, QLight I-Switch Model IS22, Advance Product Information, `www.alphion.com`

[23] Taken online from: `www.go4fiber.com`, Product Datasheets.

2 Broadband access networks: current and future directions

Abu (Sayeem) Reaz, Lei Shi, and Biswanath Mukherjee
University of California-Davis, USA

Abstract: Internet users and their emerging applications require high-data-rate access networks. Today's broadband access technologies – particularly in US – are Digital Subscriber Line (DSL) and Cable Modem (CM). But their limited capacity is insufficient for some emerging services such as IPTV. This is creating the demand for Fiber-to-the-X (FTTX) networks – typically employing Passive Optical Network (PON) – to bring the high capacity of fiber closer to the user. Long-Reach PON can reduce the cost of FTTX by extending the PON coverage using Optical Amplifier and Wavelength-Division-Multiplexing (WDM) technologies. Since Internet users want to be untethered (and also mobile), whenever possible, wireless access technologies also need to be considered. Thus, to exploit the reliability, robustness, and high capacity of optical network and the flexibility, mobility, and cost savings of wireless networks, the Wireless-Optical Broadband Access Network (WOBAN) is proposed. These topics are reviewed in this chapter.

2.1   Introduction

An access network connects its end-users to their immediate service providers and the core network. The growing customer demands for bandwidth-intensive services are accelerating the need to design an efficient "last mile" access network in a cost-effective manner. Traditional "quad-play" applications, which include a bundle of services with voice, video, Internet, and wireless, need to be delivered over the access network to the end-users in a satisfactory and economical way. High-data-rate Internet access, known as *broadband access*, is therefore essential to support today's and emerging application demands.

2.1.1   Current broadband access solutions

The most widely deployed broadband access technologies in the US today are Digital Subscriber Line (DSL) and Cable Modem (CM). DSL uses the same

Next-Generation Internet Architectures and Protocols, ed. Byrav Ramamurthy, George Rouskas, and Krishna M. Sivalingam. Published by Cambridge University Press. © Cambridge University Press 2011.

twisted pair wiring as telephone lines and requires a DSL modem at the customer premises and Digital Subscriber Line Access Multiplexer (DSLAM) in the telecom Central Office (CO). Basic DSL is designed with Integrated Services Data Network (ISDN) compatibility and has 160 kbps symmetric capacity. Asymmetric Digital Subscriber Line (ADSL) is the most widely deployed flavor of DSL and typically provides up to 8 Mbps of downstream bandwidth and 512 kbps of upstream bandwidth. On the other hand, CM uses the Community Antenna Television (CATV) networks which were originally designed to deliver analogue broadcast TV signals to subscriber TV sets. CATV networks provide Internet services by dedicating some Radio Frequency (RF) channels in the coaxial cable for data transmission.

Although the bandwidth provided by these networks is improving significantly compared to the 56 kbps dial-up lines, such low bandwidth has been criticized by customers for a long time, and because of the low bandwidth in both upstream and downstream directions, access networks are the major bottleneck for providing broadband services such as Video-on-Demand (VoD), interactive gaming, and two-way video conferencing to end-users. Although some variations of the traditional DSL, such as very-high-bit-rate DSL (VDSL), can support a much higher bandwidth (up to 50 Mbps of downstream bandwidth in case of VDSL), these technologies also introduce other drawbacks such as short distance between CO and end-users that prevents them from large-scale deployment.

The explosive demand for bandwidth is leading to new access network architectures which are bringing the high capacity of the optical fiber closer to residential homes and small businesses [1]. The FTTX model – Fiber to the Home (FTTH), Fiber to the Curb (FTTC), Fiber to the Premises (FTTP), etc. – offers the potential for unprecedented access bandwidth to end-users (up to 100 Mbps per user). These technologies aim at providing fiber directly to the home, or very near the home, from where technologies such as VDSL or wireless can take over.

2.1.2   Passive Optical Network (PON)

FTTX solutions are mainly based on the Passive Optical Network (PON) [2]. A PON is a point-to-multipoint fiber network in which unpowered passive optical splitters are used to enable a single optical fiber to serve multiple premises, typically 16–32. PONs are designed for local-loop transmission rather than long-distance transmission and they bring the fiber closer to the customer to provide higher speed. PON is a high-performance cost-effective broadband access solution because active elements are only used in end clients while providing high data-rate to end-users because of the vast bandwidth of the fiber. Developments in PON in recent years include Ethernet PON (EPON) [3], ATM PON (APON) [4], Broadband PON (BPON) [5] (which is a standard based on APON), Gigabit PON (GPON) [5] (which is also based on ATM switching), and wavelength-division-multiplexing PON (WDM-PON) [6].

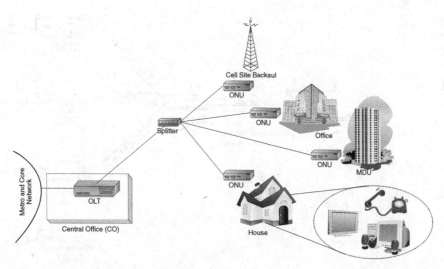

Figure 2.1 Passive Optical Network (PON).

A PON consists of an Optical Line Terminal (OLT) at the service provider's central office and a number of Optical Network Units (ONUs) near end-users, as shown in Figure 2.1.

The typical distance between ONUs and the OLT is 10 km to 20 km. In the downstream direction (from the OLT to ONUs), a PON is a point-to-multipoint network. The OLT typically has the entire downstream bandwidth available to it at all times. In the upstream direction, a PON is a multipoint-to-point network: i.e., multiple ONUs transmit towards one OLT, as shown in Figure 2.2 [3]. The directional properties of a passive splitter/combiner are such that an ONU transmission cannot be detected by other ONUs. However, data streams from different ONUs transmitted simultaneously may still collide. Thus, in the upstream direction (from user to CO), a PON should employ some channel-separation mechanism to avoid data collisions and fairly share the channel capacity and resources.

One possible way of separating the ONUs' upstream channels is to use Wavelength-Division Multiplexing (WDM), in which each ONU operates on a different wavelength. Several alternative solutions based on WDM-PON have been proposed, namely, Wavelength-Routed PON (WRPON) [6] which uses an Arrayed Waveguide Grating (AWG) instead of a wavelength-independent optical splitter/combiner. Time-Division Multiplexing PON (TDM-PON) is another solution for channel separation. In a TDM-PON, simultaneous transmissions from several ONUs will collide when they reach the combiner if no proper control mechanism is employed. To avoid data collisions, each ONU must transmit in its own transmission window (time slot). One of the major advantages of a TDM-PON is that all ONUs can operate on the same wavelength and they can be identical componentwise. Several architectures for TDM-PON-based access networks have been standardized. ITU-T G.983, the first PON standard, described APON and BPON. ITU-T G.984 standardized GPON. IEEE 802.3ah

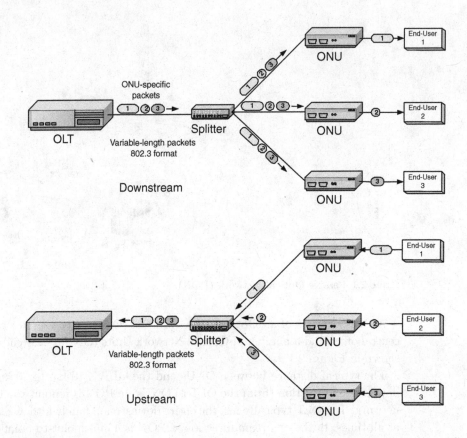

Figure 2.2 Downstream and upstream traffic flows in a typical PON.

standardized EPON or GEPON, and IEEE 802.3aw standardized 10-Gigabit Ethernet PON (10G-EPON), which will also be WDM-PON compatible.

2.1.3   Extending the reach: Long-Reach PON (LR-PON)

To extend the reach of fiber, Long-Reach Passive Optical Network (LR-PON) [7] was proposed as a more cost-effective solution for the broadband optical access network. LR-PON extends the coverage span of PONs mentioned above from the traditional 20 km range to 100 km and beyond by exploiting Optical Amplifier and WDM technologies. A general LR-PON architecture is composed of an extended shared fiber connecting the CO and the local user exchange, and optical splitter connecting users to the shared fiber. Compared with traditional PON, LR-PON consolidates the multiple OLTs and COs where they are located, thus significantly reducing the corresponding Operational Expenditure (OpEx) of the network. By providing extended geographic coverage, LR-PON combines optical access and metro into an integrated system. Thus, cost savings are also achieved by replacing the Synchronous Digital Hierarchy (SDH) with a shared optical fiber. In general, the LR-PON can simplify the network, reducing the

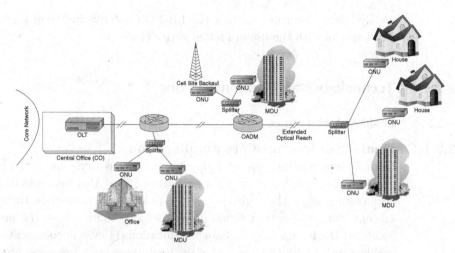

Figure 2.3 Long-Reach PON (LR-PON).

number of equipment interfaces, network elements, and even nodes [7]. Figure 2.3 shows the architecture of an LR-PON.

Although the idea of extending the reach of a PON has been around for quite a while, it is emphasized recently because the optical access is penetrating quickly into residential and small-business markets, and the simplification of telecom networks requires an architecture to combine the metro and access networks. Figure 2.4 shows how LR-PON simplifies the telecom network. The traditional telecom network consists of the access network, the metropolitan-area network, and the backbone network (also called long-haul or core network). However, with the maturing of technologies for long-reach broadband access, the traditional metro network is getting absorbed in access. As a result, the telecom network hierarchy can be simplified with the access head-end, the CO, close to the backbone network. Thus, the network's Capital Expenditure (CapEx) and OpEx can be significantly reduced, due to the need for managing fewer control units.

Note that the term LR-PON is not fully accurate since not all of its components between the OLT and ONUs are passive; but this term has been used quite

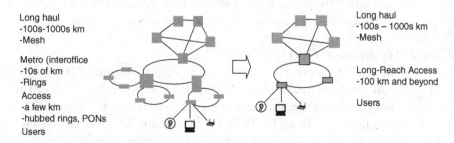

Figure 2.4 LR-PON simplifies the telecom network [8, 9].

widely in the literature to show the LR-PON's derivation from the traditional PON system with the use of limited active elements.

2.2   Technologies and demonstrations

2.2.1   Enabling technologies

2.2.1.1   Dual-stage intermediate amplification

Optical amplifiers introduce Amplified Spontaneous Emission (ASE) [10], a side effect of the optical amplification mechanism. The ASE may affect the system performance since the high split of LR-PON would attenuate the signal significantly and the optical signal power could be quite low at the input of the amplifier. Dual-stage intermediate amplification [11] was introduced to solve this problem. In this scheme, the first stage is composed of a low-noise pre-amplifier, which produces a high Signal-to-Noise Ratio (SNR) by maintaining its ASE at a low level; and the second stage consists of amplifiers to amplify the optical signal with enough power, in order to counter the large attenuation in the fiber between OLT and the local splitter (100 km and beyond).

2.2.1.2   EDFA gain control/SOA

Erbium-Doped Fiber Amplifier (EDFA) [12] is used to boost the intensity of optical signals in a fiber. EDFA features a low noise figure, a high power gain, and a wide working bandwidth, which enable it to be advantageous in an LR-PON employing WDM. But the relatively slow speed in adjusting its gain makes it disadvantageous due to the bursty nature of upstream Time-Division Multiple Access (TDMA) traffic in an LR-PON. Gain control using optical gain clamping or pump power variation is thus introduced. An example could be an auxiliary wavelength (a shorter, additional wavelength used to improve the gain of EDFA) that senses the payload wavelength (the wavelength that carries data) and is adjusted relative to the transmitted upstream packet so that the total input power at the EDFA remains constant. Hence, the gain of the EDFA remains constant for the burst duration. Researchers have also investigated the Semiconductor Optical Amplifier (SOA) [10, 13], which can be adjusted faster and offers the potential for monolithic or hybrid array integration [14] with the other optical components which makes it more cost competitive.

2.2.1.3   Reflective ONU

In order to lower the CapEx and OpEx, a standard PON may choose lower-cost uncooled transmitters in the ONU; however, the uncooled transmitter is temperature dependent, and in turn transmits a wavelength with a possible drift of 20 nm [15]. In an LR-PON which exploits WDM to satisfy the huge amount of traffic, the wavelength drift becomes crucial, especially for components such as optical filters. A possible technology is called Reflective ONU (R-ONU) [16], which generates the upstream signal from the optical carrier feeding from outside (could be the downstream optical carrier or a shared optical source at the local exchange), using a Reflective SOA (RSOA) modulator.

2.2.1.4 Burst-mode receiver

The different ONU OLT distances mean different propagation attenuations for signals from ONUs to the OLT, which in turn may result in varied DC levels of bursty packets from the ONUs at the OLT. A burst-mode receiver is designed for this purpose. The design of a burst-mode receiver at 1 Gbps has been addressed in the context of IEEE 802.3ah EPON networks, and the one at 10 Gbps or 10-G downstream/1-G upstream hybrid is currently discussed in IEEE 802.3av 10-Gbps EPON. Efforts have been made by researchers, such as a new 10 Gbps burst-mode receiver [17] that uses a multi-stage feed-forward architecture to reduce DC offsets, and a high-sensitivity Avalanche Photodiode (APD) burst-mode receiver [18] for 10-Gbps TDM-PON systems.

2.2.2   Demonstrations of LR-PON

There have been several demonstrations of LR-PON. The ACTS-PLANET (Advanced Communication Technologies and Services – Photonic Local Access NETwork) [19] is an EU-funded project. This project investigated possible upgrades of a G.983-like APON system in the aspect of network coverage, splitting factor, number of supported ONUs, and transmission rates. Installed in the first quarter of 2000, the implemented system supports a total of 2048 ONUs and achieves a span of 100 km. British Telecom has demonstrated its Long-Reach PON, which is characterized by a 1024-way split, 100-km reach, and 10-Gbps transmission rate for upstream and downstream directions [20]. The 1024-way split is made up of a cascade of two N:16 and one N:4 splitters in the drop section. The system includes a 90-km feeder section between the OLT and the local exchange, and a 10-km drop section between the local exchange and end-users. The demonstration of a hybrid WDM-TDM LR-PON is reported in [16, 21] by the Photonic System Group of University College Cork, Ireland. This work supports multiple wavelengths, and each wavelength pair (up and down stream) can support a PON segment with a long distance (100 km) and a large split ratio (256 users). Another demonstration is by ETRI, a Korean government-funded research institute, which has developed a hybrid LR-PON solution, called WE-PON (WDM E-PON) [22]. In WE-PON, 16 wavelengths are transmitted on a ring, and they can be added and dropped to local PON segments through the remote node (RN) on a ring. A possible design of the RN includes OADM and optical amplifiers. As the split ratio of the splitter is 1:32, the system can accommodate 512 users.

2.3   Research challenges in LR-PON

2.3.1   Low-cost devices: colorless ONU

In LR-PON development, a WDM transmitter, especially on the subscriber side, is considered to be the most critical component because the subscriber's transmitter should be precisely aligned with an associated WDM channel. Traditional solutions usually resort to lasers having well-defined wavelengths such as Distributed Feedback (DFB) lasers, which, on the other hand, mean it is necessary to select lasers with a right wavelength for each wavelength channel of the LR-PON system. This wavelength selectivity inherent in the source requires network operators to stock spare wavelength sources for each wavelength channel, increasing operation and maintenance cost of the network [23].

One way to ensure that costs are minimized is to use a colorless ONU, which allows the same physical unit to be used in all ONUs on the subscriber side. Several colorless remodulation schemes for ONUs have been proposed, all of which remodulate the lasers from the OLT. For example, the spectrum-sliced Light-Emitting Diode (LED) [24], the wavelength-seeded RSOA [25], and the ASE-injected Fabry-Perot Laser Diode (FP-LD) [26]. Since the spectrum-sliced LED and RSOA suffer from low power and high packaging costs, respectively, researchers have been focusing mainly on the ASE-injected FP-LD, which satisfies the requirement of the low-cost wavelength-selection-free transmitter and could possibly open the possibility of low-cost LR-WDM-PONs without any wavelength alignment.

2.3.2   Resource allocation: DBA with Multi-Thread Polling

As multiple ONUs may share the same upstream channel, Dynamic Bandwidth Allocation (DBA) is necessary among ONUs. Considering the LR-PON's benefits in CapEx and OpEx, as well as its derivation from the traditional PON, the upstream bandwidth allocation is controlled and implemented by the OLT.

Two kinds of bandwidth allocation mechanisms are used in PON: status-reporting mechanism and non-status-reporting mechanism. Although the non-status-reporting mechanism has the advantage of imposing no requirements on an ONU and no need for the control loop between OLT and ONU, there is no way for the OLT to know how best to assign bandwidth across several ONUs that need more bandwidth.

To support the status-reporting mechanism and DBA arbitration in the OLT, the proposed DBA algorithms in LR-PON are based on the Multi-Point Control Protocol (MPCP) specified in the IEEE 802.3ah standard. The proposed DBA algorithms work in conjunction with MPCP. The OLT has to first receive all ONU REPORT messages before it imparts GATE messages to ONUs to notify them about their allocated time slot [27]. As a result, the upstream channel will remain idle between the last packet from the last ONU transmission in a polling

cycle (time duration between two requests) k and the first packet from the first
ONU transmission in polling cycle k+1.

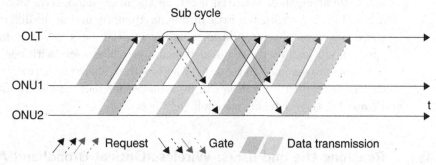

Figure 2.5 An example of Multi-Thread Polling [28].

In order to combat the detrimental effect of the increased Round-Trip Time
(RTT), the work in [28] proposed the Multi-Thread Polling algorithm, in which
several polling processes (threads) are running in parallel, and each of the threads
is compatible with the proposed DBA algorithms in a traditional PON. Figure
2.5 shows an example of Multi-Thread Polling (two threads are shown in the
example) and compares it with traditional DBA algorithms (so-called one-thread
polling with stop). As shown in Figure 2.5, the idle time in one-thread polling
[27] is eliminated because, when ONUs wait for GATE messages from OLT in the
current thread which incurs idle time in one-thread polling, they can transmit
their upstream packets which are scheduled in another thread simultaneously.
This applies to all the ONUs that share the same upstream wavelength.

2.3.3   Traffic management: behavior-aware user assignment

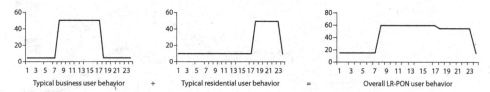

Figure 2.6 High efficiency achieved (right) by assigning a typical business user (left)
and a typical residential user (middle) on a same wavelength. (The x-axis shows the
hour of the day (1–24) and the y-axis shows the bandwidth need at a given time.)

Suppose we have N users and M channels where $N \gg M$ typically. To reduce
cost, users are equipped with fixed transceivers, so User Assignment is a planning
problem to determine which user is served on which channel.

User assignment is a trivial problem if all users behave identically, so we can
allocate to each user the same amount of bandwidth. But this method is inefficient since different users use the network in different ways, especially when

we are considering Long-Reach PON, which has a relatively large span and a lot of users. Business users, for example, require high bandwidth during daytime and little at night. Residential users, on the other hand, have large bandwidth demand in the evening but little during daytime. By noting the different network usage behaviors of typical business and residential users, the work in [29] developed an efficient user-assignment scheme in which users with complementary behaviors are assigned to the same wavelength to share the network resources so that high channel utilization can be achieved for most of the day, as shown in Figure 2.6, thus reducing channels (and cost).

2.4 Reaching the end-users: Wireless-Optical Broadband Access Network (WOBAN)

Combining wireless access technologies with a PON or an LR-PON extends the coverage of broadband access network and gives users "anytime-anywhere" access and mobility. Recent industry research [30] shows that wired and wireless networking technology must be treated as an integrated entity to create a flexible, service-centric network architecture. Hence, a Wireless-Optical Broadband Access Network (WOBAN) [31] captures the best of both worlds: (1) the reliability, robustness, and high capacity of wireline optical communication and (2) the flexibility and cost savings of a wireless network.

2.4.1 WOBAN architecture

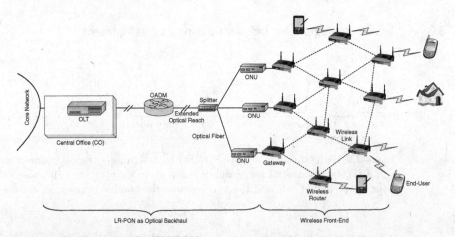

Figure 2.7 Architecture of a WOBAN.

A WOBAN consists of a wireless network at the front-end, and it is supported by an optical network at the back-end (see Figure 2.7) [31]. Its optical backhaul enables it to support high capacity, while its wireless front-end enables its users to have untethered access. This back-end can be a PON or an LR-PON. In a standard PON architecture, ONUs usually serve the end-users. However, for WOBAN, the ONUs connect to wireless Base Stations (BS) for the wireless portion of the WOBAN. The wireless BSs that are directly connected to the ONUs are known as wireless "gateway routers," because they are the gateways of both the optical and the wireless worlds. Besides these gateways, the wireless front-end of a WOBAN consists of other wireless routers/BSs to efficiently manage the network. Thus, the front-end of a WOBAN is essentially a multi-hop wireless mesh network with several wireless routers and a few gateways (to connect to the ONUs and, consequently, to the rest of the Internet through OLTs/CO). The wireless portion of the WOBAN may employ standard technologies such as WiFi or WiMax. Since the ONUs will be located far away from the CO, efficient spectrum reuse can be expected across the BSs with much smaller range but with much higher bandwidth; thus, WOBAN can potentially support a much larger user base with high bandwidth needs [31].

2.4.2   Motivation of WOBAN

The advantages of a WOBAN over wireline optical and wireless networks have made the research and deployment of this type of network more attractive [31].

- End-users in WOBAN are covered by wireless. Hence, a WOBAN is cost effective compared to a wired network. WOBAN architecture (see Figure 2.7) demonstrates that expensive FTTH connectivity is not needed to cover the end-users, because installing and maintaining the fiber all the way to each user could be quite costly. Instead, fiber can reach close to the premises and low-cost wireless solutions take over from there and connect the end-users.
- WOBAN is a flexible broadband access network because its wireless part allows the users inside the WOBAN to seamlessly connect to any wireless BS. This provides "anytime-anywhere" connectivity for the end-users while supporting mobility within WOBAN.
- WOBAN is more robust than the traditional wireline network because of its "self-organization" property. In a traditional PON, if a fiber connecting the splitter to an ONU breaks (see Figure 2.1), that ONU becomes unavailable. Even worse, if a fiber from the OLT to the splitter breaks, all the ONUs (along with the users served by the ONUs) fail. But in a WOBAN, since users have the ability to form a multi-hop mesh topology, the wireless connectivity adapts itself so that users find a neighboring ONU which is alive, communicates with the ONU, and in turn, communicates with another OLT in the CO [32].
- A WOBAN is more robust than a traditional wireless network as well because its users can access the optical backbone using any of the ONUs. This "any-cast" property of WOBAN [33] enables users to maintain connectivity even when a part of the wireless network fails, as long as it can connect to a BS.
- WOBAN is suitable for quick and easy deployment. In many parts of the world, deploying fiber to the end-users may not be feasible because of difficult

terrain and higher cost. WOBAN allows fiber to reach close to user premises using PON or LR-PON. Then, it extends the coverage to the end-users using low-cost, easily deployable wireless solutions.

2.4.3   Research challenges in WOBAN

2.4.3.1   Integrated routing

Wireless BSs of WOBAN collect traffic from end-users and carry them to the optical part of a WOBAN, possibly using multiple hops, but the traffic also experiences delay at each wireless BS. The finite radio capacity at each wireless BS limits the capacity on each outgoing link from a wireless node of a WOBAN. Thus, delay and capacity limitation in the wireless front-end of a WOBAN are major constraints [33].

To gain desirable performance from a WOBAN, an integrated routing for wireless and optical parts of a WOBAN, called Capacity and Delay Aware Routing (CaDAR) [33] is proposed. In CaDAR, each wireless BS advertises the wireless link states (Link-State Advertisement or LSA) periodically. Based on the LSA information, it optimally distributes the radio capacity of a wireless BS among its outgoing links. Then it selects the shortest-delay path through the entire WOBAN to route packets by calculating delays on wireless and optical links. This integrated routing enables CaDAR to support higher load in the network and minimize packet delay.

2.4.3.2   Fault tolerance

Risk-and-Delay Aware Routing Algorithm (RADAR) [32] is proposed to manage fault tolerance for WOBAN. RADAR minimizes packet delay in the wireless front-end of WOBAN and reduces packet loss for multiple failure scenarios: gateway failure, ONU failure, and OLT failure. It finds the shortest-delay path between a wireless BS and a gateway using LSA information of the wireless links and sends packets via this path. It maintains a Risk List (RL) table in each wireless BS. If a failure occurs, RL is updated accordingly and the subsequent packets are rerouted.

2.4.3.3   Directionality

Directionality As Needed (DAN) [34] is a simple connectivity algorithm in a WMN of WOBAN that uses directional antennas. It strikes a good balance between the dual goals of minimum network design cost and minimum interference among different links. Given the locations of wireless BSs along with the connectivity that is desired, DAN provides connectivity using only minimal wired infrastructure. It assumes that each wireless BS is equipped with a single radio capable of beam switching. It determines the beam widths for each node as well as power levels and specific beam index to be used for each link.

2.5   Conclusion

Increasing demands for bandwidth-intense applications by end-users are making broadband access networks essential. FTTH is the only technology that delivers high enough bandwidth reliably to meet the demands of next-generation applications. PON has become a popular low-cost, high-bandwidth solution for broadband access because of its passive infrastructure, point-to-multipoint communication over shared fiber, and fast, non-disruptive provisioning. Enabling technologies, such as efficient amplifiers and receiver and reflective ONUs have made it possible to extend the reach of PON and to create LR-PON. It has the benefits of PON with high cost-effectiveness and reduced CapEx and OpEx, and it connects the access network directly to the core network. LR-PON requires low-cost devices, efficient resource allocation, and intelligent user management to successfully cover a larger area and a higher number of users.

Today's end-users want to access their applications "on-the-go." Extending the reach of optical fiber to the users over wireless using WOBAN gives users mobility and tetherless, flexible access. Integrated routing over optical and wireless links, integrated fault management of both optical and wireless devices and link failures, and efficient resource management of wireless via directionality make WOBAN an excellent choice to deliver broadband access to users.

References

[1] Wegleitner, M. (2007). Maximizing the impact of optical technology, *Keynote Address at the IEEE/OSA Optical Fiber Communication Conference.*

[2] Telcordia (2006).   *Passive Optical Network Testing and Consulting.* www.telcordia.com/services/testing/integrated-access/pon/.

[3] IEEE 802.3ah (2004).   *EPON – Ethernet Passive Optical Network.* www.infocellar.com/networks/new-tech/EPON/EPON.htm.

[4] Yano, M., Yamaguchi, K., and Yamashita, H. (1999). Global optical access systems based on ATM-PON, *Fujitsu Science and Technology Journal,* **35**:1, 56–70.

[5] Nakanishi, K., Otaka, A., and Maeda, Y. (2008). Standardization activities on broadband access systems, *IEICE Transactions on Communication,* **E91B**:8, 2454–2461.

[6] Banerjee, A., Park, Y., Clarke, F., *et al.* (2005). Wavelength-division multiplexed passive optical network (WDM-PON) technologies for broadband access: a review [invited], *OSA Journal of Optical Networking,* 4:11, 737–758.

[7] Song, H., Kim, B., and Mukherjee, B. (2009). Long-reach optical access, in *Broadband Access Networks,* ed. Shami, A., Maier, M., and Assi, C., pp. 219–235. Springer.

[8] Mukherjee, B. (2006). *Optical WDM Networks.* Springer.

[9] Gerstel, O. (2000). Optical networking: a practical perspective, *Tutorial at the IEEE Hot Interconnects 8.*

[10] Keiser, G. (2000). *Optical Fiber Communications.* McGraw-Hill.

[11] Deventer, M., Angelopoulos, J., Binsma, H., *et al.* (1996). Architecture for 100 km 2048 split bidirectional SuperPONs from ACTS-PLANET, *Proceedings of the Society of Photo-Optical Instrumentation Engineers (SPIE)*, **2919**, 245–251.

[12] Encyclopedia of Laser Physics and Technology (2008). *Erbium-doped Fiber Amplifiers.* www.rp-photonics.com/encyclopedia.html.

[13] Suzuki, K., Fukada, Y., Nesset, D., and Davey, R. (2007). Amplified gigabit PON systems [Invited], *OSA Journal of Optical Networking*, **6**:5, 422–433.

[14] Paniccia, M., Morse, M., and Salib, M. (2004). Integrated photonics, in *Silicon Photonics*, ed. Pavesi, L. and Lockwood, D., pp. 51–121. Springer.

[15] Shea, D. and Mitchell, J. (2007). Long-reach optical access technologies, *IEEE Network*, **21**:5, 5–11.

[16] Talli, G. and Townsend, P. (2006). Hybrid DWDM-TDM long-reach PON for next-generation optical access, *IEEE/OSA Journal of Lightwave Technology*, **24**:7, 2827–2834.

[17] Talli, G., Chow, C., Townsend, P., *et al.* (2007). Integrated metro and access network: PIEMAN, *Proceedings of the 12th European Conf. Networks and Opt. Comm.*, 493–500.

[18] Nakanishi, T., Suzuki, K., Fukada, Y., *et al.* (2007). High sensitivity APD burst-mode receiver for 10 Gbit/s TDM-PON system, *IEICE Electronics Express*, **4**:10, 588–592.

[19] Voorde, I., Martin, C., Vandewege, J., and Qiu, X. (2000). The superPON demonstrator: an exploration of possible evolution paths for optical access networks, *IEEE Communication Magazine*, **38**:2, 74–82.

[20] Shea, D. and Mitchell, J. (2007). A 10 Gb/s 1024-way split 100-km long reach optical access network, *IEEE/OSA Journal of Lightwave Technology*, **25**:3, 685–693.

[21] Talli, G. and Townsend, P. (2005). Feasibility demonstration of 100 km reach DWDM SuperPON with upstream bitrates of 2.5 Gb/s and 10 Gb/s, *Proceedings of the IEEE/OSA Optical Fiber Communication Conference*, OFI1.

[22] ETRI (2007). *WDM E-PON (WE-PON).* Working Document.

[23] Shin, D., Jung, D., Shin, H., *et al.* (2005). Hybrid WDM/TDM-PON with wavelength-selection-free transmitters, *IEEE/OSA Journal of Lightwave Technology*, **23**:1, 187–195.

[24] Jung, D., Kim, H., Han, K., and Chung, Y. (2001). Spectrum-sliced bidirectional passive optical network for simultaneous transmission of WDM and digital broadcast video signals, *IEE Electronics Letters*, **37**, 308–309.

[25] Healey, P., Townsend, P., Ford, C., *et al.* (2001). Spectral slicing WDM-PON using wavelength-seeded reflective SOAs, *IEE Electronics Letters*, **37**, 1181–1182.

[26] Kim, H., Kang, S., and Lee, C. (2000). A low-cost WDM source with an ASE injected Fabry–Pérot semiconductor laser, *IEEE Photonics Technology Letters*, **12**:8, 1067–1069.

[27] Kramer, G. (2005). *Ethernet Passive Optical Networks*. McGraw-Hill Professional.

[28] Song, H., Banerjee, A., Kim, B., and Mukherjee, B. (2007). Multi-thread polling: a dynamic bandwidth distribution scheme in long-reach PON, *Proceedings of the IEEE Globecom*, 2450–2454.

[29] Shi, L. and Song, H. (2009). Behavior-aware user-assignment in hybrid PON planning, *Proceedings of the IEEE/OSA Optical Fiber Communication Conference*, JThA72.

[30] Butler Group (2007), *Application Delivery: Creating a Flexible, Service-centric Network Architecture*. www.mindbranch.com/Application-Delivery-Creating-R663-21/.

[31] Sarkar, S., Chowdhury, P., Dixit, S., and Mukherjee, B. (2009). Hybrid wireless-optical broadband access network (WOBAN), in *Broadband Access Networks*, ed. Shami, A., Maier, M., and Assi, C., pp. 321–336. Springer.

[32] Sarkar, S., Mukherjee, B., and Dixit, S. (2007). RADAR: risk-and-delay aware routing algorithm in a hybrid wireless-optical broadband access network (WOBAN), *Proceedings of the IEEE/OSA Optical Fiber Communication Conference*, OThM4.

[33] Reaz, A., Ramamurthi, V., Sarkar, S., *et al.* (2009). CaDAR: an efficient routing algorithm for wireless-optical broadband access network (WOBAN), *IEEE/OSA Journal of Optical Communications and Networking*, 1:5, 392–403.

[34] Ramamurthi, V., Reaz, A., Dixit, S., and Mukherjee, B. (2008). Directionality as needed – achieving connectivity in wireless mesh networks, *Proceedings of the IEEE ICC*, 3055–3059.

3 The optical control plane and a novel unified control plane architecture for IP/WDM networks

Georgios Ellinas[†], Antonis Hadjiantonis[⊕], Ahmad Khalil[‡],
Neophytos Antoniades[‡], and Mohamed A. Ali[‡]

[†]University of Cyprus, Cyprus   [⊕]University of Nicosia, Cyprus   [‡]City University of New York, USA

An effective optical control plane is crucial in the design and deployment of a transport network as it provides the means for intelligently provisioning, restoring, and managing network resources, leading in turn to their more efficient use. This chapter provides an overview of current protocols utilized for the control plane in optical networks and then delves into a new unified control plane architecture for IP-over-WDM networks that manages both routers and optical switches. Provisioning, routing, and signaling protocols for this control model are also presented, together with its benefits, including the support of interdomain routing/signaling and the support of restoration at any granularity.

3.1 Introduction

In the last two decades optical communications have evolved from not only providing transmission capacities to higher transport levels, such as inter-router connectivity in an IP-centric infrastructure, to providing the intelligence required for efficient point-and-click provisioning services, as well as resilience against potential fiber or node failures. This is possible due to the emergence of optical network elements that carry the intelligence required to efficiently manage such networks. Current deployments of wavelength division multiplexed (WDM)-based optical transport networks have met the challenge of accommodating the phenomenal growth of IP data traffic while providing novel services such as rapid provisioning and restoration of very high bandwidth circuits, and bandwidth on demand.

One of the most important considerations of a carrier in designing and deploying its transport network is the design and implementation of an effective control

Next-Generation Internet Architectures and Protocols, ed. Byrav Ramamurthy, George Rouskas, and Krishna M. Sivalingam. Published by Cambridge University Press. © Cambridge University Press 2011.

plane that can be utilized to build a dynamic, scalable, and manageable backbone network, which can in turn support cost-efficient and reliable services to its customers. The optical control plane is responsible for supporting functionalities such as point-and-click provisioning for a connection, protection/restoration against faults, and traffic engineering. These are functionalities that optimize network performance, reduce operational costs, and allow the introduction of novel applications and services, as well as multi-vendor interoperability across different carrier networks.

This chapter provides an overview of current protocols utilized for the control plane in optical networks and then delves into a new unified control plane architecture for IP-over-WDM networks. Section 3.2 presents existing link management, routing, and signaling protocols for the control of optical networks by presenting an overview of the GMPLS framework that supports the control plane functionalities currently present or proposed for optical networks. The reader should note that this is an evolving subject area with new recommendations and proposals continuously being developed by standards bodies such as the Internet Engineering Task Force (IETF), the Optical Interworking Forum (OIF), and the International Telecommunication Union (ITU). Thus, the aim of the first part of this chapter is to provide the basic methodology and ideas utilized for the optical control plane. For a more detailed description of protocols and standards related to the optical control plane, the reader is referred to books exclusively dedicated to this subject [11, 21]. Section 3.3 provides an introduction to IP/WDM interconnection and presents the different control paradigms proposed in the literature for such an architecture. Section 3.4 presents a unified optical layer-based control plane for such an infrastructure that manages both routers and optical switches and explains its advantages over the traditional approaches. Provisioning, routing, and signaling protocols for this control model are also presented in Section 3.4, together with its benefits, such as the support of interdomain routing/signaling and the support of restoration at any granularity. The chapter ends with Section 3.5 that offers some concluding remarks.

3.2   Overview of optical control plane design

For IP networks, distributed management schemes, such as Multi-Protocol Label Switching (MPLS), are used to provide the control plane necessary to ensure automated provisioning, maintain connections, and manage the network resources (including providing Quality of Service (QoS) and Traffic Engineering (TE)). In recent years, industry organizations like the OIF and the IETF have been continuously working on extending the MPLS-framework to support not only devices that perform packet switching, but also those that perform switching in time, wavelength, and space (Generalized-MPLS (GMPLS)) [4, 40]. Thus, GMPLS can now be applied as the control plane for wavelength-routed optical networks. GMPLS includes extensions of signaling and routing protocols developed for MPLS traffic engineering [35], and also supports the new feature of link management.

3.2.1   Link Management Protocol

The Link Management Protocol (LMP) in GMPLS is responsible primarily for neighbor discovery (automated determination of the connectivity in the network) that is subsequently used for up-to-date topology and resource discovery in the network (used for opaque, as well as transparent network architectures) [39]. Other tasks performed by LMP include management of the control channel, link bundling, and link fault isolation. Information that is collected by the neighbor discovery protocol includes the physical properties of a fiber link interconnecting two nodes (length, available bandwidth, etc.), node, port, and link identification parameters, etc. Information exchange concerning these parameters can be achieved via an out-of-band control channel, via in-band signaling, or via a completely separate out-of-band data communications network (DCN). When the network topology is ascertained, this information is either kept at a central location (central network manager) or distributed to the network nodes (distributed network control) to be subsequently used by the routing protocol to determine the routing paths for the various network connections (see Section 3.2.2).

The Link Management Protocol uses the periodic exchange of "Hello" and "Configuration" messages between neighboring nodes so that each node obtains the required information about its neighbors. These messages are also utilized to monitor the health of the communication channel used for the LMP sessions. To minimize the information exchange between neighboring nodes that are linked with a large number of fiber-optic links, a "link bundling" technique is used that bundles together a number of these links that have the same characteristics for routing purposes [36]. This group of links is then called a "TE link" (with a corresponding TE link ID). An additional mechanism, namely, link verification, is then utilized to separate the component links used for different connections. As LMP has information on the physical adjacencies between neighboring nodes, it can also be used to isolate a fault in the case of a network failure (fiber link cut, laser failure, etc.). In the event of a fault that has propagated downstream, a simple "backtracking mechanism" is used in the upstream direction to determine the location of the fault.

3.2.2   GMPLS routing protocol

To implement the GMPLS routing protocol in optical networks, extensions to the routing approaches used for MPLS, such as the Open Shortest Path First protocol with Traffic Engineering extensions (OSPF-TE), are utilized [14, 29, 31, 37, 41, 42, 47]. OSPF is a distributed, link-state shortest path routing algorithm (based on Dijkstra's algorithm [18] which computes a shortest path tree from a node to all other nodes) that uses a table in each node for routing purposes. These tables contain the complete network topology, as well as other link parameters (such as link costs). This information is created by exchanging information on the state of links between the nodes via Link State Advertisements (LSAs) [31, 41, 42]. In these types of networks, information on the links is required in order to successfully route an optical connection. For example, optical transmission impairments and wavelength continuity (for transparent connections) [16, 17, 46], bit-rates and modulation formats (for opaque connections), bandwidth availability, etc., must be taken into account when routing an optical connection. The topology information at each node is modified when a change occurs in the network (i.e., a new node or link is added to the network), or is updated periodically (e.g., every 30 minutes), in order to ensure that the topology information is correct and up-to-date.

OSPF also supports hierarchical routing in the case where the network is divided in multiple areas with a hierarchical structure. If routing takes place across multiple areas, no information, complete information, or summary information can flow between different areas. For example, in the case of summary information distribution, summarized TE LSAs can be distributed to the entire Autonomous System (AS) [14, 35, 47].

3.2.2.1 Extensions of OSPF-TE

While MPLS routing is implemented on a hop-by-hop basis, in optical networks routing is source-based (explicit from the source to the destination node). Furthermore, there are specific requirements needed to route optical connections (impairments, wavelength continuity, link types, etc.) and the number of links between different nodes may be quite large. Thus, several extensions to the OSPF-TE utilized in MPLS are required for the case of implementing a GMPLS routing protocol in optical networks. The most important extensions to OSPF-TE are as follows:

- **Dissemination of link state information**: Additional information is advertised utilizing opaque LSAs that is either optical network specific or is needed for protection purposes [37]. Such information includes a link's
  - Protection Type
  - Encoding Type
  - Bandwidth Parameter
  - Cost Metric
  - Interface Switching Capability Descriptor
  - Shared Risk Link Group (SRLG).
- **Link bundling**: As previously discussed, link bundling is a technique used to combine several parallel links having the same properties for purposes of routing, into a single logical group, called a TE link [36]. When link bundling is used, explicit routing takes into account only the TE links and not the individual links in the bundle. The specific link used to route the connection is only decided locally during the signaling process.

- **Nested label switched paths (LSPs)**: A hierarchy in the LSPs is created by introducing "optical LSPs" that are groups of LSPs that have the same source and destination nodes [6, 37, 38]. A number of LSPs (with different bandwidth values) can now use this optical LSP provided that the bandwidth of the optical LSP can support all of them.

3.2.3   GMPLS signaling protocol

Two types of signaling are in place in optical networks: signaling between the client and the transport network (that takes place at the User–Network Interface (UNI) [48]), and signaling between intermediate network nodes (that takes place at the Network–Network Interface (NNI)). Signaling at the UNI is used so that the clients can request connections across the transport network, specifying such parameters as the bandwidth of the connection, the Class of Service (CoS) requirements, and the protection type for the connection. After a route is determined, signaling is required to establish, maintain and teardown a connection. Furthermore, in the event of a fault, signaling is also employed to restore connections. An enhanced Resource Reservation Protocol with Traffic Engineering extensions (RSVP-TE) is a possible signaling protocol that can be used for optical networks [8].

3.2.3.1   Extensions of RSVP-TE for GMPLS

RSVP is a signaling protocol that utilizes the path computed by the routing protocol (e.g., OSPF) to reserve the necessary network resources for the establishment of a session (supports both point-to-point and multicast traffic in IP networks with QoS requirements) [12]. In RSVP, a *Path* message, containing information on the traffic characteristics of the session, is sent downstream from the source to the destination and is being processed by each intermediate node. The destination node then sends back a *Resv* message along the path, which allocates resources on the downstream link. When the reservation is complete, data can flow from source-to-destination adhering to the QoS requirements specified. RSVP also utilizes other message types that are used for error notification (*PathErr* and *ResvErr*), for connection establishment confirmation purposes (*ResvConf*), and for deleting reservations (*PathTear* and *ResvTear*).

RSVP with traffic engineering extensions (RSVP-TE) supports the establishment and management of LSP tunnels (including specifying an explicit route and specifying traffic characteristics and attributes of the LSP). It also supports fault detection capabilities by introducing a "Hello" protocol between adjacent label switched routers (LSRs) [5]. When RSVP is extended for GMPLS support, it is adapted for circuit-switched rather than packet-switched connections and it accommodates the independence between the control and data planes. To this end, new label formats are defined in RSVP-TE in order to support a variety of switching and multiplexing types (such as wavelength switching, waveband switching, fiber switching, etc.) [7, 9]. Thus, several new objects are defined in GMPLS RSVP-TE, such as (a) the Generalized Label Request object that includes LSP encoding type, switching type, generalized protocol ID (the type of payload), source and destination endpoints, and connection bandwidth, (b) the Upstream Label object that is used for bi-directional connections (not supported in MPLS RSVP-TE), and (c) the Interface Identification object that identifies the data link on which labels are being assigned. This is essential for optical networks where control and data planes are separate.

3.2.3.2   Signaling for lightpath establishment

As in MPLS networks, *Path* and *Resv* are again used to establish lightpaths in optical networks utilizing GMPLS RSVP-TE [30]. The *Path* message now carries extra information concerning the connection to be provisioned (LSP tunnel information, explicit route information, etc.) and is again forwarded downstream from the source to the destination node, and is being processed at intermediate nodes. When the message reaches the destination node, a *Resv* message is created that is forwarded upstream. At each intermediate node processing the *Resv* message the cross-connects are now set in order to establish a bi-directional connection. Note that, in contrast to signaling in IP networks, in optical networks the DCN does not necessarily use the data transfer links.

3.2.3.3   Signaling for protection/restoration

In networks where protection is offered, GMPLS signaling can be used to provision secondary protection paths during the connection provisioning phase. In the case of reactive protection (in which case alternative paths are precomputed anticipating a failure), when a failure occurs it is detected by the source (with the help of a Notify message that contains the failed connection information) and GMPLS RSVP-TE is used to activate the precomputed protection path in the same way that a regular connection is provisioned in an optical network [7].

3.3   IP-over-WDM networking architecture

The phenomenal growth in Internet traffic along with the abundance of transmission capacity offered by WDM has signaled the beginning of a new networking paradigm where the vision of integrating data and optical networking seems to be indispensable. A simplified, IP/WDM two-tiered networking architecture combining the strengths of both IP and optical transport in which IP can be implemented directly over WDM, bypassing all of the intermediate layer technologies, is the key to realizing such a vision [1, 22, 44, 45]. Advances in WDM optical networking technology and of data-centric equipment, such as IP/MPLS routers, have further increased the attention being paid to the topic of how these layers will be optimally interacting.

The notion of supporting "data directly over optics" has been fueled by the promise that elimination of unnecessary network layers will lead to a vast reduction in the cost and complexity of the network. This has led to the migration from the multilayer networking architectures (e.g., IP/ATM/SONET/WDM) to the two-layer IP/WDM networking architecture, where high-performance routers are interfaced directly to the optical transport network (OTN), thus diminishing the role of Synchronous Digital Hierarchy/Synchronous Optical Network (SDH/SONET). Even though some of the layers are eliminated, the important functionalities provided by the intermediate layers (traffic engineering in ATM, multiplexing and fast restoration in SONET) must be retained in future IP/WDM networks. This can be accomplished in one of three ways: (a) by distributing the functionalities between the IP and optical layer [51]; (b) by moving them down to the optical layer; and (c) by moving them up to the IP layer [10, 23, 50].

Several architectural options have been proposed on how the client layer (IP routers) must interact with the optical layer to achieve end-to-end connectivity. Of these, the predominant are the overlay, the peer-to-peer, and the augmented models [4, 13, 40, 45].

3.3.1   The overlay model

The overlay model is the simplest interaction model between client and transport layers that treats the two layers (optical and IP layers) completely separately. The client routers request high-bandwidth connections from the optical network, via some User-to-Network Interface (UNI), and the OTN provides lightpath services to the client IP/MPLS layer. The collection of lightpaths in the optical layer therefore defines the topology of the virtual network interconnecting IP/MPLS routers and the routers have no knowledge of the optical network topology or resources. The overlay model is best suited for multidomain networks where different domains are under different administrative control and there is no exchange of topology and resource information between the domains. In this case, the need to maintain topology and control isolation between the optical transport and the client layers is dictated by the network infrastructure itself. Even though this model is simple, scalable, and most suited for current telecommunications network infrastructures, its simplicity also results in inefficient use of network resources as there is no exchange of state and control information between the layers.

3.3.2   The peer and augmented models

In the peer model the network elements in the transport and client layers (i.e., optical cross-connects (OXCs) and IP/MPLS routers) act as peers using a unified control plane to establish paths that could traverse any number of routers and OXCs with complete knowledge of network resources. Thus, from a routing and signaling point of view, there is no distinction between the UNI, NNI, and router-to-router interface; all network elements are direct peers and fully aware of the network topology and resources. Using an enhanced Interior Gateway Protocol (IGP) such as OSPF supported by GMPLS (described in Section 3.2.2), the edge routers can collect the resource usage information at both layers. The peer model supports an integrated routing approach where the combined knowledge of resource and topology information at both layers is taken into account [15]. Compared to the overlay model, the peer model is more efficient in terms of network resource usage and it improves the coordination and management of failures among network elements with different technologies. However, the peer model is not as scalable as the overlay model, as the network elements have to process a large amount of control information. This model is also not as practical for near-term deployment, as it is highly improbable that service providers of the transport and service infrastructures will grant each other full access to their topology and resources, as required for this model implementation.

Finally, the augmented approach combines the peer and the overlay models with some of the client network elements being peers to the transport network elements while others are isolated and communicate via the UNI.

3.4   A new approach to optical control plane design: an optical layer-based unified control plane architecture

Ideally, the control plane functionalities should be integrated and implemented at one and only one layer. While there is a school of thought that favors moving the network intelligence up to the IP layer (favoring the intelligence of routers over optical switches) [22, 23, 50, 51], the model described in this section demonstrates that moving the networking functionality and intelligence down to the optical layer (favoring the intelligence of optical switches over routers), is more compelling in terms of simplicity, scalability, overall cost savings, and the feasibility for near-term deployment. Specifically, this model utilizes an optical layer-based control plane that manages the network elements in both client and transport layers (analogous to the peer model), while still retaining the complete separation between the layers (analogous to the overlay model). This is to say that the unified model retains the advantages of these two models while avoiding their limitations [32, 33].

Under this architecture, the optical core can be thought of as an AS whose members (OXC controllers) are hidden completely from the outside domains. In other words, in this architecture, both the logical and physical layers belong to a single administrative domain and all of the networking intelligence belongs to the optical layer. In this model, GMPLS can support the unified control plane to provide full-lambda and sub-lambda routing, signaling, and survivability functionalities. Utilizing this unified control model, all network resources are visible and thus more efficient resource utilization can be achieved in provisioning, restoring, and managing connections. This is because the model supports an integrated routing/signaling approach where the combined knowledge of resource and topology information at both the IP and optical layers are taken into account (see Sections 3.4.2.1 and 3.4.2.2). Adapting this unified interconnection model also opens up new avenues for implementing several significant novel applications (see Section 3.4.2.3) that can drastically enhance and transform the vision of the next generation optical Internet such as:

- The network is now capable of being managed end-to-end (from the access network through the metro and core networks to another access network), across multiple ASs.
- The physical layer can restore both link/OXC and router failures at any granularity (sub-lambdas and/or full lambdas).
- Optical Ethernet (GigE/WDM) network infrastructures where native Ethernet frames are mapped directly over WDM can now be developed, by combining the simplicity and cost effectiveness of Ethernet technology with the ultimate intelligence of the WDM-based optical transport layer.

3.4.1   Node architecture for the unified control plane

Figure 3.1 depicts the optical node architecture for the unified control plane design. As it can be seen, this architecture is composed of three components. A backbone IP/MPLS router used as an Electronic Cross-Connections (EXC), shown as the sub-wavelength (sub-λ) switch, an OXC switch (the λ optical switch), and an IP/MPLS-aware, non-traffic bearing OXC controller module. The backbone IP/MPLS router is attached to the optical switch and can generate and terminate the traffic to/from a lightpath, while the OXC performs the wavelength switching function. An incoming wavelength-multiplexed signal on a fiber is first demultiplexed and then, based on its wavelength channel, is either optically switched (i.e. continue riding the same lightpath if this is not terminated at this optical node), or dropped to the backbone IP/MPLS switch for electronic processing. From there, the signal can be either switched and added to a new lightpath (in the case of multi-hop logical routing), or dropped to the client network (i.e., to the local IP/MPLS router, operating in the client layer) if this is the egress optical node.

The main intelligence component of this model is the OXC controller. It is responsible for creating, maintaining, and updating both the physical and logical connectivity and with it, the optical node is capable of provisioning on-demand lightpaths (full wavelength channels), as well as low-speed (sub-lambda) connection requests. The client IP/MPLS router's responsibility is, then, simply to request a service from the OTN via a predefined UNI that will encompass a Service Level Agreement (SLA) to govern the specifics of the service requests. Note that this is in contrast to the peer model implementation, where it is the traffic-bearing edge router that performs such functions. Supported by GMPLS, with an IGP such as OSPF (suitably enhanced to support WDM networks) running among the OXCs, the resource usage information at both layers can be collected

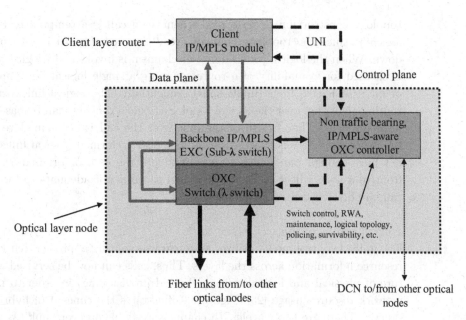

Figure 3.1 Node architecture of the unified control plane model.

by the OXC controllers and the OXC controllers can communicate with each other over a DCN.

3.4.2   Optical layer-based provisioning

In this model, the OTN is responsible for providing this service, but in whichever way it deems optimum. For instance, it can (a) open up one or more lightpaths to service the new traffic, (b) use one or more existing lightpaths to multiplex the new traffic on, or (c) use a combination of existing lightpath(s), together with setting up new ones to serve the new traffic (hybrid provisioning).

Physical provisioning

Physical provisioning refers to solving the Routing and Wavelength Assignment (RWA) problem. As the name implies, this is made up of two sub-problems, which deal with path discovery (routing), and assigning wavelength channel(s) along its fiber links. Many ways of solving the RWA problems have been proposed including treating the sub-problems separately, or trying to solve them simultaneously. When RWA is attempted to be solved jointly, a dynamic routing and wavelength assignment algorithm can be used by means of dynamic routing over multi-layered graphs where each layer represents a wavelength.

Logical provisioning

Logical provisioning is achieved by considering established lightpaths as directional logical links that comprise the logical (virtual) topology. The logical topology construction is performed whenever a call is attempted to be served logically, since the topology changes every time a lightpath is set up or torn down. When the logical provisioning mechanism is invoked, the logical topology is checked for availability of a route spanning a single logical link (single-hop) or multiple logical links (multi-hops) sequentially. The logical link cost for the routing algorithm over the logical topology (logical routing) can be based on the normalized used bandwidth of the link after the call is accommodated. When the logical topology is checked, *pruning* is also performed (logical links that do not have enough bandwidth to accommodate the call, or originate/terminate from/to a node whose IP module residual speed is not adequate to forward the call, are deleted from the topology).

Hybrid provisioning

The unified control model utilizes an integrated routing approach that combines resource information across the layers. Thus, calls can now be serviced as a mixture of physical and logical routing (hybrid provisioning) in order to minimize network resource usage [25]. Figure 3.2 illustrates the concept of hybrid provisioning. There are two wavelength channels available on every link (λ_1 and λ_2), and every call is a sub-lambda request. Assuming that call 1 from node A to node E arrives at the network first, the lightpath between nodes A and E is set up on wavelength λ_1 (by invoking RWA) to service it. Then, call 2 arrives at the network requesting service from node B to node C. Since there is no logical connectivity (previously established lightpath) from node B to node C, this call can only be serviced by the RWA on a new lightpath (on wavelength λ_2, as shown in the figure). Suppose now that call 3 arrives at the network and requests service from node A to node F. Since the link between nodes B and C is completely utilized at the wavelength level, and no logical connectivity exists between nodes A and F, traditionally this call would have been blocked. Under a hybrid provisioning approach, however, the optical core is aware of both the logical and physical topologies, and can thus utilize the already established lightpath between nodes A and E and simply try to complete the remaining portion from node E to node F on wavelength λ_2 by means of the RWA.

Note that the concept of hybrid provisioning was implicitly introduced in [49], where the authors proposed an integrated model to groom the client traffic onto a dynamic logical topology over optical networks. Another integrated routing approach (based on the max-flow, min-cut theorem) was proposed in [34], in which, again, the concept of the hybrid provisioning scheme was also implicitly introduced. Both these approaches suffer from the well-known scalability problem due to the significant amount of state and control information that has to be exchanged between the IP and optical layers. This has forced the authors of [34] to consider only a limited number of network nodes as potential ingress-egress pairs, and the authors of [49] to also consider fewer number of network nodes by introducing the concept of Collaboration Groups (CGs). (The CG of a client node is defined as a list of client nodes located within a

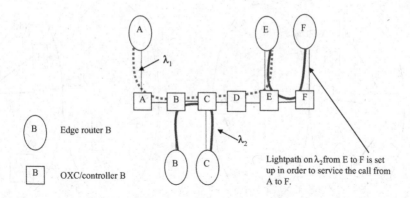

Figure 3.2 Concept of hybrid provisioning.

certain number of physical links.) The integrated routing approach presented in Section 3.4.2.1 below, however, requires no exchange of information between the boundaries of the two layers except for that of the simple UNI. Thus, it shifts most of the intelligence and burden from the IP layer (traffic-bearing edge routers) to the optical layer (IP/MPLS-aware, non-traffic bearing OXC controllers). This renders the unified optical layer-based model simpler, less expensive, and most importantly partially alleviates the scalability problem. Several algorithms can be used to implement the hybrid provisioning scheme. Three of these algorithms are described below [25]. The performance analysis of these techniques is also presented in this section.

(A) Shortest Path Exhaustive Search (SPES) algorithm

The goal of the hybrid provisioning approach is to find an intermediate node (like node E in Figure 3.2 above) that splits the source-destination path into a logical segment (obtained from the existing logical topology) and a physical segment (to be created using RWA). The SPES algorithm is a greedy algorithm that finds the shortest path (minimum number of hops) between the source (S) and destination node (D) and then for every intermediate node (I) on that path it tries to find lightpaths from S to I or from I to D that have adequate residual bandwidth to multiplex the new connection request on them. If the search is successful, it multiplexes the new connection on the found lightpath and for the remainder of the path it tries to set up a new lightpath (RWA – physical provisioning). If the RWA is successful, the algorithm stores the hybrid path found and repeats this process until all intermediate nodes have been examined. Finally, the algorithm chooses the hybrid path that has the smallest hop-count for the new lightpath to be established.

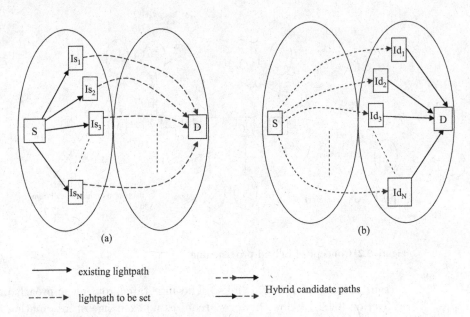

Figure 3.3 NETWES topology partitioning. (From [25], Figure 2. Copyright 2004 IEEE. Used by permission of The Institute of Electrical and Electronics Engineers Inc.)

(B) Network-Wide Exhaustive Search (NETWES) algorithm

This algorithm is a generalization of SPES that searches over the entire network for intermediate nodes that are associated with lightpaths that can be used to multiplex the new connection request on them. It is easier to visualize this by partitioning the network into two regions: one containing nodes that node S has an existing lightpath to (excluding node D, even if there exists a lightpath to node D), and one with the remaining nodes (including node D). This is shown in Figure 3.3(a). The algorithm stores the hybrid paths found for the nodes in the first partition and repeats this process for a second partition where the candidate intermediate nodes are the ones that have existing paths to node D, as illustrated in Figure 3.3(b). Finally, the algorithm chooses the hybrid path, over all stored hybrid paths, that has the smallest hop-count for the new lightpath to be established.

(C) First-Fit on Shortest Path (FFSP) algorithm

In addition to NETWES and SPES, another option called First-Fit on Shortest Path (FFSP) can be used. This is an algorithm similar to SPES and NETWES that terminates when the first successful hybrid path is found on the shortest path between S and D. Thus, in this case there is no forced optimization.

Performance evaluation

Some results on the performance of the unified control plane model, in terms of hybrid provisioning, are presented in this section. A mesh-based NSF network

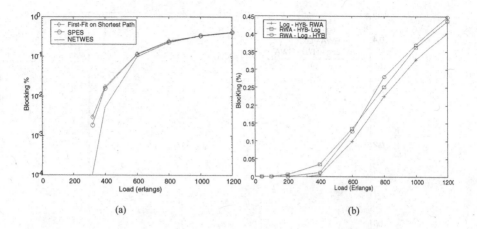

(a)   (b)

Figure 3.4 (a) Performance evaluation of the NETWES-SPES-FFSP algorithms. (b) Effect of order changing for sequential provisioning. (From [25], Figures 5 and 3 respectively. Copyright 2004 IEEE. Used by permission of The Institute of Electrical and Electronics Engineers Inc.)

consisting of 14 nodes and 21 bi-directional links (with 4 fibers (2 in each direction) and 4 wavelengths per fiber) was used for the simulations, together with a dynamic traffic model in which call requests arrive at each node according to a Poisson process and the session holding times are exponentially distributed. The wavelength channel capacity was OC-48 ($\approx$ 2.5 Gb/s) and the sub-lambda requests had bit-rate demands that were normally distributed around 400 Mbps with a standard deviation of 200 Mbps, in multiples of 50 Mbps. Sequential provisioning was the approach used to service a given request. For instance, LOG-HYB-RWA means that the logical search is performed first and if it fails, the hybrid algorithm is invoked, and if that fails the RWA algorithm is invoked. A connection request (call) is blocked if all three approaches fail sequentially.

Figure 3.4(a) compares the performance of the FFSP, NETWES, and SPES algorithms for the hybrid approach [25]. It is clear that the network-wide search (NETWES) yields better results, as it optimizes the choice of the intermediate node based on exhaustive search. Figure 3.4(b) presents the effect of changing the sequential search order for provisioning a call [25]. As it can be seen from the figure, the search order LOG-HYB-RWA exhibits the best performance. For figure clarity, only the best three orders out of the possible six are illustrated. From these it is clear that it is important to have the hybrid approach before the RWA, for it enriches the logical topology connectivity.

Figure 3.5 compares the performance of the conventional sequential search (LOG-RWA or RWA-LOG) [25] with a search that includes hybrid provisioning (LOG-HYB-RWA). These results demonstrate that the presence of the hybrid approach improve significantly the performance of the network, thus further

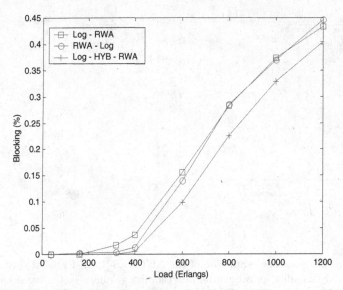

Figure 3.5 Effect of introducing hybrid provisioning. (From [25], Figure 4. Copyright 2004 IEEE. Used by permission of The Institute of Electrical and Electronics Engineers Inc.)

validating the need for the unified control plane model that can support hybrid provisioning.

3.4.2.1 The Integrated Routing algorithm

An integrated dynamic routing algorithm is required in order to facilitate the hybrid provisioning described in the previous section. This algorithm incorporates information about the network topology and resources for both the optical and IP layers that are kept at the OXC controllers. Unlike the peer model and other integrated approaches [2, 15], where a significant amount of information is exchanged between the layers, in this integrated routing approach only the information exchange required by the simple UNI is needed, thus making this unified model simpler, and alleviating the scalability problem associated with the peer model. An integrated directed layered-graph is used to model the network, and each layer corresponds to a wavelength channel [34]. A lightpath on a specific wavelength from source to destination is modeled by a "cut-through path" that connects these nodes (this is a logical link in the IP layer and all other layers must also be aware of this lightpath). The physical links corresponding to this path, on the layer corresponding to the wavelength used, are then removed from the graph. Figure 3.6 shows an example of a four-node network with two wavelengths per fiber (thus a two-layer model is created). If a call request arrives for a connection from node A to node C requiring m units of bandwidth ($m < W$; W is the full wavelength capacity) it is established using λ_1 on path A–B–C. These links are then removed from the graph and a logical link is created for

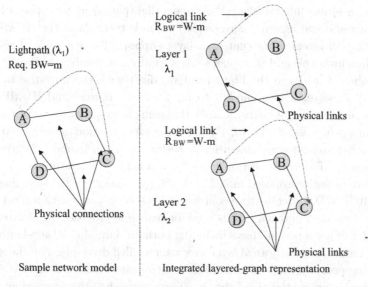

Figure 3.6 Integrated layered-graph representation. (From [32], Figure 2. Copyright 2003 IEEE. Used by permission of The Institute of Electrical and Electronics Engineers Inc.)

this layer between nodes A and C having (remaining) bandwidth $(W - m)$. In this algorithm, the logical link is deleted and the corresponding physical links are reinstated when the logical link is back to full capacity (due to disconnection events).

Routing in this integrated graph can be implemented using any routing algorithm (e.g., Dijkstra's shortest path algorithm) and the routing path can traverse both physical and logical links. The link costs can be a function of a number of metrics such as number of hops, residual (available) bandwidth capacities, etc. In general, the total cost of a path, C_{path}, will be the combined total cost of the logical and the physical links. Different ways for assigning the link costs will result in different routing results in the integrated graph. As an example, if the routing is performed so as to minimize the residual bandwidth, physical links will always be preferred over logical links as they offer the maximum available bandwidth. The goal of the cost metrics should be to obtain the best performance results, without constraining the order of the search (use of physical versus logical links) [32].

3.4.2.2   Integrated signaling component

Provisioning of connections requires algorithms for path selection (Section 3.4.2.1), and signaling mechanisms to request and establish connectivity within the network along a chosen path. Most of the MPLS/GMPLS-based routing and signaling algorithms and protocols which have been reported by standards bodies, as well as by research communities, were developed to provision either full wavelength channels (lightpaths) at the optical layer only (find a route and assign a wavelength) or packets/LSPs at the IP/MPLS layer only (logical layer). Note that each layer supports its own suite of routing and signaling protocols and that each suite is totally independent of the other. While a number of works in the literature have dealt with the initiative of a unified control plane (e.g., peer model) to integrate the optical and IP/MPLS layers into a single administrative domain that runs a single instance of routing and signaling protocols [32, 33, 40, 45, 52], the implementation of integrated signaling protocols that can simultaneously provision both full-lambda (lightpaths) and sub-lambda (LSPs) connection requests at a single domain in a unified manner was only recently proposed in [26] and [28] ([28] describes the signaling protocol in a GigE/WDM architecture context). This section deals with a short overview of the signaling approach that is used for real-time provisioning of diverse traffic granularity (on a per-call basis including both full-lambda and sub-lambda traffic flows) entirely on the optical layer's terms. Detailed description on the integrated signaling protocol used for the unified control plane can be found in [27, 28].

The main characteristic of the signaling approach is the prepending of n electronic labels in a label stack to each packet of a sub-lambda flow, where each label in the stack corresponds to a given lightpath, provided that the flow is to be routed over n lightpaths from source to destination. The label stack is organized in a last-in/first-out order; forwarding decisions are based solely on the top label in the stack. The electronic label has meaning only to the IP/MPLS modules that perform packet-by-packet switching based on it (the data plane). Since the OXC controllers handle the entire control plane, it is their responsibility to perform label assignment and label distribution by executing the appropriate protocols (the control plane). Signaling to accommodate calls is achieved by registering an electronic label at the source and destination IP/MPLS modules for every successfully set-up lightpath. This is in addition to the "physical" label (wavelength) that bears meaning only to the OXC controllers and links input ports to output ports based on the wavelength (i.e., $<$[input port]$_i$ on λ_j, [output port]$_k$ on $\lambda_j>$).

A. Signaling for pure physical provisioning

This section describes the signaling mechanism for a connection that will be served using one lightpath that is to be set up for this connection. When a call request is received through UNI at the OXC controller, the provisioning algorithm (RWA) is run and a path is returned. Then, GMPLS-like signaling algorithms can be used (forward-probing, backward reserving schemes to conform with GMPLS-based signaling standards) to set up the lightpath (a lightpath is identified by a sequence of physical links and an available wavelength on them). Intermediate OXC switches that the path traverses configure their input and output port connections (based on the chosen wavelength) and the controllers assign a unique electronic label to the lightpath, which they communicate to the ingress and egress IP/MPLS modules. The ingress module then attaches the electronic label to each packet in the flow. At the egress side, the IP/MPLS module strips off this electronic label and, since there are no other electronic labels in the stack, it drops the packets to the local network.

B. Signaling for pure logical provisioning

This section describes the signaling mechanism for a connection that will be served using time multiplexing on one or more existing lightpaths. Since existing lightpaths already have electronic labels communicated to their ingress and egress modules, once the controllers decide that the selected route will be using one or more existing lightpaths, then simple reservation along these lightpaths is required. If the reservation is successful, the source controller instructs its module to stack the n electronic labels to every packet of the flow, which correspond to the n lightpaths this call will use, beginning with the last lightpath label. An intermediate module strips off the "outmost" electronic label it receives; the next label in the stack (which corresponds to a lightpath whose ingress module is this intermediate module) instructs it to add the traffic to the next lightpath. This process continues up to the egress node, where the IP/MPLS module strips off the last electronic label and then drops the packets to the local network.

C. Signaling for hybrid provisioning

The signaling for the portions of the returned new lightpath that need to be set up is performed as in A above, where each new lightpath is registering an electronic label to its ingress and egress modules. The existing portions of the path need mere reservation for the bandwidth along them. Then, as in B, the ingress node of the call stacks the electronic labels, beginning with the last lightpath label, and intermediate modules strip off these labels one by one until the destination strips off the last electronic label and then drops the packets to the local network.

D. An illustrative example

Consider the simple network of Figure 3.7 with a call request from node B to node D (call 1). The controllers, after performing the provisioning algorithm, return an RWA path through node C on wavelength λ_1 (pure physical provisioning). The intermediate OXC switch at node C is then configured and an electronic label (called l_1) is communicated to the modules B and D. Module B then needs to attach the l_1 label to all the packets that are destined to node D and are to use the λ_1 lightpath (note that a new flow, even if it is destined to the same destination, will require a fresh UNI request to the OXC controller).

Now suppose that as the λ_1 lightpath has traffic on it, a new request arrives at node A and requests (through UNI) service to node D (call 2). The controllers, after performing the provisioning algorithm, decide that this new request is best served on the hybrid path $(A–B–C–D)$. This necessitates setting up a new $A–B$ lightpath on wavelength λ_2 and reusing the existing λ_1 lightpath on $B–C–D$ (hybrid provisioning), provided that the λ_1 lightpath has enough

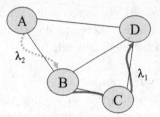

Figure 3.7 A test network to illustrate the concept of the sub-lambda signaling scheme.

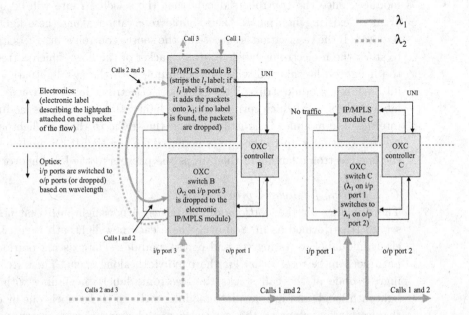

Figure 3.8 Processing of the sub-lambda signaling at nodes B and C.

residual bandwidth to accommodate the call 2 request. Then, the λ_2 lightpath is set up and the l_2 electronic label is communicated to its ingress and egress modules (namely, module A and module B, respectively). Furthermore, controller A instructs module A to stack labels l_1 and l_2 to all the packets destined to node D. Then, intermediate module B strips off the l_2 label and reads the l_1 label, which instructs it automatically to add the traffic to the outgoing lightpath associated with the l_1 label (λ_1 lightpath).

Last, a call arrives at the network and requests service from node A to node B (call 3). The controllers, after performing the provisioning algorithm (and checking for enough residual bandwidth), return a logical path that reuses the λ_2 lightpath from node A to node B (pure logical provisioning). After reservation on the λ_2 lightpath, controller A instructs module A to attach only the l_2 label for this new traffic. At node B, this label is stripped, and, since no other electronic labels are read, the traffic is dropped. This process is depicted at nodes C and B in Figure 3.8. In Figure 3.9, the above scenario is depicted at the packet level.

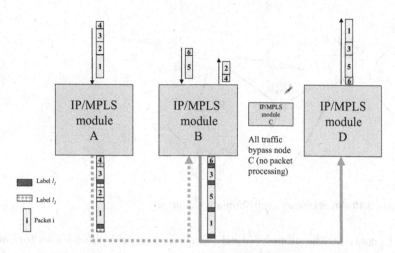

Figure 3.9 Processing the sub-lambda signaling at the packet level (IP/MPLS module).

In this figure, packets 1 and 3 belong to call 2, packets 2 and 4 belong to call 3, and packets 5 and 6 belong to call 1.

3.4.2.3   Benefits of the unified optical layer-based control model

A. End-to-end on-line inter-domain routing/signaling
For the most part, the Internet today is a set of autonomous, interconnected and inter-operating networks that ride on top of a physical telecommunications infrastructure that was largely designed to carry voice telephony. The Internet is a hierarchical structure comprised of backbone providers, enterprise networks, and regional Internet Service Providers (ISPs). Local ISPs, that ultimately provide access to the end-user, connect to the regional ISPs. This hierarchical structure leads to congestion on the Internet at points where the sub-networks must interconnect. The decentralized nature of the Internet means that there is only limited coordination between network providers, a fact that exacerbates problems related to network performance. These problems are further manifested by the so-called "hot-potato routing," where backbone providers place traffic destined for another backbone as soon as possible at the nearest traffic exchange point, usually a public network access point (NAP) [3]. This creates traffic flows that are asymmetrical and causes congestion at the public NAPs, resulting in performance degradation. This practice also limits the level of control a provider has over end-to-end service quality. Compounding the problem, is the fact that most of the GMPLS-based routing and signaling techniques discussed above [19, 20, 40, 45], which were developed to address the problem of real-time provisioning in IP-over-optical networks, were developed to provision connection requests at the full wavelength capacity and run only within the boundaries of a single AS, namely "the OTN."

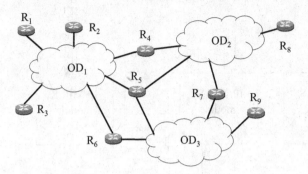

Figure 3.10 An arbitrary multi-domain topology.

In contrast, the unified optical layer-based control plane allows for the design of a network capable of being managed end-to-end (from the access network through the metro and core networks to another access network), across multiple Optical Domains (ODs) that belong to different administrative domains. To illustrate the concept of this architecture, we use the arbitrary network topology shown in Figure 3.10, which comprises of three optical networks that belong to different administrative domains. Each OD consists of multiple OXCs interconnected via WDM links in a general mesh topology. As previously stated, the intelligence of the OXCs lies with their controllers that are capable of performing routing and wavelength assignment (RWA), as well as maintaining a database with existing lightpaths (intra-domain logical connectivity). Attached to the ODs are high-performance backbone IP routers from and to which traffic requests are generated and terminated respectively. Routers interconnecting multiple optical cores are defined as Gateway Routers (GRs) (routers R_4, R_5, R_6, and R_7 in Figure 3.10). Two steps are required in order to achieve end-to-end connectivity across these ODs, namely, the initialization phase and the provisioning phase.

A.1 Initialization phase

During the initialization phase, the OXCs disseminate information about the presence of edge routers attached to them. It is thus possible for routers to identify reachable routers attached to the same OD through an intra-domain connectivity table. In this case, each edge router runs a limited reachability protocol with the corresponding edge OXC and obtains the address of every other edge router belonging to the same OD. Using this information, an initial set of IP routing adjacencies is established between edge routers. The edge routers then run an IP routing protocol amongst themselves to determine all the IP destinations reachable over the OD that they are attached to. (It is important to note that GRs are "seen" by all ODs they are connected to.) The next step in the initialization phase is for the GRs to advertise their intra-domain connectivity tables from one OD to all other ODs they are connected to, declaring themselves as the gateway points. It is possible then to define a Global

Table 3.1. Global topology database for example in
Figure 3.10

ODs are servicing:	GRs connecting:
OD_1: R_1, R_2, R_3, R_4, R_5, R_6.	R_4: OD_1, OD_2.
OD_2: R_4, R_5, R_7, R_8.	R_5: OD_1, OD_2, OD_3.
OD_3: R_5, R_6, R_7, R_9.	R_6: OD_1, OD_3.
	R_7: OD_2, OD_3.

Topology (GT), which considers the ODs as black-box switches interconnecting all routers. For example, with respect to Figure 3.10, after completion of the initialization phase, the routers keep the global topology database shown in Table 3.1.

A.2 Provisioning phase

When a request arrives at a router (the request may originate at this router, or at another router within the global topology; in the latter case, the router is a GR), the router checks whether the destination belongs to its own OD:

- If yes, then it requests service from its OD via a defined UNI. The OD will then provision the request by setting up a lightpath, using already established lightpath(s), or by combining the two. In the case that no optical resources can be allocated, the call is blocked.
- If no, then the router consults its global topology table, finds out all gateways that can connect it to the destination OD, and simultaneously requests (using different instances of signaling) service to them from its OD. These gateway routers then act as the source for the path and the above steps are repeated.

All ODs traversed have to return additive costs associated with the paths they computed (as opposed to returning the paths). Note that information about the so-far traversed ODs is kept in the signaling message, so that an OD can only be traversed once. When an instance of the signaling reaches the destination router, the latter sends it back to the source router (using the instance-specific route the signaling information carries). The source router, then, after collecting all the instances of its signaling, chooses the lowest-cost path, and initiates a reservation protocol to reserve the resources along the global path.

A.3 An illustrative example

If router R_1 gets a request to send data to router R_9 (Figure 3.10), it will send three parallel signaling instances requesting service from OD_1 to routers R_4, R_5, and R_6. OD_1 will decide the best routes to these destinations and will propagate the three signaling instances (note that whenever an OD performs route calculations from a source to a destination, it attaches the path-cost and its ID to the signaling information) to them:

- When router R_4 receives the signaling information it will request paths to routers R_5 and R_7 from OD_2, which will then request service to the destination from OD_3.
- When router R_5 receives the signaling information it will request a path to the destination from OD_3.
- When router R_6 receives the signaling information it will request a path to the destination from OD_3.

In this specific example, a total of four possible paths are examined: (a) R_1–R_4–R_5–R_9, (b) R_1–R_4–R_7–R_9, (c) R_1–R_5–R_9, and (d) R_1–R_6–R_9. Global logical links in these paths (i.e. from R_x to R_y) have additive costs associated with them (which the corresponding OD that created them assigned). The least-cost path is then chosen by the source router R_1, which will then use a reservation protocol to reserve the resources.

B. Optical layer-based restoration at any granularity

One of the most important considerations of a carrier in designing and deploying its transport network is the reliability offered by the network to the services and customers it supports. Service reliability considerations are profoundly critical when high-capacity WDM transport technologies are involved, since certain single WDM transport system failures may affect thousands of connections. Restoration may be provided at the IP layer and/or the optical layer. With optical layer protection, a significant number of failure scenarios including fiber cable breaks and transmission equipment outages can be detected and recovered much quicker than IP rerouting (e.g., 50 ms compared with tens of seconds). These failures can be restored transparently (core routers are not notified of the fault) and therefore do not cause IP routing reconvergence [15, 19]. When router failures are taken into account, however, the situation changes significantly as practical reasons favor the use of the IP layer to restore network failures:

- Since the optical layer operates independently and has no awareness of a router failure, router failures and certain other failures cannot be recovered by optical protection/restoration. This means that protection against router failures must be provided directly at the IP layer, and that IP network operators have to provide extra link capacity to recover from such failures. This extra capacity (provided via the IP layer) can also be used to protect against optical layer failures (e.g., a fiber cut).
- Although link failures occur relatively frequently due to cable breaks, in practice this may only be a small fraction of the overall number of faults causing reconvergence. Router hardware and software failures may constitute the majority of faults in many IP backbone networks (routers have

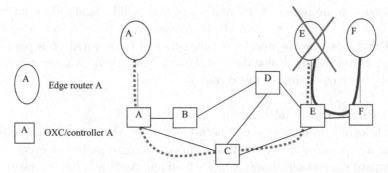

Figure 3.11 Edge router failure. (From [24], Figure 1. Copyright 2003 IEEE. Used by permission of The Institute of Electrical and Electronics Engineers Inc.)

historically had yearly downtimes that are orders of magnitude higher than legacy TDM/SONET equipment [20]). In such cases the IP-layer reconvergence would still be required to provide resilience.

- Different classes of emerging IP services need varying degrees of resilience requirements. However, the coarser granularity of optical layer protection/restoration would lead to costly and inflexible resilient network architectures.

Adapting the unified control model with an optical layer capable of provisioning on a per-call basis makes it now possible for the optical layer to restore all disrupted traffic (both full lambda and/or sub-lambda) independently in the case of a link/node and router failures. The all-optical resilience strategy described in this section is based on fast restoration techniques where the recovery path is established dynamically after the detection of a failure [24]. Since the optical layer is capable of restoring on a per-call basis, different levels of restoration (differentiated resilience) for different classes of service can be provided.

B.1 Edge router failure

In the case of an edge router failure, the OXC controller attached to the failed router detects the failure and floods the network with a message identifying the failed router. Note that the optical layer cannot restore the traffic originated from or terminated at the failed router. Conventionally, the traffic traversing the failed router would need to wait for the IP restoration mechanisms to take over in order to be restored. In the unified model approach, however, the affected traffic traversing the failed router can be restored by rerouting the individual affected calls. In addition, all lightpaths originating or terminating at the failed router can be immediately released so that the resources can be made available for future connections.

Figure 3.11 shows a failure of router E, illustrating that the traffic that originates and terminates at that router is lost and the two lightpaths ("dotted" lightpath from router A to router E and "solid" lightpath from router E to router F) are released. Multi-hop connections from router A to router F utilizing these two lightpaths sequentially can be restored on a per-call basis, (e.g., by setting up a new direct lightpath from router A to router F on path A–C–E–F bypassing the failed router).

B.2 Physical link failure

Traditionally, in an end-to-end path-based protection scheme, the optical layer typically provides node and link disjoint alternate paths for every restorable lightpath connection. However, in such an approach, a failure to restore a single lightpath will result in the loss of a potentially large number of sub-lambda connections contained in it. Thus, the coarse granularity of restoration leads to inflexible and inefficient resiliency. In the unified control model approach, the optical layer can provide fine granularity of restoration. Thus, in the case of a fiber cut, all affected calls appear to the network as new calls and are individually reprovisioned following the provisioning techniques previously discussed (physical, logical, and hybrid).

B.2.1 Path-based lightpath restoration

After a failure event, the optical layer restores the affected lightpaths sequentially, starting from the lightpath with the highest bandwidth allocation. For a call to be restored all lightpaths servicing it must be restored. Subsequently, if a lightpath cannot be restored, all the calls utilizing that lightpath are lost.

B.2.2 Sub-lambda restoration

After a link failure, the restoration algorithm restores all affected calls sequentially starting from the one with the highest bandwidth. Any provisioning order (e.g., LOG-HYB-PHY) can be utilized to reprovision the affected calls.

B.3 Performance evaluation

Some results on the performance of the unified control plane model, in terms of sub-lambda restoration, are presented in this section. The network parameters used for this analysis are the same as the ones used for the evaluation of the hybrid provisioning techniques in Section 3.4.2.

Figure 3.12 compares the optical layer-based restoration capability of the unified model (restoring sub-lambdas using the LOG-HYB-RWA sequence) with that of the conventional path-based restoration (restoring full lambdas) [24]. The fault is a link cut that brings down all four fibers in the trunk (a maximum of 16 wavelength channels can fail). The results in Figure 3.12 show the percentage of restored bandwidth for the two approaches, and demonstrate that when the granularity of the restoration is on a per-call basis, much more percentage of the bandwidth affected can be restored. It is interesting to note the oscillations that the conventional path-based lightpath restoration exhibits when the network operates at higher loads. This is because the load increase is a feedback factor

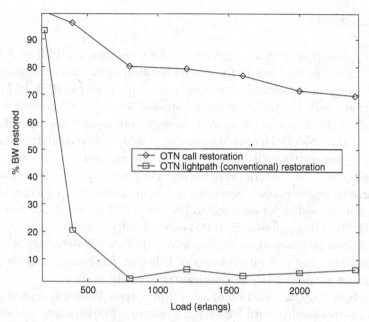

Figure 3.12 Optical layer-based sub-lambda restoration. (From [24], Figure 2. Copyright 2003 IEEE. Used by permission of The Institute of Electrical and Electronics Engineers Inc.)

in the restorability of the network. When the network is highly loaded, then the fiber cut causes more lightpaths to fail, leading to more network resources being freed. These resources could, in turn, be reused to restore some of the affected lightpaths. This can effectively, in some cases, provide a network with higher restoration capability at higher loads.

C. GigE/WDM network infrastructure

A further advantage is that this unified model is not confined to IP-over-WDM networks but could also be applied, for example, to GigE/WDM network infrastructures where native Ethernet frames are mapped directly over WDM [28]. Utilizing the unified optical layer-based control plane to manage both GigE switches and OXCs can allow for the transport of native Ethernet (layer-2 MAC frame-based) end-to-end (from the access network through the metro and core networks to another access network), bypassing unnecessary translation of the Ethernet data traffic (the majority of enterprise data traffic) to some other protocol and then back to Ethernet at its destination. Such a network will diminish the role of SDH/SONET, and will combine the simplicity and cost effectiveness of Ethernet technology with the ultimate intelligence of WDM-based optical transport layer, to provide a seamless global transport infrastructure for end-to-end transmission of native Ethernet frames.

3.5   Conclusions

This chapter presented an overview of the Generalized MPLS that has been widely viewed as the basis for the control plane for optical networks and then presented a new unified optical layer-based control plane for the IP/WDM networks. This model uses IP/MPLS-aware, non-traffic bearing OXC controller modules located within the optical domain to manage both routers and optical switches. Utilizing this model, the optical layer can provide lambda/sub-lambda routing, signaling, and survivability functionalities, and can also enable the end-to-end management of the network (from the access network through the metro and core networks to another access network), across multiple ASs. A further advantage is that this model is not confined to IP-over-WDM networks but could also be applied, for example, for GigE/WDM network infrastructures where native Ethernet frames are mapped directly over WDM. This infrastructure will combine the simplicity and cost effectiveness of Ethernet technology with the ultimate intelligence of the WDM-based optical transport layer.

Clearly, any implementation of an optical control plane will depend primarily on its interoperability with legacy infrastructure. Furthermore, ease of use and reliability is critical for network operators who manage these types of networks. The optical control plane implementation will depend on the choice of a design that can show improved decisions for provisioning, restoration, and overall efficient management of network resources, which will ultimately lead to reduced network operational expenses [43].

References

[1] M. A. Ali, A. Shami, C. Assi, Y. Ye, and R. Kurtz. Architecture Options for Next-Generation Networking Paradigm: Is Optical Internet the Answer? *Springer Journal of Photonic Network Communications*, **3**:1/2 (2001), 7–21.

[2] J. Armitage, O. Crochat, and J.-Y. Le Boudec. Design of a Survivable WDM Photonic Network, *Proc. IEEE Conference on Computer Communications (Infocom)*, Kobe, Japan, April 1997, pp. 244–252.

[3] C. Assi, *et al.* Optical Networking and Real-time Provisioning; An Integrated Vision for the Next-generation Internet, *IEEE Network Magazine*, **15**:4 (2001), 36–45.

[4] D. Awduche and Y. Rekhter. Multi-Protocol Lambda Switching: Combining MPLS Traffic Engineering Control with Optical Crossconnects, *IEEE Communications Magazine*, **39**:3 (2001), 111–116.

[5] D. Awduche, *et al.* RSVP-TE: Extensions to RSVP for LSP Tunnels, *Internet Engineering Task Force (IETF) RFC 3209*, (December 2001).

[6] A. Banerjee, *et al.* Generalized Multiprotocol Label Switching: An Overview of Routing and Management Enhancements, *IEEE Communications Magazine*, **39**:1 (2001), 144–150.

[7] A. Banerjee, *et al.* Generalized Multiprotocol Label Switching: An Overview of Signaling Enhancements and Recovery Techniques, *IEEE Communications Magazine*, **39**:7 (2001), 144–151.

[8] L. Berger (editor). Generalized MPLS-Signaling Functional Description, *Internet Engineering Task Force (IETF) RFC 3471*, (January 2003).

[9] L. Berger (editor). Generalized MPLS Signaling – RSVP-TE Extensions, *Internet Engineering Task Force (IETF) RFC 3473*, (January 2003).

[10] G. Bernstein, J. Yates, and D. Saha. IP-Centric Control and Management of Optical Transport Networks, *IEEE Communications Magazine*, **38**:10 (2000), 161–167.

[11] G. Bernstein, B. Rajagopalan, and D. Saha. *Optical Network Control: Architectures, Protocols, and Standards*, (Addison Wesley, 2004).

[12] R. Braden (editor). Resource Reservation Protocol (RSVP), *Internet Engineering Task Force (IETF) RFC 2205*, (September 1997).

[13] N. Chandhok *et al.* IP over Optical Networks: A Summary of Issues, *Internet Engineering Task Force (IETF) Internet Draft draft-osu-ipo-mpls-issues-00.txt*, (July 2000).

[14] D. Cheng. OSPF Extensions to Support Multi-Area Traffic Engineering, *Internet Engineering Task Force (IETF) Internet Draft draft-cheng-ccamp-ospf-multiarea-te-extensions-01.txt*, (February 2003).

[15] A. Chiu and J. Strand. Joint IP/Optical Layer Restoration after a Router Failure, *Proc. IEEE/OSA Optical Fiber Communications Conference (OFC)*, Anaheim, CA, March 2001.

[16] A. Chiu and J. Strand. Control Plane Considerations for All-Optical and Multi-domain Optical Networks and their Status in OIF and IETF, *SPIE Optical Networks Magazine*, **1**:4 (2003), 26–34.

[17] A. Chiu and J. Strand (editors). Impairments and Other Constraints on Optical Layer Routing, *Internet Engineering Task Force (IETF) RFC 4054*, (May 2005).

[18] E. W. Dijkstra. A Note on Two Problems in Connection with Graphs, *Numerische Mathematik*, **1**(1959), 269–271.

[19] R. Doverspike, S. Phillips, and J. Westbrook. Transport Network Architecture in an IP World, *Proc. IEEE Conference on Computer Communications (Infocom)*, **1**, Tel Aviv, Israel, March 2000, pp. 305–314.

[20] R. Doverspike and J. Strand. Robust Restoration in Optical Cross-connects, *Proc. IEEE/OSA Optical Fiber Communications Conference (OFC)*, Anaheim, CA, March 2001.

[21] A. Farrel and I. Bryskin. *GMPLS: Architecture and Applications*, (Morgan Kaufmann, 2006).

[22] N. Ghani *et al.* On IP-over-WDM Integration, *IEEE Communications Magazine*, **38**:3 (2000), 72–84.

[23] A. Greenberg, G. Hjalmtysson, and J. Yates. Smart Routers – Simple Optics: A Network Architecture for IP over WDM, *Proc. IEEE/OSA Opti-*

cal Fiber Communications Conference (OFC), paper ThU3, Baltimore, MD, March 2000.

[24] A. Hadjiantonis, A. Khalil, G. Ellinas, and M. Ali. A Novel Restoration Scheme for Next Generation WDM-Based IP Backbone Networks, *Proc. IEEE Laser Electro-Optic Society (LEOS) Annual Meeting*, Tucson, AZ, October 2003.

[25] A. Hadjiantonis, *et al.* A Hybrid Approach for Provisioning Sub-Wavelength Requests in IP-over-WDM Networks, *Proc. IEEE Canadian Conference on Electrical and Computer Engineering (CCECE)*, Niagara Falls, May 2004.

[26] A. Hadjiantonis, *et al.* On the Implementation of Traffic-Engineering in an All-Ethernet Global Multi-Service Infrastructure, *Proc. IEEE Conference on Computer Communications (Infocom)*, Barcelona, Spain, April 2006.

[27] A. Hadjiantonis. A Framework for Traffic Engineering of Diverse Traffic Granularity Entirely on the Optical Layer Terms, Ph.D. Thesis, City University of New York, (2006).

[28] A. Hadjiantonis, *et al.* Evolution to a Converged Layer 1, 2 in a Hybrid Native Ethernet-Over WDM-Based Optical Networking Model, *IEEE Journal on Selected Areas in Communications*, **25**:5 (2007), 1048–1058.

[29] K. Ishiguro, *et al.* Traffic Engineering Extensions to OSPF Version 3, *Internet Engineering Task Force (IETF) RFC 5329*, (September 2008).

[30] Distributed Call and Connection Management Mechanism using GMPLS RSVP-TE, *ITU, Recommendation G.7713.2*, (2003).

[31] D. Katz, K. Kompella, and D. Yeung. Traffic Engineering (TE) Extensions to OSPF version 2, *Internet Engineering Task Force (IETF) RFC 3630*, (September 2003).

[32] A. Khalil, *et al.* A Novel IP-Over-Optical Network Interconnection Model for the Next-Generation Optical Internet, *Proc. IEEE Global Communications Conference (GLOBECOM)*, San Francisco, CA, December 2003.

[33] A. Khalil, *et al.* Optical Layer-Based Unified Control Plane for Emerging IP/MPLS Over WDM Networking Architecture, *Proc. IEEE LEOS/OSA European Conference on Optical Communications (ECOC)*, Rimini, Italy, September 2003.

[34] M. Kodialam and T. V. Lakshman. Integrated Dynamic IP and Wavelength Routing in IP over WDM Networks, *Proc. IEEE Conference on Computer Communications (Infocom)*, Anchorage, AK, April 2001, pp. 358–366.

[35] K. Kompella, *et al.* Multi-area MPLS Traffic Engineering, *Internet Engineering Task Force (IETF) Internet Draft draft-kompella-mpls-multiarea-te-04.txt*, (June 2003).

[36] K. Kompella, Y. Rehkter, and L. Berger. Link Bundling in MPLS Traffic Engineering, *Internet Engineering Task Force (IETF) RFC 4201*, (October 2005).

[37] K. Kompella and Y. Rehkter (editors). OSPF Extensions in Support of Generalized MPLS, *Internet Engineering Task Force (IETF) RFC 4203*, (October 2005).

[38] K. Kompella and Y. Rehkter (editors). LSP Hierarchy with Generalized MPLS TE, *Internet Engineering Task Force (IETF) RFC 4206*, (October 2005).

[39] J. P. Lang (editor). Link Management Protocol (LMP), *Internet Engineering Task Force (IETF) RFC 4204*, (October 2005).

[40] E. Mannie (editor). Generalized Multi-Protocol Label Switching (GMPLS) Architecture, *Internet Engineering Task Force (IETF) RFC 3945*, (October 2004).

[41] J. Moy. *OSPF: Anatomy of an Internet Routing Protocol*, (Addison Wesley Longman, 1998).

[42] J. Moy. OSPF Version 2, *Internet Engineering Task Force (IETF) RFC 2328*, (April 1998).

[43] S. Pasqualini, *et al.* Influence of GMPLS on Network Provider's Operational Expenditures: A Quantitative Study, *IEEE Communications Magazine*, **43**:7 (2005), 28–38.

[44] B. Rajagopalan, *et al.* IP over Optical Networks: Architecture Aspects, *IEEE Communications Magazine*, **38**:9 (2001), 94–102.

[45] B. Rajagopalan, J. Luciani, and D. Awduche. IP over Optical Networks: A Framework, *Internet Engineering Task Force (IETF) RFC 3717*, (March 2004).

[46] A. Saleh, L. Benmohamed, and J. Simmons. Proposed Extensions to the UNI for Interfacing to a Configurable All-Optical Network, *Optical Interworking Forum (OIF) Contribution oif2000.278*, (November 2000).

[47] P. Srisuresh and P. Joseph. OSPF-xTE: An Experimental Extension to OSPF for Traffic Engineering, *Internet Engineering Task Force (IETF) RFC 4973*, (July 2007).

[48] User Network Interface (UNI) v1.0 Signaling Specification, *Optical Interworking Forum (OIF), OIF Contribution*, (December 2001).

[49] C. Xin, *et al.* An Integrated Lightpath Provisioning Approach in Mesh Optical Networks, *Proc. IEEE/OSA Optical Fiber Communications Conference (OFC)*, Anaheim, CA, March 2002.

[50] J. Yates, *et al.* IP Control of Optical Networks: Design and Experimentation, *Proc. IEEE/OSA Optical Fiber Communications Conference (OFC)*, Anaheim, CA, March 2001.

[51] Y. Ye, *et al.* A Simple Dynamic Integrated Provisioning/Protection Scheme in IP over WDM Networks, *IEEE Communications Magazine*, **39**:11 (2001), 174–182.

[52] H. Zhu, H. Zang, K. Zhu, and B. Mukherjee. A Novel, Generic Graph Model for Traffic Grooming in Heterogeneous WDM Mesh Networks, *IEEE/ACM Transactions on Networking*, **11**:2 (2003), 285–299.

4 Cognitive routing protocols and architecture

Suyang Ju and Joseph B. Evans

University of Kansas, USA

4.1 Introduction

Nowadays, there are many routing protocols [1-4] available for mobile ad-hoc networks. They mainly use instantaneous parameters rather than the predicted parameters to perform the routing functions. They are not aware of the parameter history. For example, AODV, DSDV, and DSR use the hop counts as the metric to construct the network topology. The value of hop counts is measured by the route control packets. Current physical topology is used to construct the network topology. If the future physical topology is predicted, a better network topology might be constructed by avoiding the potential link failure or finding a data path with high transmission data rate.

Most traditional routing protocols do not consider the channel conditions and link load. In this case, it is assumed that the channel conditions for all links are the same and the load levels for all links are the same. Unlike the wired networks, the channel conditions and the link load in a wireless network tend to vary significantly because of the node mobility or environment changes. Therefore, the nodes in a wireless network should be able to differentiate the links with different channel conditions or load levels to have a general view of the network. In this way, the routing functions can be better performed. Further, the network performance might be increased.

In recent years, cognitive techniques are increasingly common in wireless networks. Most research focuses on the solutions that modify the PHY layer and MAC layer. Few papers propose the cognitive routing protocols which use cognitive techniques. Compared to the traditional routing protocols, the main advantage of the cognitive routing protocols is that the network topology can be better constructed, because the routing functions are performed based on the predicted parameters. The predicted parameters are implied from the parameter history. With the predicted parameters, the nodes can have a general view which reflects the history and the future rather than an instantaneous view of the network

Next-Generation Internet Architectures and Protocols, ed. Byrav Ramamurthy, George Rouskas, and Krishna M. Sivalingam. Published by Cambridge University Press. © Cambridge University Press 2011.

topology. Consequently, cognitive routing protocols should be able to increase the network performance.

In optimum situations, the network topology should be adaptive and stable. To make the network topology adaptive, the routing updates should be triggered frequently to accommodate the physical topology changes which might incur lots of overhead. On the other hand, to make the network topology stable, the routing updates should be triggered infrequently to minimize the overhead. Therefore, there is a tradeoff between the accuracy and overhead. Cognitive routing protocols should be able to adjust the tradeoff to maximize the network performance by learning the history and predicting the future.

This chapter illustrates some theories on cognitive routing protocols and their corresponding protocol architectures. The protocol architectures used by the cognitive routing protocols are evolved from the protocol architectures used by the cross-layer optimization routing protocols, since most of the cognitive routing protocols are cross-layer optimized routing protocols in nature. In Section 4.2, mobility-aware routing protocol (MARP) is presented. The mobility-aware routing protocol is aware of node mobility. With MARP, the nodes are able to trigger the routing updates before the link breaks. In Section 4.3, spectrum-aware routing protocol (SARP) is presented. With SARP, the nodes are able to select an appropriate frequency for each link and select an appropriate path to route application packets. In Section 4.4, we conclude the chapter.

4.2   Mobility-aware routing protocol

4.2.1   Background

Most traditional routing protocols [5–13], such as AODV, DSDV, and DSR, trigger the routing updates after the corresponding routing table entry is removed. Mostly, this happens when the link breaks and the routing table entry timer has expired. In other words, the network topology is reactively optimized. There are two main problems for reactive optimization. The first one is that the routes might be frequently temporarily unavailable. The source nodes have to wait until a new route is found after a link breaks to resume transmission. In mobile networks, link failure happens frequently. The second one is that the packet transmission might frequently undergo low transmission data rate. In wireless networks, if the auto rate fallback is enabled, the transmission data rate is mainly impacted by the received signal strength. With strong received signal strength, high transmission data rate can be adopted. On the other hand, with weak received signal strength, low transmission data rate has to be adopted. Received signal strength is mainly impacted by the transmission distance. Mostly, when the link is about to break, the received signal strength tends to be weak. Therefore, the packet transmission has to undergo low transmission data rate. Consequently, the network performance might not be good. Therefore, because of the two main drawbacks for reactive optimization, the network performance for most traditional routing protocols is not very good.

Especially, for the routing protocols that use hop counts as the metric to construct the network topology, the problem might be serious. Mostly, in order to gain small hop counts, the distances between the neighboring nodes have to be big. In this case, the average received signal strength tends to be small because it is heavily impacted by the transmission distance. Consequently, the low transmission data rate has be adopted and the probability for link failure might be high. Further, it might incur a large amount of overhead to repair the frequent link failure. Therefore, a new metric for path selection is needed to overcome this issue.

If the nodes trigger the routing updates before the link breaks, the node might find an alternative next-hop node to the destination nodes. If the distance between the node and the alternative next-hop node is small, the channel condition might be good because the average received signal strength might be high. In order to gain better performance, the existing corresponding routing table entry should be preempted by the alternate next-hop node.

The related work includes:

- Adaptive distance vector [14]: This is able to adjust the frequency and the size of the routing updates based on mobility velocity. The frequency of the routing updates increases as the mobility velocity increases. The main issue is that the mobility velocity might not determine the actual physical topology changes.
- Preemptive AODV [15]: This uses the received signal strength as the sign to determine whether the link is about to break. The nodes trigger the routing updates if the link might break. The main issue is that the received signal strength tends to have large variance. As a result, a significant amount of overhead might be incurred.

From the related work [14–17], it is obvious that if the nodes are aware of mobility, the network performance can be increased. The link failure might be avoided. Consequently, the temporary packet transmission interruption might be avoided. Further, if the channel condition between the node and the alternative next-hop node is better than the channel condition between the node and the current next-hop node, the throughput might be increased.

4.2.2   Approach

This section introduces a novel mobility-aware routing protocol (MARP) [18]. The mobility-aware routing protocol has two new functions. First, it uses the throughput increment along a path instead of the hop counts as the metric to select the path. The throughput increment is defined as the predicted future throughput after a new application joins minus the current throughput. It is the throughput increment that determines the future overall throughput. The paths selected by MARP tend to be stable. The new metric overcomes the issues of the traditional routing protocols. Second, it uses the changes of the slope of the throughput as the sign to determine whether the link is about to break and triggers the routing updates on-demand. The alternative next-hop nodes preempt the existing next-hop nodes, if the alternative next-hop nodes are better based on the new metric for MARP. Potential link failures are avoided. Routing updates are triggered on-demand and unnecessary overhead is minimized.

Figure 4.1 shows the average throughput as a function of load. For high transmission data rate, the slope of the throughput is high. On the other hand, for low transmission data rate, the slope of the throughput is small. When the link is saturated, the slope of the throughput is almost zero. As a result, the slope of the throughput is a clear indicator of the quality of the links. With the new metric for path selection, MARP is able to easily select the links which can bear high transmission data rate to gain better performance. Interestingly, for low application data rates, the throughputs for links with different transmission data rates are almost the same, since the links are not saturated. The nodes might not able to differentiate the links if throughput is used as the sign. However, the loads of the links with different transmission data rates are quite different. Therefore, MARP uses the slope of the throughput as the sign to differentiate the links.

Mostly, the channel condition of a link is getting worse before the link breaks. The link breaks when the received signal is too weak to be detected when the channel condition is rather bad. The transmission data rate decreases as the channel condition is getting worse. Therefore, if the slope of the throughput decreases, it implies that the channel condition is getting worse. In this case, the nodes should trigger the routing updates to find an alternative next-hop node to preempt the existing next-hop node. With the new indicator for predicting the link failure, MARP is able to optimize the network topology before the link breaks.

1. Estimating the slope of the throughput

Equation (4.1) indicates how the slope of the throughput is calculated. Here S is the slope of the throughput, FT is the predicted future throughput, CT

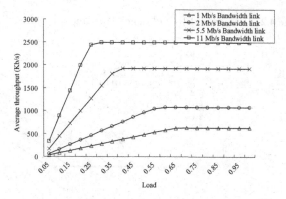

Figure 4.1 Average throughput as a function of load.

is the current throughput, FL is denoted as the future load, and CL is the current load. Throughput increment is defined as the difference between the current throughput and the future throughput when a new application joins. Throughput increment can be estimated based on the types of channels which have different transmission data rate, current load level of a link and the future load level of a link after an application joins. If the throughput increment is calculated, the slope of the throughput can be calculated:

$$S = (FT - CT)/(FL - CL). \qquad (4.1)$$

2. Estimating the types and load level of links

Figure 4.2 shows the average end-to-end delay as a function of load. Figure 4.3 shows the average packet loss rate as a function of load. From these two figures, end-to-end delay and packet loss rate can be estimated based on the types of the channels which have different transmission data rates and the load level of a link. Consequently, the nodes should be able to have some knowledge on the types of the channels and the load level of a link if they know the average end-to-end delay and the packet loss rate. End-to-end delay and packet loss rate are observable parameters for the nodes. By learning or estimating the observable parameters, the nodes are able to gain some knowledge on the predicted parameters, such as the throughput increment along a

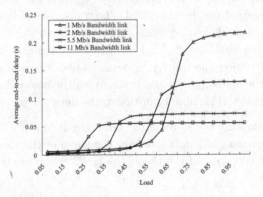

Figure 4.2 Average end-to-end delay as a function of load.

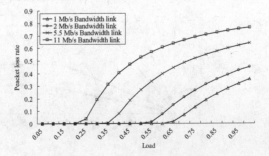

Figure 4.3 Average packet loss rate as a function of load.

path and the load level of a link. These predicted parameters are not readily available for the nodes. However, they are very useful for routing functions. The nodes use cognitive techniques to estimate the predicted parameters.

Estimating the types of links is a nonlinear problem, since the end-to-end delay as a function of load is nonlinear separable. Therefore, a three-layer neural network machine learning method should be used. The end-to-end delay and packet loss rate are the inputs of the neural network. The type of links is the output of the neural network. In the simulation, we train one neural network for estimating the types of the links. On the other hand, estimating the load level of links is a linear problem. We still use the neural network machine learning method. The end-to-end delay and packets loss rate are the inputs of neural networks. The load level of links is the output of the neural network. In the simulation, we trained four neural networks for estimating the load levels of link which are responsible for different types of the links. The neural network machine learning method is used to perform the off-line learning. From the simulation results, the accuracy of estimating the type of links is 84% and the accuracies of estimating the load level of links for each type of link are 75%, 74.2%, 82.2%, and 86.4%.

Based on the method shown above, current throughput and current load level of the link can be predicted. If the initial application data rate is predefined, the future load level of the links and the future throughput can be predicted. Therefore, the throughput increment and the slope of the throughput can be predicted.

4.2.3   Benefits

With MARP, the nodes perform the local optimization when triggering the routing updates. When a node predicts that the link is about to break, it informs the upstream node. The upstream node floods the route request packets. In many cases, the source node is transparent to the local optimization. In other words, the packet transmission is not interrupted. The source node does not need to worry about the local optimization or the link failure. Therefore, MARP performs seamless handover.

With MARP, the network topology is relatively adaptive and stable. The slope of the throughput reflects the channel condition and the load level. It has relatively small variance compared to received signal strength. The nodes can predict the slope of the throughput with confidence. As long as the slope of the throughput deceases, the local optimization is performed. So the network topology is adaptive. On the other hand, routing updates are triggered on-demand. Mobility-aware routing protocol combines the best of the proactive and reactive routing protocols. When the physical topology changes slowly, few routing updates will be triggered. In this case, MARP behaves as the reactive routing protocols. Unnecessary overhead is minimized. When the physical topology

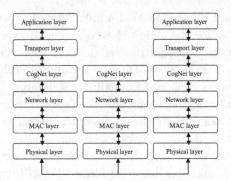

Figure 4.4 Protocol architecture.

changes rapidly, many routing updates will be triggered. In this case, MARP behaves as the proactive routing protocols. Network topology adapts fast to the physical topology changes.

Mobility-aware routing protocol uses the predicted throughput increment as the metric to select the path. Unlike in traditional routing protocols, the predicted throughput increment is updated along the path. The destination node selects the path with the biggest predicted throughput increment. In this way, MARP overcomes the issues of traditional routing protocols which use hop counts as the metric to select the path in terms of network topology construction. The new metric of throughput increment is used to determine whether the current next-hop node should be preempted if an alternative next-hop node is found.

The network performance can be increased significantly at the cost of increased overhead. The key factor that enables MARP to improve the network performance is that the nodes are able to find a better alternative next-hop node to preempt the existing corresponding routing table entry. In some cases, it might be difficult to find a better alternative next-hop node. The network performance might be even worse, because of the large amount of overhead incurred. On the other hand, if it is easy to find a better alternative next-hop node, it is worth triggering the routing updates to preempt the existing corresponding routing table entry. Therefore, the performance depends on the density of the nodes in the network.

4.2.4   Protocol architecture

Figure 4.4 shows the protocol architecture used to implement MARP. The CogNet layer is inserted between the transport layer and the network layer. This new layer is responsible for maintaining the end-to-end delay, the packet loss rate and the number of packets sent on the links for the destination nodes, and predicting the type of links for the neighbors, and predicting the load level of links for the neighbors. The CogNet layer header is inserted for each packet to transfer the information needed. The CogNet layer serves as an interface for the network layer to use the cognitive engine; MARP only uses this layer for mobility-aware purpose. This new layer provides other opportunities for protocol adaptation (e.g., network layer selection utilizing HIP).

This protocol architecture is evolved from the cross-layer optimized protocol architecture. Mobility-aware routing protocol is a cross-layer optimized routing protocol in nature. It uses the information from lower layers to optimize the routing functions. However, a big difference between MARP and traditional cross-layer optimized routing protocols is that MARP uses the observable parameters from lower layers to estimate the predicted parameters for routing protocols by cognitive engine.

The routing table entry is modified to track and manage the end-to-end delay, the packet loss rate, the incremental throughput and the number of packets sent.

The receiver monitors the slope of the throughput for the link. It learns the history for two seconds before it determines the changes of the slope of the throughput. In this way, the variance of the slope of the throughput might be decreased to increase the confidence of the prediction. The receiver sends a warning signal to the upstream node, if it is predicted that the link is about to break. The warning signal is flooded for only one hop. It has little impact on the neighboring nodes. When the upstream node receives the warning signal, it triggers the routing updates to perform local optimization. The TTL of the routing updates is the same as the existing corresponding routing table entry. The nodes should not retry the routing updates to perform local optimization. Otherwise, lots of unnecessary overhead might be incurred.

4.3   Spectrum-aware routing protocol

4.3.1   Background

A multi-channel capability is increasingly common. With multiple available frequencies, the network performance can be increased significantly, since the interference from the neighboring nodes might be reduced or avoided and the network load might be released by allocating different frequencies for links.

In recent years, the cost of the 802.11 interface has been decreasing, which make it feasible for the wireless nodes to equip multiple 802.11 interfaces. However, most research [19–24] focuses on solutions that modify the PHY layer or MAC layer. A few papers [25, 26] consider the interface assignment problem when nodes have multiple interfaces. When nodes have multiple interfaces, it is the responsibility of the network layer to assign an appropriate interface to a route.

Further, the use of the Global Positioning System (GPS) is increasingly feasible. Most GPS-based routing protocols [27, 28] focus on the physical topology provided by GPS. Few protocols specifically consider the transmission distance which heavily impacts the channel condition of the links.

Efficiently allocating the spectrum to each link in a multi-channel environment is an emerging topic. Mostly, graph-coloring techniques [29–32] are adopted. However, it is an NP-hard problem. Further, channel conditions of the links and the link load are not considered. In other words, it is assumed that the channel conditions of the links are the same and the network load for the links are the same.

A multi-channel multi-interface routing protocol is proposed. The concepts of the fixed interface and switchable interface are introduced. The nodes are able to utilize most available channels even when the number of interfaces is smaller than the number of available frequencies. Channel diversity and channel switching delay are considered to perform spectrum allocation. With the proposed approach, the interference from the neighboring nodes might be reduced or avoided. The network load of the links might be released by allocating different frequencies for the links. However, network load of the links and channel condition of the links are not considered when the nodes perform the spectrum allocation.

4.3.2   Approach

Spectrum-aware routing protocol (SARP) [33] consists of two parts. One is the intelligent multi-interface selection function (MISF). The other one is the intelligent multi-path selection function (MPSF). With multi-channel capability, the nodes should be able to assign an appropriate interface to the links and select an appropriate path to route application packets.

4.3.2.1   Intelligent multi-interface selection function

It is assumed that different interfaces are fixed on different frequencies. Unlike the nodes in wired networks, the wireless nodes are able to use any one or two interfaces which are fixed on different frequencies to relay the application packets because of the broadcast nature of wireless communication. Therefore, frequency allocation in wireless networks is flexible. In order to increase the network performance, the channel diversity should be increased as much as possible. In other words, the links should use most available frequencies and the interference from the neighboring nodes should be minimized by allocating different frequencies to the neighboring links.

The purpose of MISF is to let SARP assign an appropriate interface which is fixed on a specific frequency to the links to increase the channel diversity or minimize the interference from the neighboring nodes. It uses the delay of route request packets as the metric to assign an appropriate interface to the links.

It is assumed that each interface has a separate queue. In this way, the delay of the route request packets is used to estimate the average packet delay. The average packet delay is used to estimate the average queuing delay. The average queuing delay is used to estimate the network load of the links and the channel capacity. Therefore, the delay of the route request packets is used to estimate the network load of the links and the channel capacity.

For wireless networks, the packet delay is mainly determined by the queuing delay. Mostly, compared to the queuing delay, the packet propagation delay and the packet transmission delay are negligible. The network load of the links impacts the queuing delay, because it impacts the average queue length. The channel capacity impacts the queuing delay, because it impacts the packet transmission time for each queued packet. Consequently, by learning the queuing delay, the nodes should be able to gain some knowledge about the network load of the links and the channel capacity. When the packets are transmitted in different frequencies, they tend to have different delays. Often, the frequency in which the packets have small delay has low network load or big channel capacity.

In many cases, the interface has to perform channel switching if it is alternately fixed on different frequencies. Channel switching delay is part of the queuing delay. Consequently, MISF considers the channel switching delay. It is able to balance the load among the available interfaces of a node.

With SARP, the network topology is relatively adaptive and stable. The delay of the route request packets is mainly determined by the queuing delay. The queuing delay is the summation of the packet transmission delays of all queued packets. As a result, it should reflect the average channel capacity of a link. It is assumed that the network load of the links varies slowly. Therefore, with SARP, the constructed network topology is accurate.

4.3.2.2 Intelligent multi-path selection function

The purpose of MPSF is to let SARP select an appropriate path to route application packets. It uses the predicted throughput increment along a path as the metric to select the path.

The predicted throughput increment along a path is defined as the predicted throughput along a path after a new application joins minus the current throughput. The predicted throughput increment determines the predicted future overall throughput. The path with the biggest predicted throughput increment should be selected to route the application packets.

Five types of frequencies which have different channel characteristics are predetermined in the simulation environment. They have different shadowing means such as 4, 6, 8, 10, and 12. The Ricean K factors for them are 16.

Mostly, the received signal strength tends to have large variance because of the large-scale and small-scale fading. It might be difficult to estimate the types of frequencies based on the instantaneous received signal strength. However, it is the environment or the mobility velocity that mainly determine the large-scale and small-scale fading. Compared to the received signal strength, it is assumed that the environment changes slowly. The nodes might be able to estimate the types of the frequencies if they observe the channel characteristics for enough time. The channel characteristics include the mean and the standard deviation of the received signal strength along with the corresponding distance.

The neural network machine learning method is used to estimate the types of the frequencies. The inputs are the mean and standard deviation of the received signal strength along with the corresponding distance. The output is the types of the frequencies.

Figures 4.5 and 4.6 show the mean and the standard deviation of the received signal strength as a function of distance. Because of the undetectable signals, compared to the type of frequency which has small shadowing mean, the type of frequency which has large shadowing mean tends to have a larger mean of the received signal strength. Based on these two figures, the neural network is trained. In the simulation, the successful rate for estimating the types of the frequencies is about 80% after several thousand of packets are received and learned.

After estimating the types of the frequencies, the nodes should estimate the channel capacity. If auto rate fallback is used, the transmission data rate tends to fluctuate because of the varying channel capacity. It might be difficult to predict the trends in the transmission data rate. Alternatively, it might be easy to predict the probabilities for each transmission data rate based on the corresponding distance and the types of the frequencies. The nodes use GPS to estimate the transmission distance. Therefore, unlike most GPS-based routing protocols, SARP uses GPS to predict the probabilities for each transmission data rate.

After estimating the types of the frequencies and the channel capacity of the frequency, the nodes should predict the throughput increment along a path. The throughput increment is predicted based on the current load and the predicted

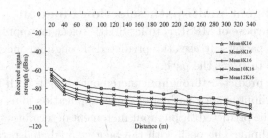

Figure 4.5 The mean of the received signal strength as a function of distance.

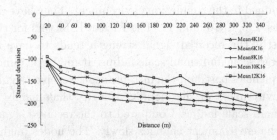

Figure 4.6 The standard deviation of the received signal strength as a function of distance.

future load after a new application joins. The current load can be calculated by dividing the application data by the transmission data rates which have different probabilities. It is assumed that the initial source application data rate is predefined. The predicted application data rate after a new application joins can be calculated based on the initial source application data rate and the packet loss rate. Further, the predicted future load can be calculated by dividing the predicted application data rate by the transmission data rates which have different probabilities. Consequently, the throughput increment along a path is averaged based on the probabilities for each transmission data rate.

4.3.3   Benefits

It is assumed that the number of interfaces determines the number of available frequencies in the simulation environment. Overhead is defined as the number of flooded RREQ packets. If there is only one interface for SARP, only MPSF works.

In order to show the benefits of SARP, a multi-channel routing protocol (MCRP) is implemented to compare with SARP. The multi-channel routing protocol is implemented as follows. There is a common control channel used for route control packets. One of the interfaces of the nodes is dedicated to the common control channel. The source node randomly selects an interface used for data transmission. The intermediate nodes have to use the same channel as the source node. The channel allocation is performed for a path rather than a link.

SARP with one interface and MCRP with two interfaces are compared. For MCRP, it means that one interface is used as control interface and the other interface is used as data interface. From the simulation results, SARP is able to increase the throughput by 77%. However, the end-to-end delay is increased by 20%. The number of packets received is increased by 77%. The overheads for the two protocols are almost the same. Therefore, MPSF increases the network performance significantly.

SARP and MCRP with the number of interfaces as 2, 4, 6, 8, and 10 are compared. In this case, both MISF and MPSF work. SARP uses all interfaces as both the control interfaces and data interfaces. If the number of interfaces is two, it means SARP has two data interfaces. However, MCRP has only one data interface. In this case, SARP increases the throughput by 250%. The end-to-end delays for the two protocols are almost the same. The overheads for the two protocols are almost the same. Therefore, SARP increases the network performance significantly. On the other hand, if the number of interfaces is ten, two routing protocols have almost the same number of data interfaces. SARP increases the throughput by 130%. End-to-end delay is decreased by 30%. The number of received packets is increased by 250%. It means that MISF increases the network performance significantly. However, the overhead is increased by 300%. The overhead of MCRP is almost constant but the overhead of SARP increases as the number of interfaces increases.

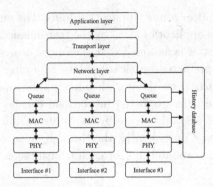

Figure 4.7 Protocol architecture.

4.3.4   Protocol architecture

Figure 4.7 shows the protocol architecture used to implement SARP. The protocol architecture uses a common database. This database is used by the lower three layers to help SARP perform the routing functions. Each node has a database. In other words, the database is allocated distributively among the nodes.

The database is used only for the nodes to perform the cross-layer optimization rather than for the network to provide some common information among the nodes. The database collects the information, such as the mean and standard deviation of the received signal strength along with the corresponding distance from the PHY layer, the application data rate from the MAC layer, and the queuing information from the queue, and keeps updating the information. When the network layer performs the routing function, it queries the database to calculate the predicted parameters. The advantage of this protocol architecture is that there is no new layer inserted. Therefore, the ISO protocol architecture is not heavily modified. Consequently, no new header is inserted in a packet. The overhead of a data packet is not increased.

The protocol architecture used by SARP is evolved from the cross-layer optimized protocol architecture, since SARP is a cross-layer optimized routing protocol in nature. It uses the common database for spectrum-aware purposes only. The protocol architecture can be easily expanded for some other purposes. It also works for some other routing protocols.

4.4   Conclusion

This chapter illustrates some theories on cognitive routing protocols and their corresponding protocol architectures. The protocol architectures used by the cognitive routing protocols are evolved from the cross-layer optimized routing protocols. Unlike most traditional routing protocols, the cognitive routing protocols perform the routing functions based on cognitive techniques. It observes the information by cross-layer optimizations, predicts the parameters by machine learning methods and performs the routing functions based on the predicted parameters. Simulation results show that cognitive routing protocols can increase the network performance significantly.

References

[1] Richard Draves, Jitendra Padhye, and Brian Zill (2006) Routing in multi-radio, multi-hop wireless mesh networks, *Proceedings of MOBICOM 2004*.

[2] Venugopalan Ramasubramanian, Zygmunt J. Haas, and Emin Gun Sirer (2003) SHARP: A hybrid adaptive routing protocol for mobile ad hoc networks, *Proceedings of the 4th ACM International Symposium on Mobile Ad Hoc Network and Computing*.

[3] Mingliang Jiang, Jinyang Li, and Y. C. Tay (1998) Cluster based routing protocol (CBRP), Internet draft.

[4] Brad Karp and H. T. Kung (2000) Greedy perimeter stateless routing for wireless networks, *ACM/IEEE MobiCom*.

[5] Elizabeth M. Belding-Royer (2003). Hierarchical routing in ad hoc mobile network, *Wireless Communications and Mobile Computing*, 515–532.

[6] Mario Joa-Ng and I-Tai Lu (1999). A peer-to-peer zone based two level link state routing for mobile ad hoc networks, *IEEE Journal on Selected Areas in Communications*.

[7] Charles E. Perkins and Elizabeth M. Royer (1997). Ad-hoc on-demand distance vector routing, *MILCOM97 panel on Ad Hoc Networks*.

[8] Charles E. Perkins and Pravin Bhagwat (1994). Highly dynamic destination sequenced distance vector routing for mobile computers, *Proceedings of the ACM SIGCOMM*.

[9] Ben Y. Zhao, Yitao Duan, and Ling Huang (2002). Brocade: Landmark routing on overlay networks, *Proceedings of 1st International Workshop on Peer-to-Peer Systems*.

[10] M. Liliana, C. Arboleda, and Nidal Nasser (2006). Cluster-based routing protocol for mobile sensor networks, *QShine'06*.

[11] Navid Nikaein, Houda Labiod, and Christian Bonnet (2000). DDR-distributed dynamic routing algorithm for mobile ad hoc networks, *MobiHOC 2000*.

[12] Navid Nikaein, Christian Bonnet, and Neda Nikaein (2001) HARP – hybrid ad hoc routing protocol, *International Symposium on Telecommunications 2001*.

[13] Atsushi Iwata, Ching-Chuan Chiang, and Guangyu Pei (1999) Scalable routing strategies for ad hoc wireless networks, *IEEE Journal on Selected Areas in Communications* Vol. 17, No. 8 August 1999.

[14] R. Boppana and S. Konduru (2001) An adaptive distance vector routing algorithm for mobile, ad hoc networks, *IEEE Infocom*.

[15] A. Boukerche and L. Zhang (2004) A performance evaluation of a pre-emptive on-demand distance vector routing protocol for mobile ad hoc networks, *Wireless Communications and Mobile Computing*.

[16] T. Goff, N. B. Abu-Ghazaleh, D. S. Phatak, and R. Kahvecioglu (2001) Preemptive routing in ad hoc networks, *ACM SIGMOBILE*.

[17] P. Srinath *et al.* (2002) Router Handoff: A preemptive route repair strategy for AODV, *IEEE International Conference*.

[18] Suyang Ju and Joseph B. Evans (2009) Mobility-Aware Routing Protocol for mobile ad-hoc networks, *CogNet Workshop 2009*.

[19] A. Nasipuri, J. Zhuang, and S. R. Das (1999) A multichannel CSMA MAC protocol for multihop wireless networks, *WCNC'99*.

[20] A. Nasipuri and S. R. Das (2000) Multichannel CSMA with signal power-based channel selection for multihop wireless networks, *VTC*.

[21] N. Jain, S. Das, and A. Nasipuri (2001) A multichannel CSMA MAC protocol with receiver-based channel selection for multihop wireless networks, *IEEE International Conference on Computer Communications and Networks (IC3N)*.

[22] Shih-Lin Wu, Chih-Yu Lin, Yu-Chee Tseng, and Jang-Ping Sheu (2000) A new multi-channel MAC protocol with on-demand channel assignment for multi-hop mobile ad hoc networks, *International Symposium on Parallel Architectures, Algorithms and Networks (ISPAN)*.

[23] Wing-Chung Hung, K. L. Eddie Law, and A. Leon-Garcia (2002) A dynamic multi-channel MAC for ad hoc LAN, *21st Biennial Symposium on Communications*.

[24] Jungmin So and Nitin H. Vaidya (2004) Multi-channel MAC for ad hoc networks: Handling multi-channel hidden terminals using a single transceiver, *Mobihoc*.

[25] Jungmin So and Nitin H. Vaidya (2004) A routing protocol for utilizing multiple channels in multi-hop wireless networks with a single transceiver, *Technical Report, UIUC*.

[26] U. Lee, S. F. Midkiff, and J. S. Park (2005) A proactive routing protocol for multi-channel wireless ad-hoc networks (DSDV-MC), *Proceedings of the International Conference on Information Technology: Coding and Computing*.

[27] S. Basagni, I. Chlamtac, V. R. Syrotiuk, and B. A. Woodward (1998) A distance routing effect algorithm for mobility (DREAM), *Proceedings of the Fourth Annual ACM/IEEE International Conference on Mobile Computing and Networking*.

[28] X. Lin, M. Lakshdisi, and I. Stojmenovic (2001) Location based localized alternate, disjoint, multi-path and component routing algorithms for wireless networks, *Proceedings of the ACM Symposium on Mobile ad-hoc Networking and Computing*.

[29] K. Leung and B.-J. Kim (2003) Frequency assignment for IEEE 802.11 wireless networks, *IEEE Vehicular Technology Conference*.

[30] P. Mahonen, J. Riihijarvi, and M. Petrova (2004) Automatic channel allocation for small wireless local area networks using graph colouring algorithm approach, *IEEE International Symposium on Personal, Indoor and Mobile Radio Communications*.

[31] A. Mishra, S. Banerjee, and W. Arbaugh (2005) Weighted coloring based channel assignment for WLANs, *Mobile Computing and Communications Review*.

[32] A. Mishra, V. Brik, S. Banerjee, A. Srinivasan, and W. Arbaugh (2006) A client-driven approach for channel management in wireless LANs, *INFOCOM*.

[33] Suyang Ju and Joseph B. Evans (2009) Spectrum-aware routing protocol for cognitive ad-hoc networks, *IEEE GlobeCom 2009*.

5   Grid networking

Anusha Ravula and Byrav Ramamurthy
University of Nebraska-Lincoln, USA

Research in Grid Computing has become popular with the growth in network technologies and high-performance computing. Grid Computing demands the transfer of large amounts of data in a timely manner.

In this chapter, we discuss Grid Computing and networking. We begin with an introduction to Grid Computing and discuss its architecture. We provide some information on Grid networks and continue with various current applications of Grid networking. The remainder of the chapter is devoted to research in Grid networks. We discuss the techniques developed by various researchers with respect to resource scheduling in Grid networks.

5.1   Introduction

Today, the demand for computational, storage, and network resources continues to grow. At the same time, a vast amount of these resources remains underused. To enable the increased utilization of these resources the tasks can be executed using shared computational and storage resources while communicating over a network. Imagine a team of researchers performing a job which contains a number of tasks. Each task demands different computational, storage, and network resources. Distributing the tasks across a network according to resource availability is called distributed computing. Grid Computing is a recent phenomenon in distributed computing. The term "The Grid" was coined in the mid 1990s to denote a proposed distributed computing infrastructure for advanced science and engineering [16].

Grid Computing enables efficient utilization of geographically distributed and heterogeneous computational resources to execute large-scale scientific computing applications [30]. Grid Computing provides an infrastructure that can realize high-performance computing. The execution of large-scale jobs is possible by using a high-performance virtual machine enabled by Grid Computing. Grid Computing demands the transfer of large amounts of data such as resource information, input data, output data and so on in a timely manner.

Next-Generation Internet Architectures and Protocols, ed. Byrav Ramamurthy, George Rouskas, and Krishna M. Sivalingam. Published by Cambridge University Press. © Cambridge University Press 2011.

In the following sections we discuss the Grid and its architecture. Grid Computing and networking are discussed in detail including some of its applications. We also present a literature review of the various scheduling schemes proposed in recent times. We discuss some studies of Grid networks based on Optical Circuit Switching and Optical Burst Switching.

5.2   The Grid

A Grid is composed of computational, network, and storage resources. Computational resources include CPU, memory, etc. The networking resources include routers and network links while the storage resources offer data generation and storage capabilities [27]. There are several applications that can benefit from the Grid infrastructure, including collaborative engineering, data exploration, high-throughput computing and, of course, distributed supercomputing. According to [9], Grid functions can be bisected into two logical grids: the computational Grid and the Access Grid. The computational Grid provides the scientists a platform, where they will be able to access virtually unlimited computing and distributed data resources. The Access Grid will provide a group collaboration environment.

5.2.1   Grid Computing

A Grid uses the resources of many separate computers, loosely connected by a network to solve large-scale computation problems. The Grid [16] will deliver high-performance computing capabilities with flexible resource sharing to dynamic virtual organizations. As mentioned earlier, Grid Computing involves coordinating and sharing computational, storage, or network resources across dynamic and geographically dispersed organizations. It enables aggregation and sharing of these resources by bringing together communities with common objectives and creating virtual organizations.

A Grid architecture represents the blueprint by which all this is possible. The architecture of the Grid is often described in terms of layers, each providing a specific function. They are (1) Network (2) Resource (3) Middleware and (4) Application and Serviceware layers. The Network layer provides the connectivity for the resources in the Grid. The Resource Layer contains all the resources that are part of the Grid, such as computers, storage systems, and specialized resources such as sensors. The Middleware Layer provides the tools so that the lower layers can participate in a unified Grid environment. The Application and Serviceware Layer includes all the applications that use the resources of the Grid to fulfill their mission. It is also called the Serviceware Layer because it includes all common services that represent mostly application-specific management functions such as billing, time logging, and others.

A network which connects the Grid resources is called a Grid network. Grid networks [27] attempt to provide an efficient way of using the excess resources. In 2004, Foster and Kesselman [16] defined the computational Grid as "a hardware and software infrastructure that provides dependable, consistent, pervasive, and inexpensive access to high-end computational capabilities."

Various grids were developed as testbeds to implement various data intensive and eScience applications. Various large-scale Grid deployments are being undertaken within the scientific community, such as the distributed data processing system being deployed internationally by "Data Grid" projects – GriPhyN (www.griphyn.org), PPDG (www.ppdg.net), EU DataGrid (http://eu-datagrid.web.cern.ch/eu-datagrid/), iVDGL (http://igoc.ivdgl.indiana.edu/), DataTAG (http://datatag.web.cern.ch/datatag/), NASA's Information Power Grid, the Distributed ASCI Supercomputer (DAS-2) system that links clusters at five Dutch universities, the DOE Open Science Grid (OSG) (www.opensciencegrid.org/) and DISCOM Grid that link systems at DOE laboratories, and the TeraGrid (www.teragrid.org/index.php) being constructed to link major US academic sites. Each of these systems integrates resources from multiple institutions, each with their own policies and mechanisms; uses open, general-purpose protocols such as Globus Toolkit protocols to negotiate and manage sharing; and addresses multiple quality of service dimensions, including security, reliability, and performance. The Open Grid Services Architecture (OGSA) modernizes and extends Globus Toolkit protocols to address emerging new requirements, while also embracing Web services. Companies such as IBM, Microsoft, Platform, Sun, Avaki, Entropia, and United Devices have all expressed strong support for OGSA.

5.2.2   Lambda Grid networks

A lambda corresponds to a wavelength which, in turn, can be used to create an end-to-end connection. A Lambda Grid is a distributed computing platform based on an optical circuit switching network, which addresses the challenging problems that originate in eScience fields. The Lambda Grid employs wavelength division multiplexing and optical paths. Optical networking plays an important role in creating an efficient infrastructure for supporting advanced Grid applications.

There are a number of noteworthy projects which are developing the general infrastructure (network provisioning solutions, middleware, protocols) required for a Lambda Grid. Some of them are Phosphorus (www.ist-phosphorus.eu/), OptIPuter (www.optiputer.net), and EGEE (www.eu-egee.org/).

Phosphorus is an FP6 IST project addressing some of the key technical challenges to enable on-demand, end-to-end network services across multiple domains. The project integrates application middleware and the optical transport network.

OptIPuter is an NSF-funded project to create an infrastructure for the coupling of high-performance computational and storage resources over parallel optical networks. The goal is to enable collaborative eScience applications which generate, process, and visualize data on a petabyte-scale.

EGEE or Enabling Grids for E-sciencE focuses on creating a production quality Grid infrastructure, and making it available to a wide range of academia and business users. A notable outcome of the project is the high-quality gLite middleware solution, which offers advanced features such as security, job monitoring, and data management in a service-oriented way.

One of the applications of Grid Computing can be seen in Cloud Computing. In the following section we discuss the research going on in Cloud Computing.

5.3   Cloud Computing

Cloud Computing [1] is aimed at applications requiring high computing power. It is enabled by a cluster of computing grids. It imparts parallelization in all the applications and hence decreases the cost and increases the horizontal scalability. This sense of distributedness and parallelization helps the users in Web-based applications. It also offers both storage and raw computing resources. The authors in [7] clarify terms and provide simple formulas to quantify comparisons between cloud and conventional computing. They also discuss how to identify the top technical and non-technical obstacles and opportunities of Cloud Computing. Many companies such as Google and IBM are offering these services to the universities by providing them with computing power [2]. Vmware, Sun Microsystems, Rackspace US, Amazon, BMC, Microsoft, and Yahoo are some of the other major Cloud Computing service providers. Some of the cloud services are discussed below.

- Amazon.com broadened access for software developers to its Elastic Compute Cloud service, which lets small software companies pay for processing power streamed from Amazon's data centers [3]. The simple storage service (Amazon S3) provides a user with storage for a very low price. S3 only costs 15 cents per Gig per month of storage.
- Using the Google App Engine, users can easily set a quota and deploy an application over the cloud. Using $Java^{TM}$ code, a user can deploy old applications onto the Cloud.
- Nimbus (www.workspace.globus.org), a Cloud Computing tool, is an open source toolkit that turns a cluster into an infrastructure-as-a-service (Iaas) cloud.
- Microsoft's cloud service, Windows Azure, supports building and deployment of cloud applications. It provides developers with on-demand compute and storage to host, scale, and manage Web applications on the Internet through Microsoft data centers.

- Sun Cloud allows a user to administer a virtual data center from a web browser. It offers an upload service to the users who wish to archive or run multiple OS/application stacks.

In the following sections we discuss the various resources considered during scheduling of jobs in a Grid network.

5.4   Resources

5.4.1   Grid network resources

There can be various Grid network resources such as network cables, switches, router, amplifiers, multiplexers, demultiplexers, etc. All the network resources can be set up differently for different networks. There are various networks in existence which are explained below.

- Switched Ethernet: An Ethernet LAN that uses switches to connect individual hosts or segments. It is an effective and convenient way to extend the bandwidth of existing Ethernets. Infrastructures such as Internet2 Network uses switched Ethernet for its advanced capabilities.
- SONET/SDH Circuit: A standard for optical transport of TDM data. SONET/SDH is operated over the optical network. Though it uses fiber as the transmission medium, it does all the switching, routing, and processing electronically.
- WDM Lightpath: Layer 1 circuit for optical transport of WDM data. It is enabled by a circuit switched network that uses wavelength routing. It is used for supporting applications that use huge bandwidth for a long time.
- WDM Wavebands: A band of wavelengths that is used to transmit data.
- Optical Burst Switching (OBS): Switching concept that operates at the sub-wavelength level and is designed to better improve the utilization of wavelengths by rapid setup and teardown of the wavelength/lightpath for incoming bursts. It is used when there is small or medium amount of data to be transmitted.

5.4.2   Optical network testbeds and projects

A network testbed gives researchers a wide range of environments to develop, debug, and evaluate their systems. Application requirements are drivers for bandwidth needs. There are various testbeds and projects being deployed on them available around the world. Some of the testbeds and projects are DRAGON (http://dragon.east.isi.edu/), UltraScienceNet (www.csm.ornl.gov/ultranet/), CHEETAH (www.ece.virginia.edu/cheetah/), ESNet (www.es.net/), HOPI (www.internet2.edu/pubs/HOPIInfosheet.pdf), NLR (www.nlr.net/), GENI (www.geni.net/), DCN (www.internet2.edu/network/dc/) to name a few. The various US-based optical network testbeds and projects are described below.

- UltraSciencenet: Developed for the needs of large-scale science applications. It provides high-speed, wide-area, end-to-end guaranteed bandwidth provisioning. It was sponsored by the US Department of Energy.
- CHEETAH: Circuit-Switched High-speed End-to-End Transport Architecture. It is developed to create a network to provide on-demand end-to-end dedicated bandwidth channels (e.g., SONET circuits) to applications. It is deployed in the US Southeast region. It is sponsored by NSF.
- DRAGON: Dynamic Resource Allocation over GMPLS Optical Networks. Provides Cyber infrastructure application support and advanced network services on an experimental infrastructure using emerging standards and technology. It also dynamically provides Authentication, Authorization and Accounting and scheduling across a heterogeneous network. This control plane is also the backbone of HOPI. It is deployed in the Washington DC region. It is sponsored by NSF.
- ESNet: Energy Sciences Network. Developed to provide a network and collaboration services in support of the agency's research missions. It was originally a packet-switched IP network. It is sponsored by the US Department of Energy. ESnet includes advanced circuit switching capabilities as well.
- HOPI: Hybrid Optical Packet Infrastructure. It is an infrastructure available at national and local level. It implements a hybrid of shared IP packet switching and dynamically provisioned optical lambdas. It uses infrastructures such as Internet2 waves and Regional Optical Networks (RONs). It is supported by Internet2 corporate participants. The HOPI testbed capabilities are provided now through Internet2 DCN (see below).
- NLR: National Lambda Rail. It was developed to bridge the gap between optical networking research and state-of-the-art applications research. It supports Dense Wavelength Division Multiplexing (DWDM). Additional bandwidths are provided to applications as needed. It also supports other testbeds such as HOPI.
- GENI: Global Environment for Network Innovations. Emerging facility to provide researchers with physical network components and software management system. GENI will provide virtual substrate that allows thousands of slices to run simultaneously. It is sponsored by NSF.
- DCN: Dynamic Circuit Network. It is a switching service for a short term between end-users that require dedicated bandwidth. It was developed to provide an automated reservation system to schedule a resource or a circuit on demand. It is supported by Internet2 architecture. The DCN service is now termed as Interoperable On-demand Network (ION).

5.4.3   Computational resources

Supercomputers today are used mainly by the military, government agencies, universities, research labs, and large companies to tackle enormously complex calculations for tasks such as simulating nuclear explosions, predicting climate change, designing airplanes, and analyzing which proteins in the body are likely to bind with potential new drugs. But these experiments take a long time. If these experiments can be executed in a distributed environment, it would reduce the execution time. To enable this feature, there are Grid testbeds available to schedule the computational resources. Portable Batch System (PBS) [4] is a queueing system to perform job scheduling. Its primary task is to allocate computational tasks among the available computing resources. Portable Batch System Pro [4] was developed to automate the process of scheduling and managing compute workload across clusters, Symmetric Multi Processor (SMP), and hybrid configurations. It thus increases productivity and decision-making capabilities. It supports on-demand computing, work load management, advanced scheduling algorithms, scalability, and availability.

The TeraGrid integrates high-performance computers, data resources and tools, and high-end experimental facilities around the USA. TeraGrid resources include more than 750 teraflops of computing capability and more than 30 petabytes of online and archival data storage, with rapid access and retrieval over high-performance networks. Researchers can also access more than 100 discipline-specific databases.

The open source Globus Toolkit (www.globus.org/toolkit/) is a fundamental enabling technology for the "Grid," letting people share computing power, databases, and other tools securely online across corporate, institutional, and geographic boundaries without sacrificing local autonomy. The Globus Resource Allocation Manager (GRAM) [5], facilitates the processing of requests for remote application execution and active job management. It manages and supports the protocol interaction between the application administration and the administration of available resources. It is the only such mechanism offered by all of TeraGrid's computational resources.

The best tools for each job depend on the individual workflow needs. Several software layers exist to simplify job submission. For local job submissions, PBS can be used to create and submit batch jobs to a large number of cluster machines. For remote job submissions, the Globus Toolkit lays on top of PBS, but includes authentication, scheduling, and resources description tools required for the remote submission. However, if there are a number of jobs to be run independently, Condor-G provides a software layer on top of Globus, allowing advanced job submission and monitoring capabilities using a single script.

Open Science Grid (OSG) is a consortium of software, service, and resource providers from universities, national laboratories, and computing centers across the USA. It is funded by the NSF and DoE. It was established to satisfy the ever-growing computing and data management applications that require high-throughput computing. OSG collaborates with TeraGrid and Internet2 to provide a better infrastructure.

Through a common set of middleware, OSG brings the computing and storage resources together from campuses and research communities into a common, shared Grid infrastructure over research networks. A combination of dedicated, scheduled, and opportunistic alternatives is used by them to offer a low-threshold access to more resources to the participating research communities than normally available. The Virtual Data Toolkit of OSG provides packaged, tested, and supported collections of software for installation on participating compute and storage nodes and a client package for end-user researchers.

5.4.4 Other resources

Apart from CPU and network bandwidth scheduling, other resources such as storage and wireless sensors can also be scheduled. For example, Planet-lab (www.planet-lab.org/) is a global network research testbed that enables sharing resources such as storage, and even network measurement tools. There are many testbeds available to share and schedule other resources such as sensor networks, telescopes, and many others. WAIL Schooner (www.schooner.wail.wisc.edu/) is a testbed managed by the University of Wisconsin that enables the end-user to schedule network hardware components such as routers, switches, and host components. There is also another testbed called EMULab (www.emulab.net/) that is hosted by the University of Utah. This testbed enables users to schedule sensor and other devices for their experiments. Similarly, DETER (www.isi.edu/deter/docs/testbed.overview.htm) and EMist (www.isi.edu/deter/emist.temp.html) are testbeds to run a variety of experiments in the area of computer security.

In the following section, we discuss scheduling in Grid networks. We also discuss some scheduling techniques developed by researchers.

5.5 Scheduling

Scheduling and management of Grid resources is an area of ongoing research and development. Effective scheduling is important in optimizing resource usage. Scheduling is the spatial and temporal assignment of the tasks of a Grid application to the required resources while satisfying their precedence constraints. The effective usage of geographically distributed computing resources has been the goal of many projects such as Globus, Condor, Legion, etc. Several open source schedulers have been developed for clusters of servers which include Maui, Condor, Catalina, Loadleveler, portable batch systems (PBS), and load sharing facility. The primary objective of the various scheduling approaches is to improve the overall system performance. The simultaneous scheduling of network and computing resources has been termed as joint scheduling with the objective to maximize the number of jobs that can be scheduled. Classical list algorithm is one of the techniques commonly used to jointly schedule the network for Grid services. ILP formulation is also used to solve the scheduling problem in the network. Recent studies [28], [21], [19], and [14] have deployed list algorithms to solve the scheduling problem in the Lambda Grid network. Some authors such as Liu *et al.* [21], Banerjee *et al.* [10], Demeyer *et al.* [14] also used the ILP to formulate the scheduling problem in the Lambda Grid network to solve the joint scheduling problem.

In [8], the co-allocation problem is formally defined and a novel scheme called synchronous queueing (SQ), which does not require advance reservation capabilities at the resources, is proposed for implementing co-allocation with quality of service (QoS) assurances in Grids. In general, for task scheduling, the Grid application can be modeled by a directed acyclic graph (DAG), where a node represents a task and an edge represents the communication between two adjacent tasks [21]. Some algorithms have been proposed previously for this kind of DAG scheduling [21], [18], [28]. However, most of them assume an ideal communication system in which the resources are fully connected and the communication between any two resources can be provisioned at any time.

A new scheduling model [22] considers both replicated data locations and maximizing processor utilizations to solve mismatch problems for scheduling. Based on the scheduling model, the scheduler implemented, called chameleon, shows performance improvements in data intensive applications that require both large number of processors and data replication mechanisms. The scheduling model is designed for job execution on one site.

Farooq *et al.* [15] presented a scalable algorithm for an NP-Hard problem of scheduling on-demand and advanced reservation of lightpaths. The proposed algorithm reschedules earlier requests, if necessary, and finds a feasible schedule if one exists, at the arrival of a new request. In addition to advance reservations, the algorithm can also schedule best-effort jobs that do not have an advance reservation and do not have a deadline to meet.

In [17], two schemes are analyzed for reducing communication contention in joint scheduling of computation and network resources in an optical Grid. The two schemes considered are adaptive routing and Grid resource selection. The adaptive routing scheme proposed detours the heavy traffic and finds an earliest start route for each edge scheduling. In the Grid resource selection scheme, a hop-bytes method is incorporated into the resource selection and a multi-level method is proposed to schedule the tasks onto the nearby resources, reducing the average data transferred across individual links.

Wang *et al.* [29] studied the accurate task scheduling in optical grids. They proposed a theoretical model and revealed the variance between the theoretical model and practical execution in an optical Grid testbed. A realistic model is proposed by investigating an optical Grid's task execution and data transfer scenarios. In the theoretical model, an optical Grid earliest finish time algorithm (OGEFT) is developed to schedule tasks (which uses list algorithms). But this algorithm does not consider an optical Grid's practical running scenarios such as lightpath establishment. The realistic model is acquired by incorporating τ_c (light establishment and data transfer time) formula into OGEFT. The realistic model demonstrated improved accuracy.

The authors of [26] proposed and developed a Grid broker that mediates access to distributed resources by discovering computational and data resources, scheduling jobs based on optimization of data resources and returning results back to the user. The broker supports a declarative and dynamic parametric programming model for creating Grid applications. The scheduler within the broker minimizes the amount of data transfer involved while executing a job by dispatching jobs to compute nodes close to the source of data. They have applied this broker in Grid-enabling a high-energy physics analysis, Belle Analysis Software Framework, on a Grid test-bed having resources distributed across Australia.

In [20], the authors proposed and compared a protection and restoration scheme in optical grids. In the protection scheme, each communication is protected by a backup lightpath link-disjoint with the working lightpath. Once an assignment of computing node with earliest finish time is decided for a given task, the lightpath with early finish time will be considered as working and the other as the backup lightpath. In the restoration scheme, a list scheduling algorithm is used to generate the initial schedule without considering the link failures. When a link failure occurs, if the backup lightpath can be established fast enough to comply with the original schedule of the destination task, then the recovery process is finished. Otherwise, the initial schedule is no longer valid and, therefore, the schedule has to be released and a new schedule is recomputed based on the available resources. The objective in both the cases is to minimize the job completion time.

The authors of [11] introduced a new procedure called traffic engineering for grids which enables Grid networks to self-adjust to fluctuations of resource availability. In order to do the self-adjusting, they monitored the network resources periodically and performed code migration accordingly.

A new Ant Colony Optimization (ACO) routing heuristic has been proposed [14] to find an optimal path between the source and the destination. The ACO heuristic simulated a real ant colony to find an optimal path. Here, with the destination node unknown and with data that cannot be sent to many destinations, the heuristic chooses a destination where the job can be successfully executed. In this work, the data packets were treated as ants which continuously explored and signaled changes in the resources and the network state. An increased performance has been determined with the proposed ACO heuristic.

In the following section we discuss various studies on Optical Grids based on Optical Circuit Switching and Optical Burst Switching.

5.6   Optical Circuit Switching and Optical Burst Switching

Several optical network architectures based on Optical Circuit Switching (OCS) or Optical Burst Switching (OBS) have been proposed, with the objective of efficiently supporting Grid services. The choice of OCS or OBS depends on bandwidth or delay requirements of the Grid applications. Optical Circuit Switching allows the user to access bandwidth at the wavelength level (e.g., 10 or 40 Gbps), while Optical Burst Switching allows bandwidth to be accessed at the sub-wavelength level.

In general, there are architectures based on either OCS (via wavelength routing) or OBS, depending on the bandwidth or delay requirement of Grid applications. In [25], an OCS-based approach (or Grid-over-OCS) was proposed for applications requiring huge bandwidth for a long period. In this approach, the Grid and optical-layer resources can be managed either separately in an overlay manner or jointly by extending the optical control plane for Grid-resource provisioning. Another type of architecture to support Grid services is based on OBS (or Grid-over-OBS), which is suitable for applications having small job sizes [13].

5.6.1   Studies on OCS-based Grids

Recently many authors such as Wang *et al.* [28], Liu *et al.* [21], Banerjee *et al.* [10], and Demeyer *et al.* [14] have proposed and developed new algorithms to jointly schedule computing and network resources by modifying the traditional list algorithm. An OCS-based approach was proposed in [25] for applications requiring huge bandwidths, where managing the Grid and optical-layer resources can be done either separately in an overlay manner or jointly by extending the optical control.

In [18] the authors defined a joint scheduling problem in the context of providing efficient support for emerging distributed computing applications in a Lambda Grid network. They focused on jointly scheduling both network and computing resources to maximize job acceptance rate and minimize total scheduling time. Various job selection heuristics and routing algorithms are proposed and tested on a 24-node NSFNet topology. The feasibility and efficiency of the proposed algorithms are evaluated on the basis of various metrics such as job blocking rate and effectiveness.

The algorithms proposed by Wang *et al.* [28] and Liu *et al.* [21] have implemented task scheduling to schedule the nodes for Grid resources. Wang *et al.* [28] has also used an adaptive routing scheme in communication scheduling to schedule the edges in the optical network along the lightpath.

Wang *et al.* [28] proposed a heuristic to minimize the completion time of jobs submitted. They proposed a modified list scheduling algorithm, Earliest Start Route First (ESRF). This algorithm has been used to map the resources from DAG to the O-Grid model extended resource system. Here the ESRF algorithm has been used to improve the accuracy of the scheduling. This algorithm determined the earliest start route for each schedule by modifying the traditional Dijkstra algorithm and reduced the total scheduling length. Better performance was identified with the ESRF algorithm, especially with average high node degree, when compared with the fixed and alternative routing algorithms. It has also been observed that the performances of all routing algorithms were identical when sufficient communication resources were available. The modified list algorithm uses the greedy approach to allocate resources which may not always be the shortest path. For this reason, the authors remark that the performance of the ESRF routing algorithm could have been improved with a better modified list algorithm.

The new algorithm [28] was implemented on a 16-Node network, the NSF-network and a mesh torus network and the algorithm of [21] was implemented on the ASAP network. The uniqueness of the work by Liu *et al.* [21] and Banerjee *et al.* [10] was that both used an integer linear programming (ILP) approach for scheduling. The authors in [10] had also implemented a greedy approach over ILP to improve the scalability of the network.

Liang *et al.* [19] proposed an optical Grid model based on the characteristics of optical network. The model presented a communication contention-aware solution to minimize the total execution time. The solution was based on the list scheduling for given tasks in an optical Grid. The Dijkstra algorithm was modified and deployed to minimize the total scheduling time. The modified Dijkstra algorithm was proved to be more feasible and efficient.

Liu *et al.* [21] present formulations to minimize the completion time and minimize the cost usage to satisfy a job. They propose an algorithm to jointly schedule network and computing resources. They use fixed routing over any adaptive algorithm to schedule the network resources. They proposed a greedy approach to schedule and execute tasks sequentially without any contention. They also propose a list scheduling approach that embeds the greedy approach in a list heuristic algorithm. To minimize the cost usage in the network, they propose a min-cost algorithm which tries to minimize the cost involved in the network along with the deadline constraint. The results showed that for a pipelined DAG, the scheduling length was less for the new list algorithm than for the traditional list algorithm. It has also been reported that for a general DAG the new list algorithm had an insignificant advantage over the traditional list algorithm. Though the network scheduling and computing node schedule is effective, it is more application-specific. Usage of an adaptive algorithm can increase the performance of the overall network.

Banerjee *et al.* [10] have considered the identification of a route for the file transfer and scheduling it over the respective circuits. The authors formulated a mathematical model and used the greedy approach to solve for the routing and scheduling on a Lambda Grid. A hybrid approach for both online and offline scheduling has been proposed. In this approach, offline scheduling solved for the route and the transfer of files. The provision for readjustment of the time in online scheduling, to transfer the entire file, reduced the total transfer time. The developed TPSP algorithm, proving its ability to optimize with MILP, was used for offline scheduling. The inappropriate scaling with the MILP made the authors use the greedy approach for TPSP. The approach chose a file and determined its earliest route and then scheduled that file transfer along that route.

They [10] also proposed two heuristics to determine the best file which was then routed and scheduled using the APT-Bandwidth scheduling or K-Random Path (KRP) algorithms. The best file was chosen with either the largest file first (LFF) or the most distant file first (MDFF) heuristics. For the chosen file, the APT algorithm computed all the time slots between the source and destination for a given duration. The bandwidth scheduling algorithm was implemented which selected the best-fit time slot. Using the KRP algorithm, the best route is chosen from K random paths. The file may be lost or transferred earlier than the finish time during the file transfer within the network. Earlier completion of the file transfer will allow the assignment of the sub-wavelength for later scheduled applications. The entire file or the lost partial file has to be retransmitted when the file is lost. Evaluating their proposed heuristics and algorithms on different networks, the authors identified that the LFF heuristics and the KRP algorithm together had better performance.

5.6.2   Studies on OBS-based Grids

In [12], the authors proposed a novel efficient and cost-effective infrastructure for Grids based on a Dual-Link-Server OBS network to improve the performance of the burst contention resolution scheme with an aim of solving the collision problem. The authors in [13] discussed an architecture which supports the Grid services based on OBS, suitable for applications having small job sizes. In [24] OBS is used for the Multi-Resource Many-cast (MRM) technique for its ability to statistically multiplex packet switching without increasing the overheads. Various heuristics are used to determine the destinations and selection of resources.

Optical Burst Switching is an alternative to jointly scheduling the network and computing resources and was proposed in [24, 25]. The work by She *et al.* [24] used many-casting over the network to perform OBS and Simeonidou *et al.* [25] have utilized an extension of the existing wavelength-switched network and also an optical burst-switching network.

Multi-Resource Many-cast over OBS networks for distributed applications was investigated in [24]. In this network, each source generates requests that required multiple resources and each destination had different computing resources. Each node was deployed with various requirements and resource availability. The objective of this paper was to determine the destination with sufficient resources and a route, to minimize the resource blocking rate. The OBS network was selected for the MRM technique for its ability to statistically multiplex packet switching without increasing the overheads. The authors have used the Closest Destination First (CDF), Most Available First (MAF), and Random Selection (RS) heuristics to determine the destinations. The Limit per Burst (LpB) and

Limit per Destination (LpD) heuristics were used as resource selection policies. It has been identified that the performance of the CDF was better than the MAF and RS heuristics on a 14-node NSF network. The authors also suggested that the use of destination selection heuristics minimized the resource blocking rate in resource selection.

Two different optical network infrastructures for Grid services have been proposed [25]. The architecture of extended wavelength-switched network facilitated the user-controlled bandwidth provisioning for applications that were data intensive. For this approach, three different solutions were provided. The first solution was to separately manage the Grid and optical layer. A Grid middleware managed the Grid resources and the optical layer managed the lightpath. The utilization of Grid middleware APIs enabled user application with the visibility of optical-network topology resources. The G-OUNI interface was the other solution that participated in resource discovery and allocation mechanism functions. The third solution utilized OBS and active router technologies. These solutions were suitable only for data-intensive applications and future Grid services.

Another approach was the use of OBS for a programmable network. This supported data-intensive and emerging Grid applications utilizing the advanced hardware solutions and a new protocol. The OBS networking scheme provided efficient bandwidth resource utilization. This proposed architecture offered a global reach of computing and storage resource using fiber infrastructure. The advantage of the OBS router was its usability in the normal network traffic and also for Grid network traffic.

Similar to the work in [25], Adami *et al.* [6] have also used a resource broker for a Grid network. It enhanced the capabilities by providing a network resource manager to manage and integrate the scheduling mechanism of network and computing resources.

In [23], the authors conducted simulation studies for various scheduling scenarios within a data Grid. Their work recommends decoupling of data replication from computation while scheduling jobs on the Grid and concludes that it is best to schedule jobs to computational resources that are closest to the data required for that job. But the scheduling and simulation studies are restricted to homogeneous nodes with a simplified First-In-First-Out (FIFO) strategy within local schedulers.

5.7 Conclusion

The introduction to Grid Computing and the ongoing research in Grid networking were presented in this chapter. The basic architecture of the Grid and Lambda Grids was introduced. Various scheduling schemes in Grid networks have been proposed under different scenarios. Some of the job scheduling schemes in Lambda Grids proposed by researchers are discussed in this chapter. The concept of cloud computing was also discussed. There is a huge scope for research and development in Grid networking.

References

[1] cloudcomputing.qrimp.com/portal.aspx.

[2] www.nsf.gov/pubs/2008/nsf08560/nsf08560.htm.

[3] www.amazon.com/gp/browse.html?node=201590011.

[4] www.pbsgridworks.com/Default.aspx.

[5] www.globus.org/toolkit/docs/2.4/gram/.

[6] Adami, D., Giordano, S., Repeti, *et al.* Design and implementation of a grid network-aware resource broker. In *Proceedings of the IASTED International Conference on Parallel and Distributed Computing and Networks, as part of the 24th IASTED International Multi-Conference on Applied Informatics* (Innsbruck, Austria, February 2006), pp. 41–46.

[7] Armbrust, M., Fox, A., Griffith, R., *et al.* Above the clouds: A Berkeley view of cloud computing.

[8] Azzedin, F., Maheswaran, M., and Arnason, N. A synchronous co-allocation mechanism for grid computing systems. *Cluster Computing 7*, 1 (2004), 39–49.

[9] Baker, M., Buyya, R., and Laforenza, D. The grid: International efforts in global computing. *International Conference on Advances in Infrastructure for Electronic Business, Science, and Education on the Internet (SSGRR 2000)* (July 2000).

[10] Banerjee, A., Feng, W., Ghosal, D., and Mukherjee, B. Algorithms for integrated routing and scheduling for aggregating data from distributed resources on a lambda grid. *IEEE Transactions on Parallel and Distributed Systems 19*, 1 (January 2008), 24–34.

[11] Batista, D., da Fonseca, N. L. S., Granelli, F., and F. Kliazovich, D. Self-adjusting grid networks. In *Communications, 2007. ICC '07. Proceedings of IEEE International Conference '07.* (Glasgow, June 2007), pp. 344–349.

[12] Chen, Y., Jingwei, H., Chi, Y., *et al.* A novel OBS-based grid architecture with dual-link-server model. In *First International Conference on Communications and Networking in China, 2006. ChinaCom '06.* (October 2006), pp. 1–5.

[13] De Leenheer, M., Thysebaert, P., Volckaert, B., *et al.* A view on enabling-consumer oriented grids through optical burst switching. *IEEE Communication Magazine 44*, 3 (2006), 124–131.

[14] Demeyer, S., Leenheur, M., Baert, J., Pickavet, M., and Demeester, P. Ant colony optimization for the routing of jobs in optical grid networks. *OSA Journal of Optical Networking 7*, 2 (February 2008), 160–172.

[15] Farooq, U., Majumdar, S., and Parsons, E. Dynamic scheduling of lightpaths in lambda grids. In *2nd International Conference on Broadband Networks, 2005.* (Boston, Massachusetts, October 2005), pp. 1463–1472.

[16] Foster, I., and Kesselman, C. *The Grid: Blueprint for a New Computing Infrastructure.* Morgan Kaufmann, 2004.

[17] Jin, Y., Wang, Y., Guo, W., Sun, W., and Hu, W. Joint scheduling of computation and network resource in optical grid. In *Proceedings of 6th*

International Conference on Information, Communications and Signal Processing, 2007 (Singapore, December 2007), pp. 1–5.

[18] Lakshmiraman, V., and Ramamurthy, B. Joint computing and network resource scheduling in a lambda grid network. In *Proceedings of IEEE International Conference on Communications (ICC 2009)* (2009).

[19] Liang, X., Lin, X., and Li, M. Adaptive task scheduling on optical grid. In *Proceedings of the IEEE Asia-Pacific Conference on Services Computing (APSCC 2006)* (Xian, China, December 2006), pp. 486–491.

[20] Liu, X., Qiao, C., and Wang, T. Survivable optical grids. In *Optical Fiber Communication Conference* (San Diego, California, February 2008).

[21] Liu, X., Wei, W., Qiao, C., Wang, T., Hu, W., Guo, W., and Wu, M. Task scheduling and lightpath establishment in optical grids. In *Proceedings of the INFOCOM 2008 Mini-Conference and held in conjuction with the 27th Conference on Computer Communication (INFOCOM 2008)* (Phoenix, Arizona, 2008).

[22] Park, S.-M., and Kim, J.-H. Chameleon: A resource scheduler in a data grid environment. In *Cluster Computing and the Grid, 2003. Proceedings. CCGrid 2003. 3rd IEEE/ACM International Symposium on Cluster Computing and the Grid, 2003.* (May 2003), pp. 258–265.

[23] Ranganathan, K., and Foster, I. Data scheduling in distributed data-intensive applications. In *Proceedings of 11th IEEE International Symposium on High Performance Distributed Computing (HPDC-11)* (July 2002).

[24] She, Q., Huang, X., Kannasoot, N., Zhang, Q., and Jue, J. Multi-resources many cast over optical burst switched networks. In *Proceedings of 16th International Conference on Computer Communications and Networks (ICCCN 2007)* (Honolulu, Hawaii, August 2007).

[25] Simeonidou, D., Nejabati, R., Zervas, G., *et al.* Dynamic optical-network architectures and technologies for existing and emerging grid services. *OSA on Journal of Lightwave Technology 23*, 10 (October 2005), 3347–3357.

[26] Venugopal, S., and Buyya, R. A grid service broker for scheduling distributed data-oriented applications on global grids. ACM Press, pp. 75–80.

[27] Volckaert, B., Thysebaert, P., De Leenheer, M., *et al.* Grid computing: The next network challenge! *Journal of The Communication Network 3*, 3 (July 2004), 159–165.

[28] Wang, Y., Jin, Y., Guo, W., *et al.* Joint scheduling for optical grid applications. *OSA Journal of Optical Networking 6*, 3 (March 2007), 304–318.

[29] Wang, Z., Guo, W., Sun, Z., *et al.* On accurate task scheduling in optical grid. In *First International Symposium on Advanced Networks and Telecommunication Systems, 2007* (Mumbai, India, December 2007), pp. 1–2.

[30] Wei, G., Yaohui, J., Weiqiang, S., *et al.* A distributed computing over optical networks. In *Optical Fiber Communication/National Fiber Optic Engineers Conference, 2008. OFC/NFOEC 2008.* (January 2008), pp. 1–3.

Part II

Network architectures

6   Host Identity Protocol (HIP): an overview

Pekka Nikander[†], Andrei Gurtov[‡], and Thomas R. Henderson[⊗]

[†]Ericsson Research, Finland   [‡]University of Oulu, Finland   [⊗]Boeing, USA

6.1   Introduction

The Host Identity Protocol (HIP) and architecture [1] is a new piece of technology that may have a profound impact on how the Internet will evolve over the coming years.[1] The original ideas were formed through discussions at a number of Internet Engineering Task Force (IETF) meetings during 1998 and 1999. Since then, HIP has been developed by a group of people from Ericsson, Boeing, HIIT, and other companies and academic institutions, first as an informal activity close to the IETF and later within the IETF HIP working group (WG) and the HIP research group (RG) of the Internet Research Task Force (IRTF), the research arm of the IETF.

From a functional point of view, HIP integrates IP-layer mobility, multi-homing and multi-access, security, NAT traversal, and IPv4/v6 interoperability in a novel way. The result is architecturally cleaner than trying to implement these functions separately, using technologies such as Mobile IP [2, 3], IPsec [4], ICE [5], and Teredo [6]. In a way, HIP can be seen as restoring the now-lost end-to-end connectivity across various IP links and technologies, this time in a way that it secure and supports mobility and multi-homing. As an additional bonus, HIP provides new tools and functions for future network needs, including the ability to securely identify previously unknown hosts and the ability to securely delegate signaling rights between hosts and from hosts to other nodes.

From a technical point of view, the basic idea of HIP is to add a new name space to the TCP/IP stack. These names are used above the IP layer (IPv4 and IPv6), in the transport layer (TCP, UDP, SCTP, etc.) and above. In this new name space, hosts (i.e., computers) are identified with new identifiers, namely Host Identifiers. The Host Identifiers (HI) are public cryptographic keys, allowing hosts to authenticate their peer hosts directly by their HI.

The Host Identity Protocol can be considered as one particular way of implementing the so-called identifier/locator split approach [7] in the stack. In the

---

[1] This chapter is based on "Host Identity Protocol (HIP): Connectivity, Mobility, Multi-homing, Security, and Privacy over IPv4 and IPv6 Networks" by P. Nikander, A. Gurtov, T. Henderson, which will appear in *IEEE Communication Surveys and Tutorials*. © 2010 IEEE.

Next-Generation Internet Architectures and Protocols, ed. Byrav Ramamurthy, George Rouskas, and Krishna M. Sivalingam. Published by Cambridge University Press. © Cambridge University Press 2011.

current IP architecture, the IP addresses assume the dual role of acting both as host identifiers and locators; in HIP, these two roles are cleanly separated. The Host Identifiers take, naturally, the host identifying role of the IP addresses while the addresses themselves preserve their locator role.

As a result of adding this new name space to the stack, when applications open connections and send packets, they no longer refer to IP addresses but to these public keys, i.e., Host Identifiers. Additionally, HIP has been designed in such a way that it is fully backwards compatible with applications and the deployed IP infrastructure. Hence, for example, when an existing, unmodified e-mail client opens a connection to the e-mail server hosting the mailbox, the e-mail client hands over a reference to the public key to the operating system, denoting that it wants the operating system to open a secure connection to the host that holds the corresponding private key, i.e., the e-mail server. The resulting connection can be kept open even if both of the hosts, i.e., the client and the server, are mobile and keep changing their location. If the hosts have multiple access links at their disposal, HIP allows these multiple links to be used for load balancing or as backups, invisible from the applications.

To deploy HIP in a limited environment, all that is required is to update the operating system of the involved hosts to support HIP. No changes are required to typical applications nor the IP routing infrastructure; all nodes will remain backwards compatible with existing systems and can continue to communicate with non-HIP hosts. For full HIP support, it is desirable to add HIP related information to the Domain Name System (DNS); additionally, a piece of new infrastructure is needed to support HIP rendezvous services. If it is impossible to upgrade the operating system of some particular host, e.g., a legacy mainframe, it is also possible to add a front-end processor to such a system. The front-end processor acts as a HIP proxy, making the legacy host appear (a set of) HIP host(s) to the rest of the network.

In this chapter, we describe the history behind HIP, the HIP architecture, the associated protocols, the potential benefits and drawbacks for prospective users, and the current status of HIP acceptance.

6.2   Fundamental problems in the Internet today

Before discussing the details of the HIP architecture and protocols, we take a brief look at a few of the most challenging problems in the contemporary Internet: loss of universal connectivity, poor support for mobility and multi-homing, unwanted traffic, and lack of authentication, privacy, and accountability. This forms a baseline for later sections, where we explain how HIP and other mechanisms relying on HIP alleviate these problems.

At the same time, it must be understood that HIP does not provide much remedy, at least not currently, to a number of other hard problems in the Internet, such as self-organizing networks and infrastructure, or intermittent connectivity.

On the other hand, there may certainly be new, still-unfound ways to utilize HIP to make those or other hard problems easier to solve.

6.2.1   Loss of universal connectivity

Compared to the original Internet, perhaps the largest of all current problems is the loss of connectivity, caused by NATs, firewalls, and dynamic IP addresses.

The HIP architecture offers a possible path for overcoming these limitations by providing a new end-to-end naming invariant and protocol mechanism that allows the IP addresses used on a wire to be used as ephemeral locators rather than host identifiers themselves.

6.2.2   Poor support for mobility and multi-homing

Effective mobility support requires a level of indirection [7] to map the mobile entity's stable name to its dynamic, changing location. Effective multi-homing support (or support for multi-access/multi-presence) requires a similar kind of indirection, allowing the unique name of a multi-accessible entity to be mapped to the multitude of locations where it is reachable.

As briefly mentioned above, HIP provides a novel approach to implementing mobility and multi-homing. It explicitly adds a new layer of indirection and a new name space, thereby adding the needed level of indirection to the architecture.

Mobility of hosts participating to multicast is a largely unsolved problem. Especially, if the multicast source changes its IP address, the whole multicast tree needs to be reconstructed. Two common approaches for multicast receiver mobility are bi-directional tunneling, based on Mobile IP [2], which tunnels multicast data to and from the home and visited networks, and remote subscription [8], in which each visiting multicast receiver joins the local multicast tree.

Some solutions to provide authentication to multicast receivers have been proposed. However, they are only able to authenticate subnetworks where the user is located, but not the host itself. Therefore, other hosts from the same subnetwork can receive the stream without authentication. Researchers have started to explore whether HIP can help also to alleviate these problems by allowing hosts to be more directly authenticated by the network [9].

6.2.3   Unwanted traffic

The various forms of unwanted traffic, including spam, distributed denial of service (DDoS), and phishing, are arguably the most annoying problems in the current Internet. Most of us receive our daily dosage of spam messages; the more lucky of us just a few of them, the more unlucky ones a few hundreds each day. Distributed denial of service attacks are an everyday problem to large ISPs, with each major website or content provider getting their share. Phishing is getting increasingly common and cunningly sophisticated.

The current unwanted traffic problem is a compound result from the following factors:

- An architectural approach where each recipient has an explicit name and where each potential sender can send packets to any recipient without the recipient's consent.
- A business structure where the marginal cost of sending some more packets (up to some usually quite high limit) is very close to zero.
- The lack of laws, international treaties, and especially enforcement structures that would allow effective punishment of those engaging in illegal activity in the Internet.
- The basic profit-seeking human nature, driving some people to unethical behavior in the hopes for easy profits.

Of course, there is nothing that one can do with the last cofactor (human nature), other than to accept it. The third one is more regulatory in nature and therefore falls beyond the scope of this chapter. For the other two, the separation of identifiers and locators can be used to create architectures where a sender must acquire the recipient's consent before it can send data beyond severely rate-limited signaling messages.

6.2.4   Lack of authentication, privacy, and accountability

The aim of authentication, privacy, and accountability is to prevent organizationally or socially undesirable things from happening. The Host Identity Protocol does not directly provide means to address the privacy and accountability problems. However, it changes the landscape in a number of ways. Firstly, it uses cryptographic host identifiers as an integral part of connectivity, thereby providing automatic identity authentication. This makes it easier to attribute a series of acts to a distinct host. Secondly, the separation of identities and locators makes it easier to hide the topological location of communicating parties. Thirdly, there are a few privacy extensions to HIP [10, 11] that allow the identities of the communicating parties to be hidden from third parties.

6.3   The HIP architecture and base exchange

In this section, we describe what HIP is in detail; after that, in the two next sections, we turn our attention to how HIP can be used as a tool in alleviating the above discussed architectural problems.

Over the years, the aims of the HIP-related work have broadened from a name space that works as a security and mobility tool, through generalizing the mobility support to cover also multi-homing, towards a general sublayer that provides interconnectivity in the same way the original IP did. In other words, the current HIP aim can be characterized as providing the lowest layer in the stack that encompasses location-independent identifiers and end-to-end connectivity. Furthermore, HIP cleanly separates host-to-host signaling and data traffic into separate planes, i.e., it can act both as an interconnectivity-level signaling protocol and as a general carrier for higher-layer signaling. Thereby it has appeal for architectures where control is separated from data, e.g., due to commercial requirements.

Starting from the architectural ideas, i.e., the desire to add a new, secure name space and a new layer of indirection, HIP has been carefully engineered towards two goals that are in partial conflict with the architectural ideas. Firstly, it is inherently designed to utilize, in an unmodified form, the current IP-based routing infrastructure (IPv4 and IPv6). At the same time, it was engineered to support the current application networking APIs with sufficient semantic compatibility. Hence, HIP is backwards compatible and deployable in parallel to the existing stack, requiring no changes to the routing infrastructure or to the typical user-level applications. Secondly, the goal has been to implement the desired new (or renewed) functionality with a minimal set of changes to the existing system. In practical terms, a HIP implementation that is integrated with an existing kernel-level TCP/IP typically requires only a few hundred lines of code modifications to the kernel; all the rest runs in user space. .

6.3.1 Basics

As mentioned above, the core of HIP lies in implementing the so-called identifier/locator split in a particular way. As already briefly discussed, in traditional IP networks each host has an IP address that serves for two different purposes: it acts both as a locator, describing the current topological location of the host in the network graph, and as a host identifier, describing the identity of the host, as seen by the upper-layer protocols [7]. Today, it has become impossible to use the same IP address for both purposes, due to the host mobility and multi-homing requirements.

A solution to this problem is to separate the identity and location information from each other. The Host Identity Protocol separates the locator and identifier roles of IP addresses by introducing a new name space, the Host Identity (HI) name space. In HIP, a Host Identity is a public cryptographic key from a public–private key-pair. A host possessing the corresponding private key can prove the ownership of the public key, i.e., its identity. As discussed briefly above, this separation of the identifiers and locators makes it also simpler and more secure to handle mobility and multi-homing than is currently possible.

Figure 6.1 shows, in approximate terms, how the new HIP sublayer is located in the current stack. On the layers above the HIP sublayer, the locator(s) of the host need not be known. Only the HI (or its 128-bit representation, a Host Identity Tag, HIT, or a 32-bit local representation, a Local Scope Identifier, LSI) are used. The Host Identity sublayer maintains mappings between identities and locators.

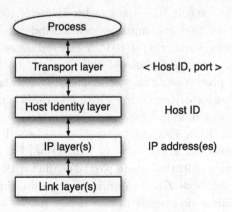

Figure 6.1 Approximate location of the HIP sublayer within the TCP/IP stack.

When a mobile host changes its location, HIP is used to transfer the information to all peer hosts. The dynamic mapping from the identifier to locators, on other hosts, is modified to contain the new locator information. Upper layers, e.g., applications, can remain unaware of this change; this leads to effective division of labor and provides for backwards compatibility.

During the connection initialization between two HIP hosts, a four-way handshake, a Base Exchange, is run between the hosts [12]. During the exchange, the hosts identify each other using public key cryptography and exchange Diffie–Hellman public values. Based on these values, a shared session key is generated. Further, the Diffie–Hellman key is used to generate keying material for other cryptographic operations, such as message integrity and confidentiality. During the Base Exchange, the hosts negotiate what cryptographic protocols to use to protect the signaling and data messages. As of today, the default option is to establish a pair of IPsec Encapsulated Security Payload (ESP) Security Associations (SA) between the hosts [13]. The ESP keys are retrieved from the generated Diffie–Hellman key and all further user data traffic is sent as protected with the ESP SAs. However, the HIP architecture is not limited to support only ESP. With suitable signaling extensions, some in preparation [14], it is possible to use HIP for protecting the user data of almost any standalone data protection protocol, such as SRTP [15] for real-time multimedia, and perhaps even with data-oriented, copy-and-forward protocols, such as S/MIME [16].

6.3.2   HITs and LSIs

When HIP is used, the Host Identity public keys are usually not written out directly. Instead, their 128-bit long representations, Host Identity Tags (HIT), are used in most contexts. According to the current specifications [17], a HIT looks like an IPv6 address with the special 28-bit prefix 2001:0010::/28, called Orchid, followed by 100 bits taken from a cryptographic hash of the public key.

It is important to note that embedding a cryptographic hash of the public key to the short identifier allows one to verify that a given HIT was derived from the given Host Identity. That is, due to the second pre-image resistance of the used hash functions, it is believed to be computationally unfeasible to construct a new Host Identity that hashes to a given, existing Host Identity Tag. Therefore, HITs are compact, secure handles to the public keys they represent.

As described in sections below, HITs are used to identify the communication parties both in the HIP protocol itself and in the legacy APIs. From the legacy API point of view, the second pre-image resistance establishes implicit channel bindings between the HIT and the underlying IPsec or other security associations. That is, if an application uses a HIT to connect a socket, it implicitly gains assurance that once the socket connects, the sent packets will be delivered to the entity identified by the corresponding Host Identity, and any received packets indeed come from that entity.

Unfortunately, in the IPv4 API the host identifiers, i.e., IP addresses, are only 32 bits long. Hence, even if all of these 32 bits were derived from the hash of the public key, there still would be occasional collisions. Hence, the approach taken by HIP is to use so-called Local Scope Identifiers (LSIs) in the IPv4 API [18]. These are assumed to be only locally unique; there are also no implicit channel bindings. However, with suitable IPsec policy expression it is still possible to create explicit channel bindings even for LSIs. Unlike HITs, LSIs are not sent on the wire within the HIP protocol.

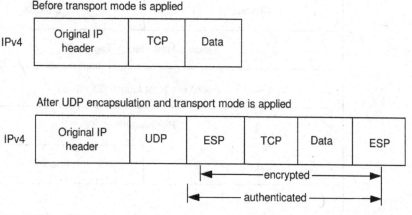

Figure 6.2 IPsec NAT traversal.

6.3.3   Protocols and packet formats

From the protocol point of view, HIP consists of a control protocol, a number of extensions to the control protocol, and any number of data protocols. The control protocol consists of the base exchange, any number of status update packets (that are typically used to convey extension protocols), and a three-message termination handshake that allows the peer hosts to cleanly terminate a protocol run [12]. For most extensions, there is some flexibility in which messages are used to carry the parameters comprising the extension. For example, multi-homing related information may be sent in three of the initial handshake messages or in update packets. With suitable extensions, HIP may be extended to use almost any data protocols. However, today the only defined data protocol is IPsec ESP [13].

By default, the HIP control protocol is carried directly in IPv4 and IPv6 packets, without any intervening TCP or UDP header. However, a larger fraction of the existing IPv4 NATs will not pass traffic with this protocol number through, or at best will allow only one host to communicate from behind the NAT. Therefore, work is being conducted to specify how HIP control messages may be carried in UDP packets [14–19]. The basic idea there is to use UDP encapsulation identical to IPsec IKE NAT traversal [20, 21]; see Figure 6.2. However, the other aspects will differ, and since mere packet encapsulation is not enough to allow NATed hosts to be contacted from the Internet, the UDP encapsulation format must be accompanied by a specification for NAT traversal details. At the time of writing, there were several competing proposals for that [19, 22, 23].

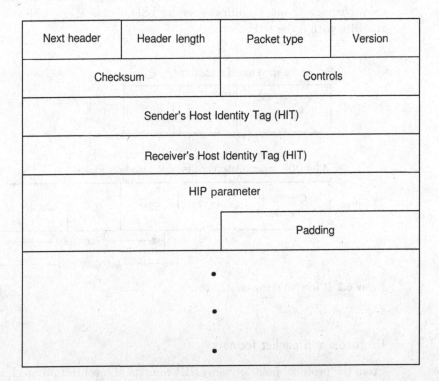

Figure 6.3 HIP control packet format.

The HIP control protocol packet format is depicted in Figure 6.3. The packet consists of a fixed header that is modeled after IPv6 extension headers. As the most important information, it also carries a packet type field and the sender's and receiver's HITs. The fixed header is followed by a variable number of parameters. The base exchange and each extension defines what parameters are needed and on what kind of HIP control messages the parameters may be carried. Most (but not all) messages also carry a cryptographic Hashed Message Authentication Code (HMAC) and a signature in the end of the packet. The former is meant for the peer host, which can use the Diffie–Hellman session keys necessary to verify the HMAC. The latter is meant for middle boxes, which typically do not have access to the Diffie–Hellman key but may well have access to the sender's public key [24]. After the base exchange, where the signature is used by the peer hosts for authentication, the signature is typically ignored by the receiver.

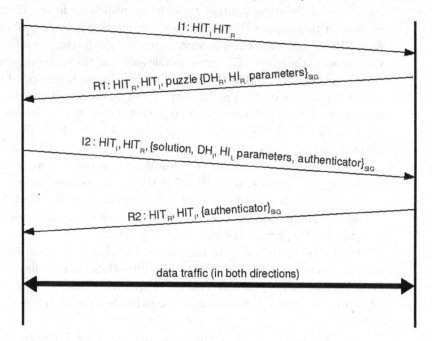

Figure 6.4 HIP base exchange.

The base exchange is depicted in Figure 6.4. It consists of four messages, named by letters and numbers. The letters denote the sender of the packet, I for initiator or R for responder. The numbers are simply sequential. Hence, the four messages are named as I1, R1, I2, and R2. The I1 message is a mere trigger. It is used by the initiator to request an R1 message from the responder. By default, any HIP host that receives an I1 packet will blindly reply with an R1 packet; that is, the responder will not remember the exchange.

Remaining stateless while responding to an I1 with an R1 protects the responder from state-space-exhausting denial-of-service attacks, i.e., attacks similar to the infamous TCP SYN one [25, 26]. However, as a side effect, it adds flexibility to the architecture. It does not need to be the responder itself that replies to an I1. Hence, if there is some other means by which the initiator may acquire a fresh R1 message, such as a directory lookup, it is perfectly fine to skip the I1/R1 exchange. As long as the host responding with an R1 has a supply of fresh R1s from the responder, it can be any node. This flexibility is used by some more advanced architecture proposals based on HIP, such as the Hi^3 proposal [27].

The R1 message contains a cryptographic puzzle, a public Diffie–Hellman key, and the responder's public Host Identity key. The Diffie–Hellman key in the R1 message allows the initiator to compute the Diffie–Hellman session key. Hence, when constructing the I2 message, the initiator already has the session key and can use keys derived from it.

In order to continue with the base exchange, the initiator has to solve the puzzle and supply the solution back to the responder in the I2 message. The purpose of this apparently resource-wasting method is to protect the responder from CPU-exhausting denial-of-service attacks by enforcing the initiator to spend CPU to solve the puzzle. Given the puzzle solution, the responder can, with very little effort, make sure that the puzzle has been recently generated by itself and that it has been, with high probability, solved by the initiator and is not a result of a puzzle posted much earlier or a puzzle generated by someone else. That is, by verifying the puzzle solution the responder knows that, with high probability, the initiator has indeed used quite a lot of CPU to solve the puzzle. This, assumedly, is enough to show the initiator's commitment to the communication, thereby warranting the forthcoming CPU cycles that the responder needs to process the rest of the I2 message. The difficulty of the puzzle can be varied depending on the load of the responder. For example, if the responder suspects an attack, it can post harder puzzles, thereby limiting its load.

The I2 message is the main message in the protocol. Along with the puzzle solution, it contains the initiator's public Diffie–Hellman key, the initiators public Host Identity key, optionally encrypted with the Diffie–Hellman key, and an authenticator showing that the I2 message has been recently constructed by the initiator.

Once the responder has verified the puzzle, it can confidently continue to construct the Diffie–Hellman session key, to decrypt the initiator's Host Identity public key (if encrypted), and to verify the authenticator. If the verification succeeds, the responder knows that there is out there a host that has access to the private key corresponding to the initiator's Host Identity public key, that the host wants to initiate a HIP association with the responder, and that the two hosts share a Diffie–Hellman session key that no other node knows (unless one of the hosts has divulged it) [28]. Given this information, the responder can consult its policy database to determine if it wants to accept the HIP association or not. If it does, the responder computes an authenticator and sends it as the R2 packet to the initiator.

The details of this relatively complex cryptographic protocol are defined in the HIP-based exchange specification [12]. From the high-level point of view, the HIP protocol can be considered as a member of the SIGMA family [29] of key exchange protocols.

6.3.4   Detailed layering

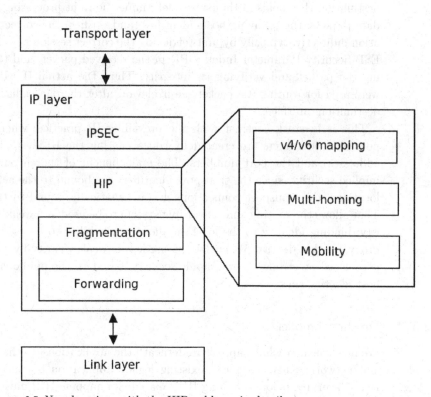

Figure 6.5 New layering, with the HIP sublayer in detail.

Let us now focus in more detail on how the new HIP sublayer is wedged into the existing stack. Figure 6.5 depicts the positioning of the new functionality in detail. The current IP layer functionality is divided into those functions that are more end-to-end (or end-to-middle) in nature, such as IPsec, and those that are more hop-by-hop in nature, such as the actual forwarding of datagrams. The Host Identity Protocol is injected between these two sets: architecturally immediately below IPsec, in practice often functionally embedded within the IPsec SA processing [30].

Now, in a communications system, the main function of in-packet identifiers is to allow demultiplexing. Forwarding nodes, such as routers, use the identifiers to determine which of the outgoing links to forward the packet to. The receiving host uses the identifiers to make sure that the packet has reached its right destination and to determine which upper layer protocol (if any) should process the packet. In the classical IP system, the IP address is used both by the routers to determine the next outgoing link and by the destination system to make sure that the packet has reached the right end host. With HIP, the separation of the location and identity information disentangles these two functions. Routers continue to use IP addresses to make their forwarding decisions.

At the hosts the behavior changes. For HIP control packets, the source and destination HIT fields in the packet determine the right processing context. For data packets, the receiving hosts identifies (and verifies) the correct HIP association indirectly, typically by first retrieving the correct session keys based on the ESP Security Parameter Index (SPI) in the received packet, and then decrypting the packet and verifying its integrity. Thus, the actual IP addresses that were used for routing the packet are irrelevant after the packet has reached the destination interface.

This is in stark contrast with the prevailing IP practice, where the transport layer identifiers are created by concatenating the IP-layer identifiers (IP addresses) and the port numbers. The main benefit of the current practice is implied security: since the transport identifiers are bound to the actual network locations, the transport connections get automatically bound to the locations. That allows the routing and forwarding system to be used as a weak form of security: binding identity to the location allows reachability to be used as a (weak) proxy for the identity. When HIP is used, this weak-security-by-concatenation is replaced by strong cryptographic security, based on the public cryptographic host identity keys.

6.3.5   Functional model

We now consider what happens underneath the applications, in the API, kernel, and network, when a typical, existing legacy application is configured to use HIP. Naturally, before anything HIP-related can happen, HIP must be installed into the system and the application(s) must be configured to use it. Today, typically the installation requires that a pre-compiled HIP package is installed; in most operating systems this requires administrator privileges. As the second step, the applications must be configured to use HIP. In most cases, there are three alternative configuration options. The simplest but least generic way is to configure the peer hosts' HITs directly into the application. For example, an IPv6-capable e-mail application can be configured to use HIP by entering the mail server's HIT into the configuration field that usually contains either a DNS name or an IP address. An IPv4-only application could be similarly configured with an LSI; however, to be secure, there should also be a corresponding IPsec policy rule in the IPsec policy database [18].

A more transparent way is to change the mapping from DNS names to IP addresses in a way that resolving a DNS name returns a HIT (or an LSI), instead of an IP address, to the application. Obviously, there are several options of how to implement that. In most UNIX-based systems the simplest way to change the mapping is to modify the local/etc./hosts file. Another, non-standard way is to store the HIT or LSI into the DNS in an AAAA or an A record. The drawback of this method is that as a result of such practice non-HIP-aware hosts may fail in non-obvious ways. Finally, the standard way is to store the HIT (and other information) into the new HIP resource record [31]. That allows both HIP and non-HIP hosts to create new connections with the target host without new difficulties. However, using the HIP resource record also means that the DNS software in the client has to be modified, so that it can give the requesting legacy application the HIT (or the LSI) instead of the IP address.

Once the application has got a HIT (or an LSI), it uses it in various socket API calls. In a typical implementation, the underlying libraries and the communication stack handles the HIT just as if it were a vanilla IPv6 address, all the way until the resulting packet is delivered to the IPsec module for processing. At the IPsec level, an IPsec policy rule is used to detect that the destination and source IP address fields in the packet contain the Orchid prefix. (For LSIs, an explicit, LSI-specific rule is typically required.) Usually, all Orchid packets are passed to IPsec ESP for processing. If there are no ESP Security Associations yet, the IPsec module requests a suitable pair of security associations from the HIP control functionality, which in turn creates the SAs by executing the HIP ESP extension [13], either as a part of a base exchange or over an existing HIP control association using update messages.

For ESP processing, typical HIP implementations use a non-standard variant of the ESP modes, called the BEET mode [30]. The BEET mode can be considered as standard transport mode that is enhanced with built-in address rewriting capabilities. To achieve HIP functionality, at the sending end the SA is configured so that the HITs in an outgoing packet are converted to IP addresses. Conversely, at the receiving end, the IP addresses in the packet are discarded and, if the packet passes integrity verification, the HITs are placed in the packet header. As a part of this processing, it is also possible to rewrite the IPv6 header used to carry the HITs (or the IPv4 header carrying LSIs) into an IPv4 header carrying IPv4 addresses (resp. IPv6 header carrying IPv6 addresses). Between the sender and the receiver, the packet looks like a standard IPsec ESP transport mode packet, with IP addresses in the header, and is handled by all HIP-unaware nodes as such.

At the receiving end, once an incoming packet has been processed by the IPsec ESP module, the packet contains the sender's HIT in its source field. Since this HIT was placed to the packet during the IPsec processing, and only if the packet passed verification, the HIP sublayer provides assurance to all of the upper layers that the packet was indeed received through an IPsec Security Association that was securely created through a cryptographic protocol where the private key corresponding to the HIT was present. In other words, the upper layers at the receiver end can trust that the Host Identity represented by the source address is indeed valid and not a result of IP source address spoofing.

6.3.6   Potential drawbacks

While HIP has generally been carefully designed to be backwards compatible with existing applications and infrastructure, obviously any change may have its drawbacks; so with HIP, too. In this section we briefly discuss the most general potential drawbacks; there certainly are others, more situation-specific ones.

Perhaps the biggest difference to present communication is that HIP introduces a slight delay, caused by the base exchange, whenever starting to communicate with a new host. This delay is mainly caused by the time taken to solve the puzzle, to process the public key signatures, and to produce the Diffie–Hellman public key. While the amount of time can be somewhat reduced by using trivial puzzles and short keys, it cannot be eliminated. One potential way to alleviate the situation is to use Lightweight HIP (LHIP), a variant of HIP proposed by Heer [32]. However, both using trivial puzzles/keys and using LHIP are clearly less secure than baseline HIP.

A second, often-mentioned drawback is the need to change the operating system kernel; many people are understandably concerned about this, even though the modifications are very small, as discussed above. One alternative, used by the Boeing implementation, is to divert the traffic to user level and process all packets there. However, this alternative is somewhat less efficient than a kernel-based implementation.

An often-cited potential drawback with HIP relates to the so-called third-party referral problem, where one host sends a name of a second host to a third host. In practice, in a third-party referral situation the names are IP addresses and there are three IP hosts, A, B, and C. Hosts A and B have an ongoing connection (such as a TCP connection); therefore A knows B's IP address. Host A now wants to initiate a connection between B and C by telling C to contact B. Obviously, it does that by sending B's IP address to C. With the IP address at hand, C is now able to open a new connection to B. Now, as the HITs are not routable, it is hard to open a new HIP association to a HIP host if all that one has is a HIT. Hence, if the third-party-referral application is a legacy one and if it chooses to use IP addresses and uses HITs instead, the referral process may fail.

We claim that the third-party referral problem is not that important nor bad in practice. Firstly, in a NATted environment, third-party referrals already fail now, indicating that the problem may be less important than sometimes claimed. That is, most commonly used applications no longer rely on third-party referral working. Secondly, using an overlay network to route HITs, it is possible to support even legacy applications that rely on third-party referrals working.

Another drawback is that HIP does impose a new management load on hosts and enterprises to manage the additional namespace. Key management needs to be more carefully considered and worked out; especially, the lack of aggregateability of flat key names for access control lists may cause problems in some environments. Also, an additional level of indirection may cause an increase in hard-to-debug network configuration errors and failures, which current implementations are only beginning to address adequately.

There are some considerations related to how HIP will work in a host where IPsec is used also for other, non-HIP purposes. For example, it remains an open question whether it is possible to first convert HITs into IP addresses, then run these IP addresses over an IPsec VPN connection. While these problems are real, they are not problems for HIP alone. The current IPsec architecture specification [4] is not too specific in explaining how a host should behave in a situation where IPsec is applied repeatedly.

Finally, some diagnostic applications, and probably a few other ones, will not work with HIP, at least not as intended. For example, the diagnostic tool ping can be used with HIP, but when given a HIT it no longer tests IP connectivity but HIP-based connectivity. Similar surprises are likely to be detected with other diagnostic applications. However, given the fact that HITs and other IP addresses are clearly distinguishable by the Orchid prefix, we doubt whether these new failure models would hamper operations in practice.

6.4   Mobility, multi-homing, and connectivity

Equipped with a basic understanding of what HIP is and how it works, we now continue to study how it can be used, together with a number of defined and prospective extensions, to address the problems discussed earlier.

6.4.1   HIP-based basic mobility and multi-homing

With HIP, packet identification and routing can be cleanly separated from each other. A host receiving a HIP control packet (other than I1) can verify its origin by verifying the packet signature; alternatively, the two end points of an active HIP association can simply verify the message authentication code. A host receiving a data packet can securely identify the sender through a three-step process: it first locates an ESP Security Association based on the Security Parameter Index (SPI) carried in the packet. As the second step, it verifies packet integrity and then decrypts the packet, if needed. Finally, as the third step, it places into the packet the source and destination HITs, as stored within the ESP BEET-mode SA. Thus, the actual IP addresses that were used for routing the packet are irrelevant after the packet has reached the destination interface.

Hence, to support mobility and multi-homing with HIP, all that is needed is the ability of controlling what IP addresses are placed in outgoing packets. As the addresses will be ignored by the recipient in any case, the sender may change the source address at will; for the destination address, however, it must know the address or addresses at which the receiver is currently being able to receive packets.

The HIP mobility and multi-homing extension [33] defines a Locator parameter that contains the current IP address(es) of the sending entity. For example, when the mobile host changes its location and therefore IP address, it generates a HIP control packet with one or more Locator parameters, protects the packet's integrity, and sends the packet to its currently active peer hosts. Note that the IP version of the locators may vary; it is even possible to use both IPv4 and IPv6 addresses simultaneously, and make a decision about the IP version used on outgoing IP packets depending on a local policy.

When the host receives a Locator parameter over an active HIP association, it needs to verify the reachability of the IP address(es) that are included in the parameter. Reachability verification is needed to avoid accepting non-functional and falsified updates. The verification can be skipped in special circumstances, for example, when the peer host knows that the network screens all address updates and passes only valid ones.

6.4.2   Facilitating rendezvous

While the mobility and multi-homing extension specifies how two hosts that already have a HIP association can exchange locator information and change the destination address in outgoing traffic, there remains another mobility-related problem: rendezvous. When another host wants to make a contact with a mobile host, when two mobile hosts have moved simultaneously and both have stale peer address information, or whenever the destination host's current IP address is unknown, e.g., since it is dynamic or kept private, there needs to be an external means to send packets to the host whose current locator is not known. In the Mobile IP world, this function is provided by the Home Agent, which forwards any messages sent to the mobile host's home address to the mobile host itself.

In the HIP mobility architecture, a similar function is provided by a Rendezvous server. Like a Mobile IP home agent, a HIP rendezvous server tracks the IP addresses at which hosts are reachable and it forwards packets received thereto. Unlike a home agent, a HIP rendezvous server forwards only HIP signaling packets (by default only the first packet of a base exchange), and the rest of the base exchange, as well as all subsequent communications, proceed over a direct path between the host and its peer. Furthermore, a HIP host may have more than one rendezvous server; it may also dynamically change the set of rendezvous servers that is able to serve it.

In HIP terms, a rendezvous server is simply a HIP host that knows another HIP host's current locator and is willing to forward I1 packets (and possibly other HIP control packets) to that host. In theory, any HIP host could act as a rendezvous server for any of its active peer hosts, as it already knows the peer host's locator or locators. However, in practical terms, it is expected that there will be a number of stationary hosts, located in the public Internet, providing rendezvous as a service.

The rendezvous service is defined by means of two HIP extensions. First, the generic service registration extension [34] is used by a rendezvous server and a prospective client to agree on the existence and usage of the service in the first place. Second, the rendezvous service extension [35] defines the terms of the specific service. Both of the specifications are quite simple and straightforward. The reason there are two documents, and not only one, is reusability: the registration extension can also be used to define other services besides the rendezvous service.

The HIP rendezvous server is very simple. When processing incoming HIP control packets, if the server receives a packet that does not contain any of its own HITs, the server consults its current HIT-to-locator mapping table. If there is a match, the packet is forwarded to the locator listed in the table. To facilitate tracking, the packet is augmented with the original addresses from the incoming packet. However, what is noteworthy here is that the IP addresses in the incoming packet are not used for any demultiplexing decisions. They are simply copied to a new HIP parameter, which is then added to the packet. Hence, the rendezvous server does not need a pool of IP addresses, in the way a Mobile IP Home Agent does. This is because each mobile host is identified by the HIT and not the address; the destination address in the incoming packet is transient.

6.4.3   Mobility between addressing realms and through NATs

As described earlier, HIP supports mobility between and within different IP address realms, i.e., both within and between IPv4 and IPv6 networks. However, all addresses in the scenarios above have been assumed to be drawn from the public address space, i.e., we have implicitly assumed that the (functional) addresses in the Locator payload are routable. As of today, the mobility and multi-homing specification [33] does not describe how one can be mobile when the host has addresses only from a private address space, i.e., when it is behind a legacy NAT device. However, there is ongoing work to define how NATs can be enhanced to understand HIP and how HIP can be enhanced to pass legacy NATs [19, 22, 23, 24].

We now turn our attention to how to make NATs an integral part of the architecture. In contrast to the legacy NATs, which can be seen as a hack as they require both UDP encapsulation and explicit support nodes at the un-NATed side of the network, integrated NATs are architecturally cleaner; they pass all data protocols, including ESP, and do not require external support nodes. One such approach is presented in the SPINAT proposal [36]. A SPINAT-based NAT device is HIP-aware and can take advantage of the passing-by HIP control packets, using them to learn the necessary details for demultiplexing data protocols. Additionally, a SPINAT device must also implement the HIP rendezvous functionality, acting as a rendezvous server for all hosts that are behind it.

In practice, a SPINAT device uses information from the HIP base exchange and UPDATE packets to determine the IP addresses, HITs, and SPI values (or other contextual identifiers) in the data protocol. With this information, the SPINAT device can do required address translations between public and private address spaces, mapping several HIP identities (and corresponding data protocols) to a single external IP address.

With SPINAT, a HIP host residing behind a NAT device learns directly the mapped address as a part of rendezvous registration. Furthermore, there is no difference between a mapped address, as identified by STUN, and an allocated address, as offered by TURN. They are both replaced by one shared external address, the data protocols being explicitly mapped with the help of the information from the HIP control messages.

6.4.4   Subnetwork mobility

So far we have only considered mobility in terms of mobile hosts. Now we turn our attention towards mobility at other granularity levels, first considering how HIP can be used to implement subnetwork mobility and then, in the next subsection, how application-level mobility can be implemented. A key feature of both of these is delegation. That is, the use of public cryptographic keys as identifiers adds cryptographically secure delegation, as a primary element, into the architecture.

The basic idea behind cryptographic delegation is simple, but the result is powerful [37]. When a principal (e.g., a host), identified by a public key, wants to delegate some rights (such as access rights) to another principal, identified by another public key, all that is needed is that the former signs a statement indicating that the latter is authorized to perform the operations needing the rights. The statement must include the delegate's public key to identify the delegate and it must be properly signed by the delegator's private key so that its validity can be verified. Furthermore, the delegator must itself possess the delegated right. Typically, but not necessarily, such delegations form an implicit loop where authority flows from a resource possessor through some intermediates to the prospective resource consumer, and from there back to the resource owner [37].

In the context of mobility and mobile sub-networks, delegation can be used to delegate the actual act of sending HIP control messages containing new locators from the individual mobile hosts to a mobile router, and further from the mobile router to some infrastructure node at the fixed network side.

Figure 6.6 illustrates the basic idea. First, individual mobile hosts in a mobile network delegate, to a mobile router, the right to inform their peer hosts about their location. At this stage the mobile router could send HIP signaling messages on behalf of all the hosts behind it, but such a practice would use quite a lot of capacity at the air interface. Hence, in the next step, the mobile router further delegates that right to a router (or another infrastructure node) within the fixed part of the network (illustrated as a cloud). Once the fixed-side router learns

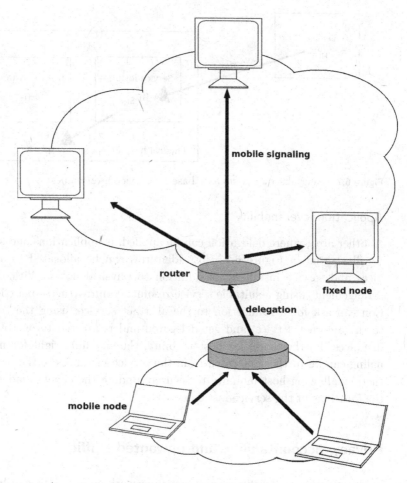

Figure 6.6 A moving network scenario.

that the mobile subnetwork has moved, it will send updates to the relevant peer hosts.

As an additional optimization, if the underlying IP-layer router mobility functionality is arranged in such a way that the fixed-side router gets directly informed whenever the mobile router changes its point of attachment, it becomes possible for the fixed-side router to send the mobility messages directly to the corresponding hosts of all mobile hosts within the mobile subnetwork, without any signaling at all over the air interface. Hence, for HIP-based mobility signaling, no HIP-related messages need to be transmitted over radio at mobility events. On the other hand, whenever a new host joins the mobile network, a few messages are needed to establish the delegation.

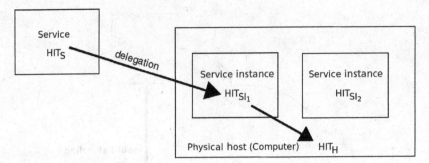

Figure 6.7 Using abstract hosts as a base for service-level names.

6.4.5   Application-level mobility

Another area, where delegation can be applied, is applications and services [38].
As illustrated in Figure 6.7, Host Identities can be allocated to abstract services and service instances, in addition to physical hosts. With that kind of
arrangement, using a suitable service resolution infrastructure, a client application can ask for a connection to the abstract service, using the HIT assigned
to the abstract service, and get delegated and redirected to one of the service
instances. Furthermore, for host mobility, the signaling right for mobility signaling can be further delegated from the service instances to the physical node,
thereby allowing host mobility to securely update the clients' understanding of
the locations of the service instances.

6.5   Privacy, accountability, and unwanted traffic

In this section, we briefly describe ongoing work on how HIP can help to solve
the privacy and accountability problem and drastically reduce some forms of
unwanted traffic. While this work is less mature than what has been discussed
above, it still appears relatively solid with a number of publications backing it.

6.5.1   Privacy and accountability

The base use of public cryptographic keys for host identification is clearly a
means that enhances accountability. At the same time, it can endanger privacy.
HIP tries to approach a balance, allowing both enhanced accountability and
privacy. Naturally, achieving such a balance is tricky, and it remains to be seen
whether the mechanisms currently built-in and proposed are sufficient.

   In HIP, the Host Identity public keys may be anonymous in the sense they do
not need to be registered anywhere. For example, a privacy conscious host could
create a multitude of key pairs and identify itself through a different key to each
website during a surfing session. Were this combined with dynamically changing
IP addresses, the websites could not correlate the activity through lower-layer

identifiers; however, anything identity-revealing in HTTP, such as cookies, could still be used, of course.

Unfortunately, while using multiple identities and changing addresses allows one to preserve privacy towards remote peer hosts, that is not sufficient to remain anonymous or pseudonymous towards the network. The network nodes close to the host could easily correlate the public keys over multiple IP addresses, determining which traffic flows belong to which host, with high probability. Encrypting the Host Identity key in the I2 packet is not sufficient to defeat such tracking, as the HITs, present in all control packets, can be easily used instead.

To alleviate the situation, a few years ago we proposed a so-called BLIND extension to HIP [10]. The basic idea is simple: instead of sending plain HITs in control packets, one hashes the HIT with a random number and sends the hash result and the random number in the initial control packets. Once the connection has been established, the actual HIT can be revealed to the responder. Alternatively, if the responder has only a limited number of HITs that it accepts connections from, it can try each of them in turn to see if the incoming connection is from a trusted peer host.

The BLIND approach was recently implemented and simultaneously enhanced by Lindqvist and Takkinen [39]. Besides using BLIND, their approach also uses identifier sequences [40]. That is, they replace all constant, easily trackable identifiers, in the HIP control protocol and in the data protocol below the encryption layer, with pseudo-random sequences of identifiers. The peer hosts derive the seeds for the pseudo-random number generator from the HIP Diffie–Hellman session key. While the exact details of their approach fall beyond the scope of this chapter, the result appears to provide a high level of privacy towards all eavesdropping third parties, while still allowing the peer hosts to communicate efficiently and securely.

Altogether, it looks like that while the exact details of a comprehensive, HIP-based privacy and accountability approach remain to be defined, the pieces of existing work clearly indicate that it is possible to simultaneously enhance both privacy and accountability through clever use of HIP-based mechanisms.

6.5.2 Reducing unwanted traffic

As discussed earlier, there are two technical aspects in unwanted traffic that allow its fundamental root causes to be affected. First, we can change the architecture so that the recipient names are either not immediately accessible to the prospective senders or require the recipient's consent before the network delivers any packet to the recipient. Second, we can attempt to raise the marginal cost of sending packets or reduce the marginal cost of receiving packets.

In the HIP context, we can consider the base exchange to employ the latter approach to defeat state-space and CPU-exhausting denial-of-service attacks. Using the former approach requires more changes; one possible way might be what we have formerly proposed in the Hi^3 overlay architecture [27]. The basic idea is to hide recipients' IP addresses and to require explicit consent from the recipients before a sender can use the addresses as destinations for sending traffic to.

From an architectural point of view, overlays similar to Hi^3 create an additional "routing" layer on top of the IP layer. Also other overlays aiming at added protection, such as SOS by Keromytis *et al.* [41] or k-anonymous overlays by Wang *et al.* [42], work basically in the same way.

An important aspect in overlay networks is that they change the naming structure. In more primitive cases, they merely replace the current IP addressing structure with another destination-oriented name space. However, at the same time, they may make denial-of-service attacks harder by dispersing the interface, i.e., instead of choking a single target host, the potential attacker must now flood the whole overlay system, which may consist of thousands or millions of nodes. That increases the overall cost of sending by introducing artificial costs to force the participating nodes to play by the rules. The more advanced overlays further change the rules by changing the naming focus from hosts and locations to pieces of information.

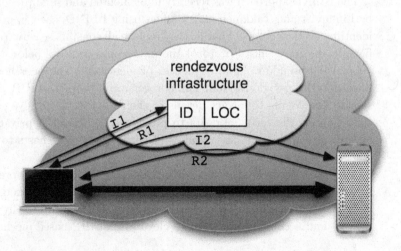

Figure 6.8 Hi^3 architecture.

The basic Hi^3 setting is illustrated in Figure 6.8. The network consists of two "planes," a data plane that is a plain IP-based router network, with HIP-enabled firewalls located at strategic points, and a control plane that is implemented as an overlay on top of the data plane. In practice, the control plane consists of enhanced HIP rendezvous servers that typically synchronize location information either partially or fully between each other.

As should be apparent by now, HIP-enabled firewalls can authenticate passing HIP base exchange and update packets, and punch holes for IPsec ESP traffic selectively [27]. In the Hi^3 architecture all servers are located behind HIP-enabled firewalls. To become accessible, the servers must register to the rendezvous infras-tructure, creating a binding between the servers' identity (ID) and the prospec-tive locators (LOC). While registering, the servers may also cache a number of pre-computed R1 packets at the rendezvous infrastructure.

When a client wants to make a contact with a server, it sends the first HIP base exchange message, I1, as usual. However, this packet cannot be sent to the server, for two reasons. First, the client will not know the server's IP address. Secondly, even if it knew one, an intervening firewall would drop the packet. Hence, the only option is to send the packet to the rendezvous infrastructure. The infrastructure looks up a cached R1 packet, if it has one, and passes it to the client (otherwise it needs to pass the packet to the server, which is undesirable). The client solves the puzzle, in the usual way, and sends an I2 packet, again to the rendezvous infrastructure. The receiving node within the infrastructure verifies that the puzzle has been correctly solved and, if so, passes the I2 packet to the server. The firewall will pass the packet as it is coming from the infrastructure. At this point, the server can verify the clients identity and determine if the client is authorized to establish a HIP association. Hence, if and only if the client can be positively identified and has proper authority, the server responds to the client with an R2 packet. The R2 packet can either be sent directly to the client if it will pass the firewalls anyway, or it may be necessary to direct it through the infrastructure, thereby indirectly triggering hole punching at the firewall. Finally, the actual data traffic traverses directly at the data plane, through the firewalls.

The result of this arrangement is that only authorized clients will ever learn the server's IP address. Furthermore, if an IP address is still revealed and used to launch a traffic-based denial-of-service attack against the server or its network, the firewall will stop most of the traffic, as the packets would be arriving with unauthorized packet identifiers. In case the firewall itself becomes congested, remaining legitimate traffic can be redirected through other firewalls, using the HIP mobility and multi-homing extension.

In summary, the Hi^3 proposal introduces one particular way to make cer-tain types of denial-of-service attacks harder than they are today, by making IP addresses less accessible, without compromising legitimate connectivity or data traffic efficiency.

6.6   Current status of HIP

This section describes the current status of HIP and its usage in the Internet today.

Maturity status

So far we have described the HIP architecture, basic design, a number of exten-sions, and discussed how they can or possibly could be used to alleviate a number of hard problems in the present Internet. We now turn our focus more to the present, and describe the current standardization and implementation status. All basic research and development for the first usable version of HIP is ready. There are three open source implementations, by Ericsson Research Nomadic Lab, Helsinki Institute for Information Technology (HIIT), and Boeing Research and Technology (the OpenHIP project). These all are mature for experimental use, differing on used platforms and supported extensions. Today, HIP is used by a few people on a daily basis, and is in daily production use at one Boeing air plane assembly factory. RFCs 5201–5207 were published during 2008 and HIP implementations were updated according to the specifications.

Usage of HIP today

Individual researchers at Ericsson, HIIT, and Boeing use HIP in their everyday life, mostly using Linux laptops to access specific HIP-enabled services, such as e-mail. Most of their traffic still flows over vanilla IPv4, though, with HIP being used only for the few services where the servers are also HIP-enabled.

The Boeing Company has been experimenting with HIP as a component of an overall Secure Mobile Architecture (SMA) [43] in its enterprise network. Boeing's SMA implementation integrates HIP with the company's public-key infrastructure (PKI) and Lightweight Directory Access Protocol (LDAP) back-end, as well as with location-enabled network services (LENS) that can identify the location of a wireless device through triangulation and signal strength measurements [44]. The SMA architecture responds to Boeing's enterprise needs to better secure a de-perimeterized network environment, as well as to Federal Aviation Administration (FAA) requirements that every step in the process of building an air plane be documented with the individual and equipment used [45]. HIP underpins the SMA architecture, and the identifiers in use in the network include machine (tool) host identifiers as well as temporary host identifiers linked to employee smart badges that are presented as part of the network log-in process.

Boeing has deployed an SMA pilot in its Everett, WA manufacturing facility. The architecture allows for network-based policy enforcement using Endboxes that can limit network connectivity using cryptographic identity rather than IP or MAC addresses. Complex trust relationships between contractors, employees, tools, and the enterprise network can be reflected in the network policy. One pilot deployment has been to secure the wireless network between Boeing's 777 "crawlers" and their "controllers" (part of the implementation of the moving assembly line for the 777 aircraft) [44].

In the IETF Peer-to-Peer SIP (P2PSIP) working group there are proposals by two independent organisations (Avaya and HIIT) to use HIP as a connectivity and security platform [46, 47]. The basic idea in these approaches is to apply HIP's ability to separate control and data traffic into different planes. This is accomplished using a distributed, peer-to-peer rendezvous service to establish HIP associations and to carry SIP packets, while running the actual data traffic directly over IP networks. The attractive features of HIP here seem to be opportunistic and leap-of-faith security [40], ability to work through NAT devices, seamless support for mobility and multi-homing, and the ability to pass control traffic through an overlay-like rendezvous structure.

Standardization situation

The HIP architecture document was published as RFC 4423 [1] in 2006. All the base protocol documents were published in 2008. The documents define the base exchange [12], using ESP transport format with HIP [13], the protocols and details for end-host mobility and multi-homing with HIP [33], the registration extension used in announcing and requesting HIP-based services [34], the rendezvous extension needed to use rendezvous services [35], HIP Domain Name System (DNS) extensions [31], and the details of using the HIP with legacy applications [18]. Additionally, a HIP Research Group at the Internet Research Task Force (IRTF) published a document specifying the issues related to NAT and firewall traversal [48].

After publication of the base RFCs, the HIP Working Group has been re-charted to focus on details for NAT traversal [19], native API [49], HIP-based overlay networks (HIP-BONE) [50], and carrying certificates in the HIP base exchange [51]. Several research drafts, such as the use of distributed hash tables (DHTs) for HIT-based lookups and a HIP experiment report [52] are being progressed at the HIP Research Group at IRTF.

6.7   Summary

In this chapter, an overview of the Host Identity Protocol was presented. The chapter discussed the need for a protocol such as HIP, the functional details of the HIP architecture and protocol messages, its potential applications and key advantages, and the current status of HIP's adoption and standardization efforts.

Acknowledgements

The authors are pleased to acknowledge the help of Mr. C. N. Sminesh, Mr. Sarang Deshpande and Professor Krishna Sivalingam of IIT Madras towards editing this document.

References

[1] Moskowitz, R., Nikander, P. Host Identity Protocol (HIP) Architecture. RFC 4423, May 2006.

[2] Perkins, C. IP Mobility Support for IPv4. RFC 3344, IETF, August 2002.

[3] Perkins, C., Calhoun P. R., Bharatia, J. Mobile IPv4 Challenge/Response Extensions (Revised). RFC 4721, IETF, January 2007.

[4] Kent, S., Seo, K. Security Architecture for the Internet Protocol. RFC 4301, IETF, December 2005.

[5] Rosenberg, J. Interactive Connectivity Establishment (ICE): A Protocol for Network Address Translator (NAT) Traversal for Offer/Answer Protocols. Work in progress, Internet Draft draft-ietf-mmusic-ice-18.txt, IETF, September 2007.

[6] Huitema, C. Teredo: Tunneling IPv6 over UDP through Network Address Translations (NATs). RFC 4380, IETF, February 2006.

[7] Chiappa, J. N. Endpoints and Endpoint Names: A Proposed Enhancement to the Internet Architecture. Unpublished Internet Draft, 1999. http://ana.lcs.mit.edu/jnc/tech/endpoints.txt

[8] Harrison, T., Williams, C., Mackrell, W., Bunt, R. Mobile Multicast (MoM) Protocol: Multicast Support for Mobile Hosts. *Proc. of the Third Annual ACM/IEEE International Conference on Computing and Networking (MOBICOM97).* pp. 151–160. ACM, 1997.

[9] Kovacshazi, Z., Vida, R. Host Identity Specific Multicast. *Proc. of the International Conference on Networking and Services* (June 19–25, 2007). ICNS07. IEEE Computer Society, Washington, DC. June 2007. DOI 10.1109/ICNS.2007.66

[10] Ylitalo, J., Nikander, P. BLIND: A Complete Identity Protection Framework for End-points. *Security Protocols, Twelfth International Workshop. Cambridge, 24–28 April 2004.* LCNS 3957, Wiley, 2006. DOI 10.1007/11861386-18

[11] Takkinen, L. Host Identity Protocol Privacy Management. Masters Thesis, Helsinki University of Technology, March 2006.

[12] Moskowitz, R., Nikander, P., Jokela, P. (ed.), Henderson TR. Host Identity Protocol. RFC 5201, IETF, April 2008.

[13] Jokela, P., Moskowitz, R., Nikander, P. Using ESP transport format with HIP. RFC 5202, IETF, April 2008.

[14] Tschofenig, H., Shanmugam, M. Using SRTP Transport Format with HIP. Work in progress, Internet Draft draft-tschofenig-hiprg-hip-srtp-02.txt, October 2006.

[15] Baugher, M., Carrara, E., McGrew, D. A., Nslund, M., Norrman, K. The Secure Real-time Transport Protocol (SRTP). RFC 3177, IETF, March 2004.

[16] Ramsdell, B. (ed.) Secure/Multipurpose Internet Mail Extensions (S/MIME) Version 3.1, Message Specification. RFC 3851, IETF, July 2004.

[17] Nikander, P., Laganier, J., Dupont, F. An IPv6 Prefix for Overlay Routable Cryptographic Hash Identifiers (Orchid). RFC 4834, IETF, April 2007.

[18] Henderson, T. R., Nikander. P., Mikka, K. Using the Host Identity Protocol with Legacy Applications. RFC 5338, IETF, September 2008.

[19] Komu, M., Henderson, T., Tschofenig, H., Melen, J., Keranen, A. Basic HIP Extensions for the Traversal of Network Address Translators. Work in progress, Internet Draft draft-ietf-hip-nat-traversal-06.txt, IETF, March 2009.

[20] Kivinen, T., Swander, B., Huttunen, A., Volpe, V. Negotiation of NAT-Traversal in the IKE. RFC 3947, IETF, January 2005.

[21] Huttunen, A., Swander, B., Volpe, V., DiBurro, L., Stenberg, M. UDP Encapsulation of IPsec ESP Packets. RFC 3948, IETF, January 2005.

[22] Tschofenig, H., Wing, D. Utilizing Interactive Connectivity Establishment (ICE) for the Host Identity Protocol (HIP). Work in progress, Internet Draft draft-tschofenig-hip-ice-00.txt, June 2007.

[23] Nikander, P., Melen, J. (ed.), Komu, M., Bagnulo, M. Mapping STUN and TURN messages on HIP. Work in progress, Internet Draft draft-manyfolks-hip-sturn-00.txt, November 2007.

[24] Tschofenig, H., Gurtov, A., Ylitalo, J., Nagarajan, A., Shanmugam, M. Traversing Middleboxes with the Host Identity Protocol. *Proc. of the Tenth Australasian Conference on Information Security and Privacy (ACISP '05)*. Brisbane, Australia, July 4–6, 2005.

[25] Schuba, C. L., Krsul, I. V., Kuhn, M. G., *et al.* Analysis of a Denial of Service Attack on TCP. *Proc. of the 1997 IEEE Symposium on Security and Privacy*, IEEE, 1997.

[26] Eddy, W. M. TCP SYN Flooding Attacks and Common Mitigations. RFC 4987, IETF, August 2007.

[27] Nikander, P., Arkko, J., Ohlman, B. Host Identity Indirection Infrastructure (Hi^3). *Proc. of the Second Swedish National Computer Networking Workshop 2004 (SNCNW2004)*. Karlstad University, Karlstad, Sweden, November 23–24, 2004.

[28] Aura, T., Nagarajan, A., Gurtov, A. Analysis of the HIP Base Exchange Protocol. *Proc. of the 10th Australasian Conference in Information Security and Privacy*. Brisbane, Australia, July 4–6, 2005, pp. 481–493, LNCS 3574, Springer, 2005.

[29] Krawczyk, H. SIGMA: the SIGn-and-MAc Approach to Authenticated Diffie–Hellman and its Use in the IKE Protocols. *Proc. of Advances in Cryptology – CRYPTO 2003, 23rd Annual International Cryptology Conference*. Santa Barbara, California, USA, August 17–21, 2003, pp. 400–425, LCNS 2729, Springer, 2003.

[30] Nikander, P., Melen, J. A Bound End-to-End Tunnel (BEET) mode for ESP. Work in progress, Internet Draft draft-nikander-esp-beet-mode-07.txt, February 2007.

[31] Nikander, P., Laganier, J. Host Identity Protocol (HIP) Domain Name System (DNS) Extensions. RFC 5205, IETF, April 2008.

[32] Heer, T. Lightweight Authentication for the Host Identifier Protocol. Masters Thesis, RWTH Aachen, August 2006.

[33] Nikander, P., Henderson, T. R., Vogt, C., Arkko, J. End-Host Mobility and Multihoming with the Host Identity Protocol. RFC 5206, April 2008.

[34] Laganier, J., Koponen, T., Eggert, L. Host Identity Protocol (HIP) Registration Extension. RFC 5203, IETF, April 2008.

[35] Laganier, J., Eggert, L. Host Identity Protocol (HIP) Rendezvous Extension. RFC 5204, IETF, April 2008.

[36] Ylitalo, J., Salmela, P., Tschofenig, H. SPINAT: Integrating IPsec into Overlay Routing. *Proc. of the First International Conference on Security and Privacy for Emerging Areas in Communication Networks (SecureComm '05)*, Athens, Greece, September 5–9, 2005.

[37] Nikander, P. An Architecture for Authorization and Delegation in Distributed Object-Oriented Agent Systems. Ph.D. Dissertation, Helsinki University of Technology, March 1999.

[38] Koponen, T., Gurtov, A., Nikander, P. Application mobility with Host Identity Protocol, Extended Abstract in *Proc. of Network and Distributed Systems Security (NDSS '05) Workshop*, Internet Society, February 2005.

[39] Lindqvist, J., Takkinen, L. Privacy Management for Secure Mobility. *Proc. of the 5th ACM Workshop on Privacy in Electronic Society*. Alexandria, Virginia, USA, October 30–30, 2006. WPES'06. pp. 63–66. ACM. DOI 10.1145/1179601.1179612

[40] Arkko, J., Nikander, P. How to Authenticate Unknown Principals without Trusted Parties. *Security Protocols, 10th International Workshop*. Cambridge, UK, April 16–19, 2002, pp. 5–16, LCNS 2845, Springer, 2003.

[41] Keromytis, A. D., Misra, V., Rubenstein, D. SOS: Secure Overlay Services. SIGCOMM *Comput. Commun. Rev.* 32:4 (October 2002), 61–72. DOI 10.1145/964725.633032

[42] Wang, P., Ning, P., Reeves, D. S. A k-anonymous Communication Protocol for Overlay Networks. *Proc. of the 2nd ACM Symposium on Information, Computer and Communications Security*. Singapore, March 20–22, 2007. Deng R and Samarati P, eds. ASIACCS '07. pp. 45–56. ACM. DOI 10.1145/1229285.1229296

[43] Estrem, B. *et al.* Secure Mobile Architecture (SMA) Vision Architecture. Technical Study E041, The Open Group, February 2004. www.opengroup.org/products/publications/catalog/e041.htm

[44] Paine, R. R. Secure Mobile Architecture (SMA) for Automation Security. ISA Conference on Wireless Solutions for Manufacturing Automation: Insights for Technology and Business Success. 22–24 July, 2007, Vancouver, CA. www.isa.org/wsummit/presentations/BoeingNGISMAAutomationSecurityVancouverISApresentationtemplates7-23-07.ppt

[45] Boeing IT Architect Pushes Secure Mobile Architecture. *Network World*, April 28, 2006. www.networkworld.com/news/2006/050106-boeing-side.html

[46] Cooper, E., Johnston, A., Matthews, P. A Distributed Transport Function in P2PSIP using HIP for Multi-Hop Overlay Routing. Work in progress, Internet Draft draft-matthews-p2psip-hip-hop-00, June 2007.

[47] Hautakorpi, J., Koskela, J. Utilizing HIP (Host Identity Protocol) for P2PSIP (Peer-to-Peer Session Initiation Protocol). Work in progress, Internet Draft draft-hautakorpi-p2psip-with-hip-00, July 2007.

[48] Stiemerling, M., Quittek, J., Eggert, L. NAT and Firewall Traversal Issues of Host Identity Protocol (HIP) Communication. RFC 5207, IRTF, April 2008.

[49] Komu, M., Henderson, T. Basic Socket Interface Extensions for Host Identity Protocol (HIP), Work in progress, draft-ietf-hip-native-api-05.txt, July 2008.

[50] Camarillo, G., Nikander, P., Hautakorpi, J., Johnston, A. HIP BONE: Host Identity Protocol (HIP) Based Overlay Networking Environment, Work in progress, draft-ietf-hip-bone-01.txt, March 2009.

[51] Heer, T., Varjonen, S. HIP Certificates, Work in Progress, draft-ietf-hip-cert-00, October 2001.

[52] Henderson, T., Gurtov, A. HIP Experiment Report, Work in Progress, draft-irtf-hip-experiment-05.txt, March 2009.

7 Contract-switching for managing inter-domain dynamics

Murat Yuksel[†], Aparna Gupta[‡], Koushik Kar[‡], and Shiv Kalyanaraman[⊗]

[†]University of Nevada – Reno, USA [‡]Rensselaer Polytechnic Institute, USA [⊗]IBM Research, India

The Internet's simple best-effort packet-switched architecture lies at the core of its tremendous success and impact. Today, the Internet is firmly a commercial medium involving numerous competitive service providers and content providers. However, current Internet architecture neither allows (i) users to indicate their *value* choices at sufficient granularity, nor (ii) providers to manage *risks* involved in investment for new innovative quality-of-service (QoS) technologies and business relationships with other providers as well as users. Currently, users can only indicate their value choices at the access/link bandwidth level, not at the routing level. End-to-end (e2e) QoS contracts are possible today via virtual private networks, but with static and long-term contracts. Further, an enterprise that needs e2e capacity contracts between two arbitrary points on the Internet for a short period of time has no way of expressing its needs.

We propose an Internet architecture that allows flexible, finer-grained, dynamic contracting over multiple providers. With such capabilities, the Internet itself will be viewed as a "contract-switched" [15] network beyond its current status as a "packet-switched" network. A contract-switched architecture will enable flexible and economically efficient management of risks and value flows in an Internet characterized by many tussle points [4]. We view "contract-switching" as a generalization of the packet-switching paradigm of the current Internet architecture. For example, *size of a packet* can be considered as a special case of the *capacity of a contract* to expire at a very short term, e.g., transmission time of a packet. Similarly, *time-to-live* of packet-switching is roughly a special case of the *contract expiration* in contract-switching. Thus, contract-switching is a more general case of packet-switching with additional flexibility in terms of its economics and carefully reduced technical flexibility due to scalability concerns particularly at the routing level.

In this chapter, we report our recent work [15, 6] on contract-switching and a sample intra-domain contracting scheme with bailouts and forwards. We describe a contract-switching network architecture for flexible value flows for the future

Next-Generation Internet Architectures and Protocols, ed. Byrav Ramamurthy, George Rouskas, and Krishna M. Sivalingam. Published by Cambridge University Press. © Cambridge University Press 2011.

Internet, and for allowing sophisticated financial engineering tools to be employed in managing the risks involved in composition of e2e QoS contracts. We concentrate on the design of our contract-switching framework in the context of multi-domain QoS contracts. Our architecture allows such contracts to be dynamically composable across space (i.e., across ISPs) and time (i.e., over longer timescales) in a fully decentralized manner. Once such elementary instruments are available and a method for determining their value is created (e.g., using secondary financial markets), ISPs can employ advanced pricing techniques for cost recovery, and financial engineering tools to manage risks in establishment of e2e contracts and performance guarantees for providers and users in specific market structures, e.g., oligopoly or monopoly. In particular, we investigate elementary QoS contracts and service abstractions at micro (i.e., *tens-of-minutes*) or macro (i.e., *hours or days*) timescales. For macro-level operation over long timescales (i.e., *several hours or days*, potentially involving contracts among ISPs and end-users), we envision a link-state like structure for computing e2e "contract routes." Similarly, to achieve micro-level operation with more flexibility over shorter timescales (i.e., *tens of minutes*, mainly involving contracts among ISPs), we envision a Border Gateway Protocol (BGP) style path-vector contract routing.

Several QoS mechanisms have been adopted within single ISP domains, while inter-domain QoS deployment has not become reality. Arguably the reasons for this include the highly fragmented nature of the ISP market and the glut in core optical capacity due to overinvestment and technological progress of the late 1990s. BGP routing convergence and routing instability issues [5] also contribute to inter-domain performance uncertainties. Recent QoS research [10] clearly identified a lack of inter-domain business models and financial settlement methods, and a need for flexible risk management mechanisms including insurance and money-back-guarantees. Specifically, attempts to incorporate economic instruments into inter-domain routing and to allow more economic inter-domain flexibility to end-users [14] have surfaced. Our work directly focuses on these issues, and also relates to the Internet pricing research [8, 16].

In Section 7.1, we define the essence of the contract-switching paradigm. Section 7.2 details architectural characteristics and challenges of contract-switching. Then, in Section 7.3, we provide the motivation and tools from financial engineering that can be used for risk management in the Internet. We conclude in Section 7.4 by summarizing the key ideas developed in this chapter.

7.1   Contract-switching paradigm

The essence of "contract-switching" is to use *contracts* as the key building block for inter-domain networking. As shown in Figure 7.1, this increases the inter-domain architecture flexibility by introducing more tussle points into the protocol design. Especially, this paradigm will allow the much-needed revolutions in the Internet protocol design: (i) inclusion of economic tools in the network layer

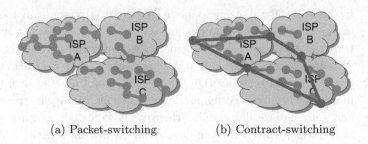

(a) Packet-switching   (b) Contract-switching

Figure 7.1 Packet-switching introduced many more tussle points into the Internet architecture by breaking the *end-to-end circuits* of circuit-switching into *routable datagrams*. Contract-switching introduces even more tussle points at the edge/peering points of domain boundaries by *overlay contracts*.

functions such as inter-domain routing (the current architecture only allows basic connectivity information exchange), and (ii) management of risks involved in QoS technology investments and participation in e2e QoS contract offerings by allowing ISPs to potentially apply financial engineering methods.

In addition to these design opportunities, the contract-switching paradigm introduces several research challenges. As the key building block, intra-domain service abstractions call for the design of (i) single-domain edge-to-edge (g2g) QoS contracts with performance guarantees, and (ii) nonlinear pricing schemes geared towards cost recovery. Moving one level up, composition of e2e inter-domain contracts poses a major research problem which we formulate as a "contract routing" problem by using single-domain contracts as "contract links." Issues to be addressed include routing scalability, contract monitoring, and verification as the inter-domain context involves large-size effects and crossing trust boundaries. Several economic tools can be used to remedy pricing, risk sharing, and money-back problems of an ISP.

7.2   Architectural issues

Inclusion of concepts such as values and risks into the design of network routing protocols poses various architectural research challenges. We propose to abstract the point-to-point QoS services provided by each ISP as a set of "overlay contracts" each being defined between peering points, i.e., ingress/egress points of the ISP. By considering these overlay contracts as "contract links," we envision a decentralized framework that composes an e2e contract from contracts at each ISP hop. This is similar to a path made up of links, except that we have "contracts" instead of links. Just like routing protocols are required for actually creating e2e paths from links, we need "contract routing" to find an inter-domain route that concatenates per-ISP contracts to compose an e2e path and an associated e2e contract bundle.

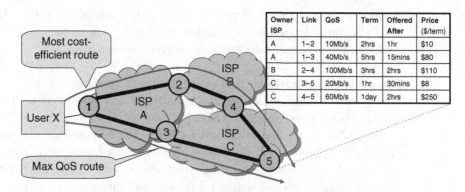

Owner ISP	Link	QoS	Term	Offered After	Price ($/term)
A	1–2	10Mb/s	2hrs	1hr	$10
A	1–3	40Mb/s	5hrs	15mins	$80
B	2–4	100Mb/s	3hrs	2hrs	$110
C	3–5	20Mb/s	1hr	30mins	$8
C	4–5	60Mb/s	1day	2hrs	$250

Figure 7.2 An illustrative scenario for link-state contract routing.

7.2.1   Dynamic contracting over peering points

We consider an ISP's domain as an abstraction of multiple g2g contracts involving buyer's traffic flowing from an ingress point to an egress point, i.e., from one edge to another. Previous work showed that this kind of g2g dynamic contracting can be done in a distributed manner with low costs [16]. Customers can only access network core by making contracts with the provider stations placed at the edge points. A key capability is that an ISP can advertise different prices for each g2g contract it offers, where locally computed prices can be advertised with information received from other stations.

Composition of better e2e paths requires flexibility in single-domain contracting capabilities. In our design, we consider point-to-point g2g contracting capabilities with unidirectional prices, which pose the question of "How should an ISP price its g2g contracts?". Though similar questions were asked in the literature [16] for simple contracts, bailout forward contracts (BFCs) require new pricing methodologies. Availability of such flexible g2g contracting provides the necessary building blocks for composing e2e QoS paths if inter-domain contracting is performed at sufficiently small timescales. This distributed contracting architecture gives more flexibility to users as well, e.g., users can potentially choose various next-hop intermediate ISPs between two peering points that are involved in users' e2e paths [14].

7.2.2   Contract routing

Given contracts between peering points of ISPs, the "contract-routing" problem involves discovering and composing e2e QoS-enabled contracts from per-ISP contracts. We consider each potential contract to be advertised as a "contract link" which can be used for e2e routing. Over such contract links, it is possible to design link-state or BGP-style path-vector routing protocols that compose e2e "contract paths." ISPs (or service overlay network providers) providing e2e services will need to be proactive in their *long-term* contracting and financial commitments and thus will need *link-state style* routing to satisfy this need. It is also possible to achieve scalable *BGP-style path-vector* routing at *short timescales* to accommodate on-demand dynamic user requests.

Macro-level operations: link-state contract routing

One version of inter-domain contract routing is link-state style with long-term (i.e., *hours or days*) contract links. For each contract link, the ISP creates a "contract-link-state" including various fields. We suggest that the major field of a contract-link-state is the forward prices in the future as predicted by the ISP now. Such contract-link-states are flooded to other ISPs. Each ISP will be responsible for its flooded contract-link-state and therefore have to be *proactively* measuring validity of its contract-link-state. This is very similar to the periodic HELLO exchanges among OSPF routers. When remote ISPs obtain the flooded contract-link-states, they can offer point-to-point and e2e contracts that may cross multiple peering points. Though link-state routing was proposed in an inter-domain context [3], our "contract links" are between peering points *within* an ISP, and not between ISPs (see Figure 7.2).

To compute the e2e "contract paths," ISPs perform a QoS-routing-like computation procedure to come up with source routes, and initiate a signaling protocol to reserve these contracts. Figure 7.2 shows a sample scenario where link-state contract routing takes place. There are three ISPs participating with five peering points. For the sake of example, a contract-link-state includes five fields: *Owner ISP*, *Link*, *Term* (i.e., the length of the offered contract link), *Offered After* (i.e., when the contract link will be available for use), and *Price* (i.e., the aggregate price of the contract link including the whole term). The ISPs have the option of advertising by flooding their contract-link-states among their peering points. Each ISP has to maintain a contract-link-state routing table as shown in the figure. Some of the contract-link-states will diminish in time, e.g., the link 1–3 offered by ISP A will be omitted from contract routing tables after 5 hrs and 15 mins. Given such a contract routing table, computation of "shortest" QoS contracts involves various financial and technical decisions. Let's say that the user X (which can be another ISP or a network entity having an immediate access to the peering point 1 of ISP A) wants to purchase a QoS contract from 1 to 5. The ISP can offer various "shortest" QoS contracts. For example, the route 1–2–4–5 is the most cost-efficient contract path (i.e., $(10\,\text{Mb/s}*2\,\text{hrs} + 100\,\text{Mb/s}*3\,\text{hrs} + 60\,\text{Mb/s}*24)/(\$10 + \$110 + \$250) = 27.2\,\text{Mb/s}*\text{hr/\$}$), while the 1–3–5 route is better in terms of QoS. The ISPs can factor in their financial goals when calculating these "shortest" QoS contract paths. The 1–2–4–5 route gives a maximum of 10 Mb/s QoS offering capability from 1 to 5, and thus the ISP will have to sell the other purchased contracts as part of other e2e contracts or negotiate with each individual contract link owner. Similarly, the user X tries to maximize its goals by selecting one of the offered QoS contracts to purchase from 1 to 5. Let's say that the ISP offers user X two options as: (i) using the route 1–2–4–5 with 10 Mb/s capacity, 2 hrs term, starting in 5 hrs with a price $15, and (ii) using the

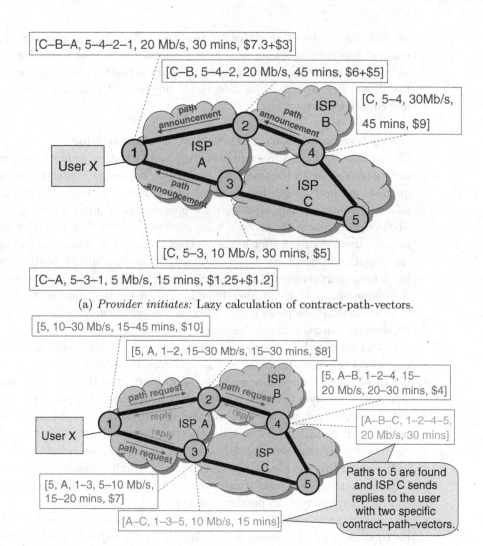

(a) *Provider initiates:* Lazy calculation of contract-path-vectors.

(b) *User initiates:* On-demand reactive calculation of contract-path-vectors.

Figure 7.3 Path-vector contract routing.

route 1–3–5 with 20 Mb/s capacity, 1 hr term, starting in 30 mins with a price $6, and user X selects the 1–3–5 route. Then, the ISP starts a signaling protocol to reserve the 1–3 and 3–5 contract links, and triggers the flooding of contract link updates indicating the changes in the contract routing tables.

One issue that will arise if an ISP participates in many peering points is the explosion in the number of "contract links," which will trigger more flooding messages into the link-state routing. However, the number of contract links can be controlled by various scaling techniques, such as focusing only on the longer-term contracts offered between the major peering points and aggregating contract-link-states "region-wise." Also, in our proposed link-state contract routing *floods only need to be performed if there is a significant change in contracting terms or in the internal ISP network conditions.* However, in traditional link-state routing, link-states are flooded periodically regardless if any change has happened.

Micro-level operations: path-vector contract routing

To provide enough flexibility in capturing more dynamic technical and economical behaviors in the network, it is possible to design contract routing that operates at short timescales, i.e., *tens of minutes.* This timescale is reasonable as current inter-domain BGP routing operates with prefix changes and route updates occurring at the order of a few minutes [11]. Further, an ISP might want to advertise a spot price for a g2g contract to a subset of other ISPs instead of flooding it to all. Similarly, a user might want to query a specific contracting capability for a short term and involving various policy factors. Such *on-demand reactive* requests cannot be adequately addressed by the link-state contract routing.

Just like BGP composes paths, e2e contract paths can be calculated in an on-demand lazy manner. In our design, each ISP has the option of initiating contract path calculations by advertising its contract links to its neighbors. Depending on various technical, financial, or policy factors, those neighbors may or may not use these contracts in composing a two-hop contract path. If they do, then they advertise a two-hop contract path to their neighbors. This path-vector composition process continues as long as there are participating ISPs in the contract paths. Users or ISPs receiving these contract paths will have the choice of using them or leaving them to invalidation by the time the contract path term expires.

Provider initiates: Figure 7.3(a) shows an example scenario where a provider initiates contract-path-vector calculation. An ISP C announces two short-term contract-path-vectors at peering points 3 and 4. The ISPs B and A decide whether or not to participate in these contract-path-vectors, possibly with additional constraints. For example, ISP B reduces the capacity of the initial path-vector to 20 Mb/s and increases its price to $11. Though each ISP can apply various price calculations, in this example ISP B adds $5 for its own contract link 2–4 on top of the price of the corresponding portion (i.e., $9*20/30 = $6) of the contract link 4–5 coming from ISP C. Similarly, ISP A constrains the two contract-path-vector announcements from ISPs B and C at peering points 2 and 3 respectively. Then, ISP A offers the two contract-path-vectors to the user X, who may choose to use the 1–5 short-term QoS path. In this path-vector computation scheme, whenever an ISP participates in a contract it will have to commit the resources needed for it, so that the users receiving the contract path announcements will have assurance in the e2e contract. Therefore, ISPs will have to decide carefully as potential security and trust issues will play a role. This game theoretic design exists in the current BGP inter-domain routing. In BGP, each ISP decides which route announcements to accept for composing its routes depending on policy, trust, and technical performance.

User initiates: Users may query for an e2e short-term contract path with specific QoS parameters which do not exist in the currently available path-vector. This kind of design can potentially allow involvement of end-users in the process depending on the application-specific needs. For example, in Figure 7.3(b), user X initiates a path-vector calculation by broadcasting a "contract-path request" to destination 5 with a capacity range 10–30 Mb/s, term range 15–45 mins with up to $10 of total cost. This contract-path request gets forwarded along the peering points where participating ISPs add more constraints to the initial constraints identified by the user X. For example, ISP B narrows the term range from 15–30 mins to 20–30 mins and the capacity range from 15–30 Mb/s to 15–20 Mb/s while deducting $4 for the 2–4 contract link of its own from the leftover budget of $8. Such participating middle ISPs have to apply various strategies in identifying the way they participate in these path-vector calculations. Once ISP C receives the contract-path requests, it sends a reply back to user X with specific contract-path-vectors. The user X then may choose to buy these contracts from 1 to 5 and necessary reservations will be done through more signaling.

7.3   A contract link: bailouts and forwards

From a services perspective, one can define a contract as an embedding of three major flexibilities [6] in addition to the contracting entities (i.e., buyer and seller): (i) *performance component*, (ii) *financial component*, and (iii) *time component*. The performance component of a contract can include QoS metrics such as delay or packet loss to be achieved. The financial component of a contract include various fields to aid financial decisions of the entities related with value and risk tradeoffs of the contract. The basic fields can be various prices, e.g., spot, forward, and usage-based. Interesting financial component fields can be designed to address financial security and viability of the contract, e.g., whether or not the contract is insured or has money-back guarantees. The time component can include operational time-stamp (e.g., contract duration, insured period when money-back guarantee expires) which are useful for both technical decisions by network protocols and economic decisions by the contracting entities. Note that all the three components operate over an aggregation of several packets instead of a single packet. For the potential scalability issues, this is the right granularity for embedding economic tools into network protocols rather than at the finer granularity of packet level, e.g., per-packet pricing.

In this section, we first formally define a forward contract and a BFC, and then present mathematical formalization for determining the price of a bailout forward contract (BFC) [6]. A bailout forward is useful for a provider since it eliminates the risk of demand for bandwidth in the future without imposing a binding obligation to meet the contract if the network cannot support it. A customer of a bailout forward contract locks in the bandwidth required in future, but obtains the bandwidth at a discount. The discount is provided since the customer shares the risk of the scenario that if the network is congested at the future time, the contracted bandwidth may not be delivered due to the bailout clause. The customer may choose not to purchase a forward, but in that case runs the risk of not being able to obtain the necessary bandwidth at the future time due to congestion or price reasons. Therefore, constructing and offering bailout forwards is beneficial for both providers and customers.

7.3.1   Bailout forward contract (BFC)

A forward contract is an obligation for delivering a (well-defined) commodity (or service) at a future time at a predetermined price – known as the 'Forward Price'. Other specifications of the contract are *Quality Specification* and *Duration* (start time – T_i, and end time – T_e, for the delivery of a timed service).

In the case of a capacitated resource underlying a forward contract, restrictions may be necessary on what can be guaranteed for delivery in future. A key factor that defines the capacity of the resource is used to define the restriction. A bailout clause added to the forward contract releases the provider from the obligation of delivering the service if the bailout clause is activated, i.e., the key factor defining the capacity rises to a level making delivery of the service infeasible. A setup is essential for the two contracting parties to transparently observe the activation of the bailout clause in order for the commoditization of the forward contract and elimination of moral hazard issues. The forward price associated with a bailout forward contract takes into account the fact that in certain scenarios the contract will cease to be obligatory.

Creation and pricing of a bailout forward contract on a capacitated resource allows for risk segmentation and management of future uncertainties in demand and supply of the resource. Contracts are written on future excess capacity at a certain price, the forward price, thus guaranteeing utilization of this capacity; however, if the capacity is unavailable at the future time, the bailout clause allows a bailout. Therefore, it hedges the precise segment of risk. The price of the bailout forward reflects this.

7.3.2   Formalization for pricing a bailout forward contract (BFC)

For pricing a bailout forward, we first need to define the price of spot contracts on which the forward is defined. The spot prices reflect present utilization of the network and price the contract using a nonlinear pricing kernel to promote utilization and cost recovery. The risks underlying the spot contract prices are key determinants for formulating the pricing framework for the bailout forward contract. Appropriate modeling abstractions are necessary.

In our "contract" abstraction of the network, a contract is defined unidirectionally between two endpoints in the network (between an ingress edge point and an egress edge point, in case of a single domain) instead of the traditional point-to-anywhere contracting scheme of the Internet. We model the time-dependent demand for the spot contract, μ_t, and the available capacity on this contract, A_t, where $\mu_t < A_t$. Price of the spot contract is a nonlinear transformation, $S_t = P(\mu_t, A_t)$. A predictive model for A_t is used as the bailout factor to define the bailout condition and to price the BFCs. Therefore, the forward price, F_t, is a function of the spot contract price, predicted future available capacity, and parameters that define the bailout term.

We model the time-dependent demand for the spot contract as follows:

$$d\mu_t = \gamma(m - \mu_t)dt + b_1\mu_t\,dW_t^1,$$ (7.1)

and the available capacity on the contract path as

$$dA_t = \beta(\overline{A} - A_t)dt + b_2A_t\,dW_t^2,$$ (7.2)

where the two Wiener processes (W_t^1 and W_t^2) are taken to be uncorrelated. Parameters in the evolution equations include the long-run mean, m and $\overline{A}$, the volatility constants, b_1 and b_2, and γ and β are rate of reversion to the long-run mean.

A specific choice of demand profile results in the spot price being the following function of μ_t and A_t:

$$S_t = P\left(\frac{\mu_t}{A_t}\right) = \int_0^{\mu_t/A_t} p(q)dq,$$ (7.3)

where $p(q)$ is the nonlinear pricing kernel, obtained as

$$p\left(\frac{\mu_t}{A_t}\right) = \frac{c + (1 - \mu_t/A_t) \times \alpha}{1 + \alpha}.$$ (7.4)

Parameters c and α are the marginal cost and the Ramsey number, respectively. The Ramsey number captures the extent of monopoly power the provider is able to exert [13], where $\alpha = 0$ corresponds to a perfect competition, and marginal price is the marginal cost. If $f(S_t, t)$ is the price of a bailout forward at some time t, then by the standard derivative pricing derivation for any derivative defined on the spot contract, S_t, gives that $f(S_t, t)$ should satisfy the following partial differential equation

$$\frac{\partial f}{\partial t} + \frac{1}{2}p^2\left(\frac{\mu_t}{A_t}\right)\left(b_1^2\frac{\mu_t^2}{A_t^2} + b_2^2A_t^2\right)\frac{\partial^2 f}{\partial S^2} + \frac{\partial f}{\partial S_t}rS_t - rf = 0,$$ (7.5)

together with the end condition

$$f(S_T, T) = (S_T - F)\mathbf{I}_{\{A_T > Th\}}.$$ (7.6)

where T is the time of delivery of service in future, F is the forward price, and $\mathbf{I}$ is the indicator function for no bailout defined in terms of a threshold level, Th [12]. The r in the partial differential equation is the short-term, risk-free interest rate. The solution of the above equation is obtained as follows:

$$f(S_0, 0) = E[e^{-rT}(S_T - F)\mathbf{I}_{\{A_T > Th\}}].$$ (7.7)

Since initially there are no payments, only the forward price is determined, we obtain the forward price, F, by equating the above equation to zero and solving for F,

$$F = \frac{1}{P(A_T > Th)} E[S_T \mathbf{I}_{\{A_T > Th\}}], \qquad (7.8)$$

where S_t in the risk-neutral world evolves by the process

$$dS_t = rS_t\, dt + p\left(\frac{\mu_t}{A_t}\right)\frac{b_1\mu_t}{A_t} dW_t^1 - p\left(\frac{\mu_t}{A_t}\right)\frac{b_2\mu_t}{A_t} dW_t^2. \qquad (7.9)$$

Multiple edge-to-edge (g2g) BFC definition and management

To capture a realistic network topology, we will need to generalize from a single g2g contract abstraction of the network to a set of g2g contracts. In the "multiple contracts" abstraction that we describe next, we model the pairwise interaction of available capacities of g2g contracts to denote the intensity of overlap between the contracts. In this abstraction, the time-dependent demand for spot contract on each contract is modeled by μ_t^i, the available capacity on each contract is modeled by A_t^i, and the price of the spot contract is a nonlinear transformation, $S_t^i = f(\mu_t^i, A_t^i)$. The key difference in models used in the single g2g contract link case is the intensity of overlap, ρ_t^{ij}, which models the correlation between the contracts. The predictive model for A_t^i, used as the bailout factor to define and price the BFCs on each g2g path of the network, utilizes the intensity of overlap as follows:

$$dA_t^i = \beta^i(\overline{A}^i - A_t^i)dt + b_2^i A_t^i\, dW_t^{2i}, \qquad (7.10)$$

where the intensity of overlap describing the correlation between available capacity on each g2g path is captured by correlation between the driving Wiener processes as

$$dW^{2i}\, dW^{2j} = \rho^{ij}\, dt, \qquad (7.11)$$

where ρ^{ij} is the **intensity of overlap** describing the shared resources between path i and path j. As before, $\overline{A}^i$ is the long-run mean, b_2^i is the volatility constant, and β^i is the rate of reversion to the long-run mean.

With the rest of the derivation as before, we obtain the forward price for contract link i, F^i, to be as follows:

$$F = \frac{1}{P(A_T^i > Th^i)} E[S_T^i \mathbf{I}_{\{A_T^i > Th^i\}}], \qquad (7.12)$$

where S_t^i in the risk-neutral world evolves by the process

$$dS_t^i = rS_t^i\, dt + p\left(\frac{\mu_t^i}{A_t^i}\right)\frac{b_1^i\mu_t^i}{A_t^i} dW_t^{1i} - p\left(\frac{\mu_t^i}{A_t^i}\right)\frac{b_2^i\mu_t^i}{A_t^i} dW_t^{2i}. \qquad (7.13)$$

Therefore, the forward price of a g2g path is modified to the extent the evolution characteristic of A_t^i is affected by variability in the available capacity in other g2g paths. To evaluate the risk involved in advertising a particular g2g contract, knowledge of the interactions among crossing flows within the underlying network is crucial. As discussed before, we develop our multiple g2g BFC terms based on the assumption that an intensity of overlap, ρ^{ij}, abstractly models the correlation between flows i and j. High correlation means that flows i and j are tightly coupled and share more of the network resources on their paths.

We model the correlation between flows i and j as

$$\rho^{ij} = U_{link} \times \left(\frac{\tau_i}{\tau_i + \tau_j} \right),$$

where τ_k is the portion of bandwidth that flow k can have according to maxmin fair share among all flows passing through the common bottleneck link, and U_{link} is the utilization of the bottleneck link. Note that this formula takes into account the *asymmetric* characteristic of overlaps arising due to the unequal amounts of traffic that flows generate. The maxmin bandwidth share can be calculated using well-known techniques like the "bottleneck" algorithm described in [2].

7.3.3   Bailout forward contract (BFC) performance evaluation

Our performance study attempts to reveal answers to the following questions:

- *Robustness of g2g BFCs:* What is the probability that a g2g BFC will break due to a link/node failure in the ISP's network?
- *Efficiency of network QoS:* There is a tradeoff between the risk undertaken to provide a better service (e.g., longer contracts with larger capacity promise) and the loss of monetary benefit due to bailouts. In comparison to simple contracting, what are the additional expected revenues, profits, or losses of the ISP due to BFCs?

Network model
In our experimental setup, we first devise a realistic network model with router-level ISP topologies obtained from Rocketfuel [9], shortest-path intra-domain routing, and a gravity-based traffic matrix estimation. We assume that the QoS metric of BFCs is the g2g capacity. We focus on developing our network model to reflect a typical ISP's backbone network. We first calculate a routing matrix R for the ISP network from the link weight information, using the shortest path algorithm. With a realistic traffic matrix T, we can then calculate the traffic load pertaining to individual links by taking the product of T and R. We use this realistic network model to identify a demand (i.e., μ) and supply (i.e., A) model, which we use to develop multiple g2g BFCs.

We obtain the topology information (except the link capacity estimation) from the Rocketfuel [9] data repository which provides router-level topology data for six ISPs: Abovenet, Ebone, Exodus, Sprintlink, Telstra, and Tiscali. We updated the original Rocketfuel topologies such that all nodes within a PoP (assuming that a city is a PoP) are connected with each other by adding links to construct at least a ring among routers in the same PoP. We estimated the capacity of the links between routers i and j as $C_{i,j} = C_{j,i} = \kappa[\max(d_i, d_j)]$ where d_i, d_j are the BFS (breadth-first-search) distances of nodes i,j, when the minimum-degree router in the topology is chosen as the root of the BFS tree, and κ is a decreasing vector of conventional link capacities.

We choose $\kappa[1] = 40\,\text{Gb/s}$, $\kappa[2] = 10\,\text{Gb/s}$, $\kappa[3] = 2.5\,\text{Gb/s}$, $\kappa[4] = 620\,\text{Mb/s}$, $\kappa[5] = 155\,\text{Mb/s}$, $\kappa[6] = 45\,\text{Mb/s}$, and $\kappa[7] = 10\,\text{Mb/s}$. The intuition behind this BFS-based method is that an ISP's network would have higher capacity and higher degree links towards the center of its topology.

To construct a reasonable traffic matrix, we first identify the edge routers from the Rocketfuel topologies by picking the routers with smaller degree or longer distance from the center of the topology. To do so, for each of the Rocketfuel topologies, we identified *Degree Threshold* and *BFS Distance Threshold* values so that the number of edge routers corresponds to 75–80 percent of the nodes in the topology. We then use gravity models [7] to construct a feasible traffic matrix composed of g2g flows. The essence of the gravity model is that the traffic between two routers should be proportional to the multiplication of the populations of the two cities where the routers are located. We used CIESIN [1] dataset to calculate the city populations, and generated the traffic matrices such that they yield a power-law behavior in the flow rates as was studied earlier [7].

Model analysis
In the Rocketfuel's Exodus topology, we used data for a total of 372 g2g paths to calibrate the mathematical model for developing the definition and pricing of BFCs. It is possible to apply the price analysis to a much larger set of g2g paths; however, we select a relatively smaller set of paths for ease of presentation. For these paths, we use summary statistics for available capacity and demanded bandwidth, such as means and standard deviations, to calibrate the models.

We begin our model-based analysis of the BFC framework by analyzing single g2g paths. For a single g2g path, we display sample paths for evolution of available capacity, demanded bandwidth, price of the spot contracts in panels (i)–(iii) of Figure 7.4. Based on a calibrated model for demand, available capacity and spot prices given by the derivation of Section 7.3.2, we determine the price of BFCs for a range of choice of the thresholds defining the bailout. The probability of bailout, plotted in panel (iv) of Figure 7.4, shows an increasing trend with an increasing threshold level, as expected. The thresholds are defined in terms of a low percentile of the distribution of available capacity.

We report price of the BFC for a sample of five g2g paths in Figure 7.5, determined within the single g2g framework of Section 7.3.2. The BFC delivers service five days in the future with the threshold for bailout set at 15th percentile of available capacity. The objective in this display is to indicate how the forward prices compare with the spot contract prices. On average the forward prices

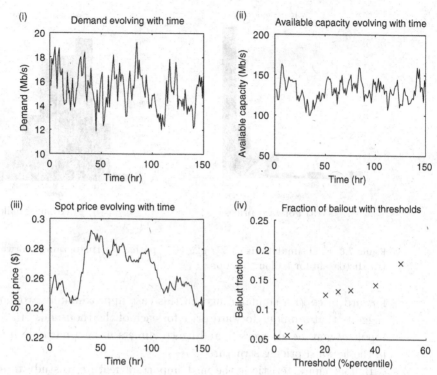

Figure 7.4 (i) One sample demand evolution for the next 5 days; (ii) Available capacity evolution for next 5 days; (iii) Price of spot contract; (iv) Probability of bailout as function of Threshold.

Link	Forward Prices	$E[S_T]$	Prob{S_T>F}	Prob{A_T<Th}
1	0.20609	0.20305	0.502	0.09
2	0.27162	0.24982	0.449	0.065
3	0.21293	0.21213	0.486	0.079
4	0.25039	0.24825	0.477	0.094
5	0.22177	0.21211	0.465	0.093

Th = 15%

Figure 7.5 Sample BFC prices for five g2g paths.

remain slightly above the spot price at maturity; however, the risk in future spot prices entails that the forward prices will be below future spot prices by a probability exceeding 45 percent (see the 4th column in Figure 7.5). We also indicate for these five g2g paths, the probability of BFCs to bailout in the last column. For these paths, the probability of bailout is well bounded by 10 percent.

We next implement the multiple g2g path framework for BFC pricing to analyze the effect of interaction between paths captured in terms of the intensity of overlap, ρ^{ij}. The forward price of a set of 372 paths is determined and plotted in a histogram in Figure 7.6(a). As the histogram suggests, although there is variability in forward prices across the set of paths, many of the paths pick a

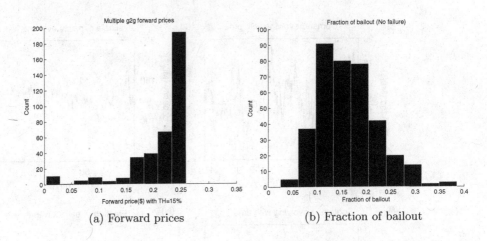

(a) Forward prices   (b) Fraction of bailout

Figure 7.6 1000 simulations of 372 g2g BFC paths: (a) Histogram of forward prices; (b) Histogram for bailout fraction.

forward price in a similar range, in this case approximately around 0.25. This suggests that a unique forward price for each of the thousands of g2g paths in a topology may be an overkill, and hence, directs us to a much desired simplicity in the forward pricing structure.

Bailout characteristic is the next important feature to study to evaluate the BFC framework. We plot the fraction of 372 g2g paths bailing out in 1000 runs of simulation in a histogram in Figure 7.6(b). The mean fraction of g2g paths bailing out from this histogram is 0.16403, or 16.4 percent. To highlight which specific paths bail out in these simulation runs, we also plot the number of times each link bails out in the 1000 runs of simulation in Figure 7.7(a). There are a few paths that clearly stand out in bailing out most frequently, marking the "skyline," while most of the paths cluster in the bottom. Another important measure of performance is how much revenue is lost when the BFC on a g2g path bails out. This is shown also by each g2g link in Figure 7.7(b). Clearly, the pattern of clusters here will be similar to Figure 7.7(a), however the height of the bars is a function of the forward price of each g2g path and how frequently it bailed out in the runs of simulation.

Network analysis

In the previous subsection we showed how our multiple g2g BFC definitions can perform when traffic demand and g2g available capacity processes change. We studied the performance under three different failure modes, each corresponding to a major link failure in the Exodus topology.

To test the viability of our BFC definitions, we evaluate the performance of our BFCs when a failure happens in the underlying network, i.e., Exodus. Specifically, we take the baseline BFC definition and identify the fraction of g2g BFCs getting invalidated (i.e., to be bailed out) due to a link failure in the underlying network

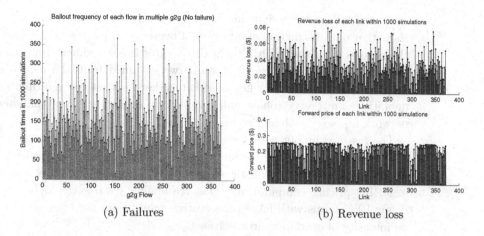

(a) Failures   (b) Revenue loss

Figure 7.7 1000 simulations of 372 g2g BFC paths: (a) The number of times each BFC path fails; (b) The amount of revenue lost for each path failure.

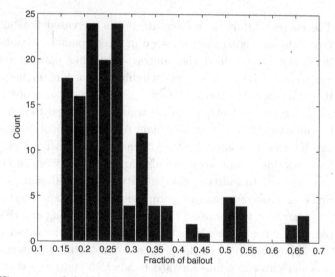

Figure 7.8 Histogram of the fraction of g2g paths bailing out after a link failure in the Exodus network. The average fraction of bailed out BFCs is 27.6 percent.

topology. Notice that this analysis *conservatively* assumes no a-priori knowledge of the failure scenarios.

To perform the analysis we take down each link of the Exodus topology one by one. After each link failure, we determine the effective g2g capacity each BFC will be able to get based on maxmin fair share (and equal share with the excess capacity). We then compare this effective g2g capacity with the bailout capacity thresholds identified for each g2g BFC. Figure 7.8 shows the distribution of the fraction of bailed out BFCs after a link failure in the Exodus network. The distribution roughly follows a similar pattern observed in Figure 7.6(b) which was obtained under a dynamic demand-capacity pattern. The results clearly show that our abstraction of intensity of overlap can be practically used to ease the process of pricing multiple g2g BFCs. This is also evident from the average bailout fraction being close to the one we obtained from the model analysis, i.e., 16.4 percent. Another key observation here is that our multiple g2g BFC definitions are pretty robust and can survive even over a very hub-and-spoke network topology such as Exodus'.

The reason why our network analysis results in an approximately 11 percent higher bailout rate is due to the fact that intensity of overlap abstraction does leave out some of the realistic situations for the sake of easing the multiple g2g BFC pricing computations. One reasonable strategy to follow can be to define BFC terms with more conservative values than the ones obtained based on intensity of overlap approximations.

7.4   Summary

The current Internet architecture needs increased flexibility in realizing value flows and managing risks involved in inter-domain relationships. To enable such flexibility, we outlined the contract-switching paradigm that promotes using contracts overlaid on packet-switching intra-domain networks. In comparison to packet-switching, contract-switching introduces more tussle points into the architecture. By routing over contracts, we showed that economic flexibility can be embedded into the inter-domain routing protocol designs and this framework can be used to compose e2e QoS-enabled contract paths. Within such a "contract routing" framework, we also argued that financial engineering techniques can be used to manage risks involved in inter-domain business relationships. Since e2e contracts can be composed by concatenating multiple single-domain g2g contracts, we discussed the fundamental pricing and risk management principles that must guide g2g contracting, including the notions of forward contracts and bailout options. In the proposed contracting mechanism, a network service provider can enter into forward bandwidth contracts with its customers, while reserving the right to bail out (for a predetermined penalty) in case capacity becomes unavailable at service delivery time. The proposed risk-neutral contract pricing mechanism allows the ISPs to appropriately manage risks in offering and managing these contracts. In the proposed architecture, providers can advertise different prices for different g2g paths, thereby providing significantly increased flexibility over the current point-to-anywhere prices. Experimentations on a Rocketfuel-based realistic topology shows that our g2g bailout contracting mechanism is quite robust to individual link failures in terms of the bailout fraction and revenue lost.

Acknowledgements

This work is supported in part by National Science Foundation awards 0721600, 0721609, 0627039, and 0831957. The authors would like to thank Weini Liu and Hasan T. Karaoglu for their assistance in research.

References

[1] (2009). The center for international earth science information network (CIESIN). www.ciesin.columbia.edu.

[2] Bertsekas D., Gallagher R. (1992). *Data Networks*. Prentice Hall.

[3] Castineyra I., Chiappa N., Steenstrup M. (1996). The nimrod routing architecture. *IETF RFC 1992*.

[4] Clark D. D., Wroclawski J., Sollins K. R., Braden R. (2002). Tussle in cyberspace: Defining tomorrow's Internet. In *Proc. of SIGCOMM*.

[5] Griffin T. G., Wilfong G. (1999). An analysis of BGP convergence properties. *ACM SIGCOMM CCR 29*, 4 (October), 277–288.

[6] Liu W., Karaoglu H. T., Gupta A., Yuksel M., Kar K. (2008). Edge-to-edge bailout forward contracts for single-domain Internet services. In *Proc. of IEEE International Workshop on Quality of Service (IWQoS)*.

[7] Medina A., Taft N., Salamatian K., Bhattacharyya S., Diot C. (2002). Traffic matrix estimation: Existing techniques and new directions. In *Proc. of ACM SIGCOMM*.

[8] Odlyzko A. M. (2001). Internet pricing and history of communications. *Computer Networks 36*, 5–6 (July).

[9] Spring N., Mahajan R., Wetherall D. (2002). Measuring ISP topologies with Rocketfuel. In *Proc. of ACM SIGCOMM*.

[10] Teitelbaum B., Shalunov S. (2003). What QoS research hasn't understood about risk. *Proc. of ACM SIGCOMM Workshops*, pp. 148–150.

[11] Teixeira R., Agarwal S., Rexford J. (2005). BGP routing changes: Merging views from two ISPs. *ACM SIGCOMM CCR*.

[12] Wilmott P., Dewynne J., Howison S. (1997). *The Mathematics of Financial Derivatives: A Student Introduction*. Cambridge University Press.

[13] Wilson R. (1993). *Nonlinear Pricing*. Oxford University Press.

[14] Yang X., Clark D., Berger A. (2007). NIRA: A new inter-domain routing architecture. *IEEE/ACM Transactions on Networking 15*, 4, 775–788.

[15] Yuksel M., Gupta A., Kalyanaraman S. (2008). Contract-switching paradigm for internet value flows and risk management. In *Proc. of IEEE Global Internet*.

[16] Yuksel M., Kalyanaraman S. (2003). Distributed dynamic capacity contracting: An overlay congestion pricing framework. *Computer Communications*, **26**(13), 1484–1503.

8   PHAROS: an architecture for next-generation core optical networks

Ilia Baldine[†], Alden W. Jackson[‡], John Jacob[⊗], Will E. Leland[‡], John H. Lowry[‡], Walker C. Milliken[‡], Partha P. Pal[‡], Subramanian Ramanathan[‡], Kristin Rauschenbach[‡], Cesar A. Santivanez[‡], and Daniel M. Wood[⊕]

[†]Renaissance Computing Institute, USA [‡]BBN Technologies, USA [⊗]BAE Systems, USA [⊕]Verizon Federal Network Systems, USA

8.1   Introduction

The last decade has seen some dramatic changes in the demands placed on core networks. Data has permanently replaced voice as the dominant traffic unit. The growth of applications like file sharing and storage area networking took many by surprise. Video distribution, a relatively old application, is now being delivered via packet technology, changing traffic profiles even for traditional services.

The shift in dominance from voice to data traffic has many consequences. In the data world, applications, hardware, and software change rapidly. We are seeing an unprecedented unpredictability and variability in traffic patterns. This means network operators must maintain an infrastructure that quickly adapts to changing subscriber demands, and contain infrastructure costs by efficiently applying network resources to meet those demands.

Current core network transport equipment supports high-capacity global-scale core networks by relying on higher speed interfaces such as 40 and 100 Gb/s. This is necessary but in and of itself not sufficient. Today, it takes considerable time and human involvement to provision a core network to accommodate new service demands or exploit new resources. Agile, autonomous, resource management is imperative for the next-generation network.

Today's core network architectures are based on static point-to-point transport infrastructure. Higher-layer services are isolated within their place in the traditional Open Systems Interconnection (OSI) network stack. While the stack has clear benefits in collecting conceptually similar functions into layers and invoking a service model between them, stovepiped management has resulted in multiple parallel networks within a single network operator's infrastructure.

Next-Generation Internet Architectures and Protocols, ed. Byrav Ramamurthy, George Rouskas, and Krishna M. Sivalingam. Published by Cambridge University Press. © Cambridge University Press 2011.

Such an architecture is expensive to build and operate, and is not well-suited to reacting quickly to variable traffic and service types. This has caused the network operators to call for "network convergence" to save operational and capital costs.

In the area of traffic engineering and provisioning, IP services now dominate core network traffic, but IP networks utilize stateless per-node forwarding – costly at high data rates, prone to jitter and packet loss, and ill-suited to global optimization. Layer 2 switching mechanisms are better by some measures but lack fast signaling. Generalized multi-protocol label switching (GMPLS) attempts layer 2 and 3 coordination but is not yet mature enough for optical layer 1 and shared protection over wide areas. Today's SONET 1+1 method of provisioning protected routes for critical services consumes excessive resources, driving down utilization, increasing cost, and limiting the use of route protection.

Thus, network operators are looking to integrate multiple L1–L2 functions to reduce cost and minimize space and power requirements. They also aim to minimize the costly equipment (router ports, transponders, etc.) in the network by maximizing bypass at the lowest layer possible. To allow maximum flexibility for unpredictable services, they require a control plane that supports dynamic resource provisioning across the layers to support scalable service rates and multiple services, e.g., Time Division Multiplexing (TDM), Storage Area Networking (SAN), and IP services. Such a control plane also enables automated service activation and dynamic bandwidth adjustments, reducing both operational and capital costs.

Surmounting these challenges requires a re-thinking of core network architectures to overcome the limitations of existing approaches, and leverage emerging technologies. In response to these challenges, the US Defense Advanced Research Projects Agency (DARPA) created the Dynamic Multi-Terabit Core Optical Networks: Architecture, Protocols, Control and Management (CORONET) program with the objective of revolutionizing the operation, performance, survivability, and security of the United States' global IP-based inter-networking infrastructure through improved architecture, protocols, and control and management software. CORONET envisions an IP (with Multi-Protocol Label Switching (MPLS)) over Wavelength Division Multiplexing (WDM) architecture on global scale. The target network includes 100 nodes, has aggregate network demands of between 20 and 100 Tb/s using up to 100 40 or 100 Gb/s wavelengths per fiber (higher demand uses higher capacity waves), and supports a mix of full wavelength and IP services.

The network is highly dynamic with very fast service setup and teardown. A key CORONET metric in this regard is very fast service setup (VFSS) in less than 50 ms + round-trip time. There are also fast services (FSS) with 2 second setup requirements, scheduled services and semi-permanent services. The IP traffic includes both best effort and guaranteed IP services with a variety of granularities as low as 10 Mb/s per flow. The network must be resilient to multiple concurrent network failures, with double- and triple-protected traffic classes in addition to singly protected and unprotected services. Restoration of services is enacted within 50 ms + round-trip time. To ensure efficiency in handling protected traffic, CORONET specifies a metric, B/W, where B is the amount of network capacity reserved for protected services and measured in wavelength-km, and W is the total working network capacity, also in wavelength-km. B/W must be less than 0.75 for the CONUS-based traffic in the CORONET target network.

In this chapter, we present PHAROS (Petabit/s Highly-Agile Robust Optical System) – an architectural framework for next-generation core networks that meets the aggressive CORONET objectives and metrics. Through its framework, optimization algorithms, and control plane protocols, the PHAROS architecture:

- substantially improves upon today's 30-day provisioning cycle with its automated systems to provide less than 50 ms + round-trip time in the fastest case;
- replaces opaque, stovepiped layer 1, 2, and 3 management systems with accessible administration;
- qualitatively improves the tradeoff between fast service setup and network efficiency; and
- assures network survivability with minimal pools of reserved (and therefore normally unused) capacity.

The CORONET program is the first program to explore control and management solutions that support services across global core network dimension with 50-ms-class setup time, and also to respond to multiple network failures in this time frame. The PHAROS architecture is designed in response to that challenge. The PHAROS architecture has been designed with awareness of current commercial core network practice and the practical constraints on core network evolution. The design of PHAROS also includes support for asymmetric demands, multicast communications, and cross-domain services, where a domain is a network or set of networks under common administrative control.

PHAROS aims to build upon recent research that highlights intelligent grooming to maximize optical bypass to reduce core network costs [1, 2, 3, 4, 5, 6, 7, 8]. The program also exploits the use of optical reconfiguration to provide bandwidth-efficient network equipment that responds gracefully to traffic changes and unexpected network outages [9, 10, 11].

While a large body of work exists on several exciting research problems in next-generation networks, our focus in this chapter is on the *system architecture* – how we can leverage individual solutions and clever breakthroughs in transport and switching from a signaling, control and management perspective in order to hasten deployment. We therefore see PHAROS as a bridge between the state of art in research and the next-generation deployed system.

Architecting any system requires selecting choices within a tradeoff space. In this chapter, we not only describe the choices we made, but in many cases, we also discuss the alternatives, their pros and cons and the reasons for our choice. We hope this gives the reader a flavor of the typical strategies in this space, and an appreciation of how requirements drive the choice.

The remainder of this chapter is organized as follows. After surveying background work, we begin with an overview of the PHAROS architecture. Following that we describe three key components of PHAROS, namely, the cross-layer resource allocation algorithm (Section 8.4), the signaling system (Section 8.5), and the core node implementation (Section 8.6). Finally, we give some preliminary results on performance estimation.

8.2   Background

We briefly survey prior work on some of the topics discussed in this chapter, namely, path computation, protection, and node architectures. Unlike IP networks, path computation in optical networks involves computation of working and protection bi-paths. Approaches can be classified by the nature of the required paths (e.g., node-disjoint, link-disjoint, k-shortest), the order for computing them (e.g., primary-then-protection vs. joint-selection), and the cost associated with each path. Some works include [12, 13]. Our approach is a hybrid one and uses the concept of joint or shared protection.

The various levels of protection defined for different traffic demands in a core optical network, along with the low-backup-capacity targets, motivate the use of shared-protection schemes for this application. Such techniques fall into broad categories of the various computational and graph-theoretic approaches: constrained shortest paths [14], cycle covers [15], and ILP formulations like the p-cycles [16]. As these techniques can guarantee only single protection for all the flows, they would have to be augmented to guarantee double or triple protection for the set of flows that require it. In this chapter, we have outlined the preliminary formulation of a shared-mesh-protection algorithm based on virtual links and jointly protected sets that deliver double- and triple-protection services.

The sophistication of optical-network-node architectures has risen as the state of the art for the optical components within these nodes has advanced. Recent advances in optical-switch reliability and functionality, along with the size of the available switch fabrics, have motivated node architectures that allow such multiple functionalities as reconfigurable add/drop, regeneration, and wavelength conversion [17]. The cost, power, size, and reliability calculations for these different implementations are highly technology-dependent and are changing rapidly as new technologies are transitioned into the commercial market. As a result of this rapidly changing trade-space, we have chosen to remain agnostic to the exact switch architecture in our nodes, a feature we discuss further in the next section.

8.3   PHAROS architecture: an overview

In designing the PHAROS system, we were guided by some high-level principles and tenets, such as technology-agnosticism, fault-tolerance, global optimizations,

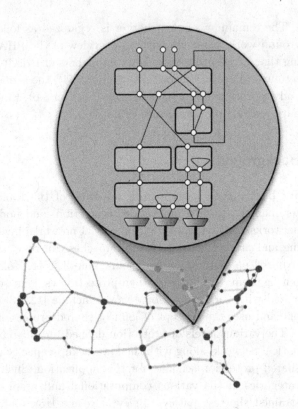

Figure 8.1 Multi-level topology abstractions make PHAROS technology-agnostic.

etc. These motivated some innovations such as topology abstractions, triangulation, etc. In this section, we first discuss these principles guiding our architecture, along with features associated with them. We then give a brief overview of the logical functional blocks and their roles.

A basic tenet of the PHAROS architecture is a *technology-agnostic* design that maximizes bypass to achieve lower cost-per-bit core network services and accommodates future generations of switch technology for long-term graceful capacity scaling. Current systems employ some degree of abstraction in managing network resources, using interface adapters that expose a suite of high-level parameters describing the functionality of a node. Such adapters, however, run the twin risks of obscuring key blocking and contention constraints for a specific node implementation, and/or tying their interfaces (and the systems resource management algorithms) too tightly to a given technology.

The PHAROS system avoids both of these problems by using *topology abstractions* – abstract topological representations for all levels of the network. The representations extend down to an abstract network model of the essential contention structure of a node, as illustrated in Figure 8.1, and extend upward to address successive (virtual) levels of functionality across the entire network.

With a uniform approach, common to all levels of resource representation and allocation, PHAROS accurately exploits the capabilities of all network elements, while remaining independent of the switching technology. At the signaling and control level, the PHAROS architecture also provides a set of common mechanisms for its own internal management functions (such as verification and failover); these mechanisms provide significant architectural immunity to changes in the technologies used in implementing specific PHAROS functional components.

The PHAROS architecture uses multi-level topological abstractions to achieve *global multi-dimensional optimization*, that is, efficient integrated resource optimization over the fundamental dimensions of network management: network extent, technology levels, route protection, and timescales. Abstraction allows a given request to be optimized across the network, simultaneously trading off costs of resources within individual network levels as well as the costs of transit between levels (such as the optical–electrical boundary). Resources of all levels can be considered, including wavelengths, timeslots, grooming ports, and IP capacity.

PHAROS optimization unites analysis of the resources needed to deliver the service with any resources required for protection against network element failures. Protection resources (at all levels) are allocated in conjunction with the resources required by other demands and their protection, achieving dramatic reductions in the total resources required for protection (the CORONET B/W metric). Our optimization design allows PHAROS to unify the handling of demand timescales, exploiting current, historical, and predicted future resource availability and consumption. Timescales are also addressed by the overall PHAROS resource management strategy, which selects mechanisms to support available time constraints: for example, PHAROS employs pre-calculation and tailored signaling strategies for very fast service setup; selects topology abstractions to perform more-extensive on-demand optimization where feasible; and evaluates long-term performance out of the critical path to enable rebalancing and improve the efficiency of the on-demand optimizations.

Finally, the PHAROS architecture achieves a high degree of *fault-tolerance* by using a design construct that combines redundancy and cross-checking in a flexible way to mitigate single point of failure and corrupt behavior in a Cross-layer Resource Allocator (CRA), a critical component of the PHAROS architecture described in Section 8.4. This design construct, which we refer to as *triangulation*, pairs up the consumer of the CRA function (typically a network element controller) with a "primary" and a "verify CRA." The verify CRA checks that the primary CRA is performing correctly, and corrupt behavior can be detected by using appropriate protocols amongst the consumer and the primary and verify CRAs.

PHAROS is a system that dynamically applies network resources to satisfy subscriber requests in an efficient and timely manner. It can be applied to a broad range of service models, topologies, and network technologies. Such broad

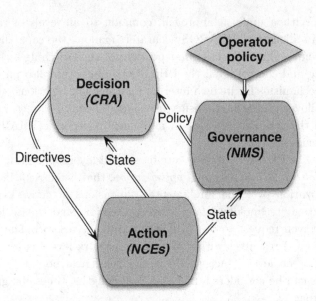

Figure 8.2 Functional components comprising the PHAROS system.

applicability is possible because our high-level architecture remains technology-agnostic; the benefit is that PHAROS can provide new capabilities and services whether they entail working with legacy infrastructures or with technologies not yet envisioned.

The PHAROS functional architecture separates governance, decision making, and action, streamlining the insertion of new services and technologies. The relationships among these roles are summarized in Figure 8.2.

The governance role is critical for correct operation but is not time-critical. Governance establishes policy and reaction on a human timescale. It is not on the critical path for service instantiations. The network management system (NMS), described further below, is the primary repository of non-volatile governance information and the primary interface between human operators and the network. For the human operator, PHAROS maintains the functionality of a single, coherent network-wide NMS; this functionality is robustly realized by an underlying multiple-agent implementation.

The decision role is the application of policy to meeting subscriber service requests, and is therefore highly time-critical. That is, the role lies in the critical path for realizing each service request on demand: the decision process is applied to each subscriber service request to create directives for control of network resources. The cross-layer-resource allocator (CRA) function, described further in the next section, is the primary owner of the decision process. Because of the critical contribution of the decision role to the network's speed, efficiency, and resilience, the CRA function is implemented by a distributed hierarchy of CRA instances. In conjunction with our mechanisms for scoping, verification, and failover, the instance hierarchy autonomously sustains our unitary strategy for resource management: each service request and each network resource within a domain is managed by exactly one CRA instance at any point in time, with a dynamic choice of the particular CRA instance. The result is globally consistent resource allocation, with consistently fast service setup.

The action role implements the decisions made in the decision role. It is a time-critical function. The responsibility of the action role is limited to implementing directives. The network element controllers (NECs), described in a later section, are the primary architectural components responsible for this role.

Our approach allows network operators to take advantage of technological improvements and emerging service models to meet the increasing requirements of their subscribers' applications. The governance function controls the behavior of the PHAROS system, establishing those actions and parameters that will be performed automatically and those that require human intervention. The decision function applies these policies to effectively allocate resources to meet real-time subscriber service requests. The action function implements the decisions quickly, reporting any changes in the state of the system.

8.4   Resource allocation

We begin with a discussion of possible resource allocation strategies and describe our approach. In Section 8.4.2 we discuss means of protecting allocated resources from failure. To be agile, we use the concept of "playbooks" described in Section 8.4.3. We conclude in Section 8.4.4 with a short description of our "grooming" approach for increasing resource utilization.

8.4.1   Resource management strategies

A key architectural decision in any communications network is the organization of the control of resources. Two of the most important aspects are whether global state or just local state is tracked, and how many nodes participate. Based on these and other choices, approaches range from "fully distributed" where each node participates using local information, to "fully centralized" where resource control is in the hands of a single node utilizing global information. We first discuss the pros and cons of several points in this spectrum and then motivate our choice.

The fully centralized or *single master* strategy entails a single processing node that receives all setup requests and makes all resource allocation decisions for the network. This approach allows, in principle, global optimization of resource allocation across all network resources. It has the further virtue of allowing highly deterministic setup times: it performs its resource calculation with full knowledge of current assignments and service demands, and has untrammeled authority to directly configure all network resources as it decides. However, it requires a single processing node with sufficient capacity for communications, processing, and memory to encompass the entire network's resources and demands. This node becomes a single point of failure, a risk typically ameliorated by having one or more additional, equally capacious standby nodes. Moreover, each service request must interact directly with the master allocator, which not only adds transit time to service requests (which may need to traverse the entire global network) but also can create traffic congestion on the signaling channel, potentially introducing unpredictable delays and so undercutting the consistency of its response time.

At the other end of the spectrum is the fully distributed or *path threading* strategy. Each node controls and allocates its local resources, and a setup request traces a route between source and destination(s). When a request reaches a given node, it reserves resources to meet the request, based on its local knowledge, and determines the next node on the request path. If a node has insufficient resources to satisfy a request, the request backtracks, undoing the resource reservations, until it fails or reaches a node willing to try sending it along a new candidate path. This strategy can yield very fast service setup, provided enough resources are available and adequately distributed in the network. There is no single point of failure; indeed, any node failure will at most render its local resource unavailable. Similarly, there is no single focus to the control traffic, reducing the potential for congestion in the signaling network. However, the strategy has significant disadvantages. Setup times can be highly variable and difficult to predict; during times of high request rates, there is an exceptionally high risk of long setup times and potential thrashing, as requests independently reserve, compete for, and release partially completed paths. Because backtracking is more likely precisely during times when there are already many requests being set up, the signaling network is at increased risk of congestive overload due to the nonlinear increase in signaling traffic with increasing request rate. The path-threading strategy is ill-suited to global optimization, as each node makes its resource allocations and next-hop decisions in isolation.

One middle-of-the-road strategy is *pre-owned resources*. In this strategy, each node "owns" some resources throughout the network. When a node receives a setup request, it allocates resources that it controls and, if they are insufficient, requests other nodes for the resources they own. This strategy has many of the strengths and weaknesses of path-threading. Setup times can be very quick, if sufficient appropriate resources are available, and there is no single point of failure nor a focus for signaling traffic. Under high network utilization or high rates of service requests, setup times are long and highly unpredictable; thrashing is also a risk. Most critically, resource use can be quite suboptimal. Not only is there the issue of local knowledge limiting global optimization, there is also an inherent inefficiency in that a node will pick routes that use resources it owns rather than ones best suited to global efficiency. In effect, every node is reserving resources for its own use that might be better employed by other nodes setting up other requests.

Which of these strategies, if any, are appropriate for the next-generation core optical network? Future core networks present some unique factors influencing our choice of PHAROS control organization. First, there is adequate signaling bandwidth and processing resources available, which allow for global tracking of resource use if necessary. Second, nodes are neither mobile nor disruption prone, again making it feasible to concentrate control functionality. Third, under high loads, efficient (preferably optimal) allocation is required. Fourth, the stringent service requirements and expectations make the user of the core optical system highly intolerant of stability issues.

We believe that these factors shift the optimum point significantly toward a centralized control for PHAROS although not completely. In essence, our approach is to move away from a single point of failure but retain the ability to use global information for resource allocation decisions resulting in a strategy that we term *unitary resource management*. The unitary strategy relies upon the previously described CRA function to determine the optimal joint resource use for a service and its protection, integrating optimization across multiple layers of technology (e.g., wavelengths, sub-wavelength grooming, and IP).

In the unitary strategy, system mechanisms autonomously sustain the following three invariants across time and across network changes: (1) the integrated CRA algorithm is sustained by a resilient hierarchy of CRA instances; (2) for each request for a given combination of service class, source, and destination(s), there is exactly one CRA instance responsible at any time; and (3) for each network resource there is exactly one CRA instance controlling its allocation at any time. Each CRA instance has an assigned scope that does not overlap with that of any other CRA instance; its scope consists of a service context and a suite of assigned resources. The service context defines the service requests for which this CRA instance will perform setup: a service context is a set of tuples, each consisting of a service class, a source node, and one or more destination nodes.

The unitary strategy allows a high degree of optimization and highly consistent setup times, as a CRA instance can execute a global optimization algorithm that takes into account all resources and all service demands within its scope. There is no backtracking or thrashing, and no risk of nonlinear increases in signaling traffic in times of high utilization or of high rates of setup requests. There is some risk of suboptimal resource decisions, but the PHAROS architecture allows for background offline measurement of the efficacy of allocation decisions and the reassignment of resources or service contexts by the NMS function. The unitary strategy uses multiple CRA instances to avoid many of the problems of the single master strategy: under the PHAROS mechanisms for scoping, failover, and mutual validation, a hierarchy of CRA instances provides load distribution, fast local decisions, and resilience against failure, partition, or attack. Moreover, by concentrating routing and resource-assignment decisions in a few computationally powerful nodes, the strategy allows for complex optimizations based on a global picture, while reducing switch cost and complexity. The CRA instances are maintained on only a small subset of the network nodes, which can be accorded higher security and a hardened physical environment if network policy so chooses.

In the case of the CORONET target network, three CRA instances are used, one CONUS based, one in Europe, and one in Asia.

8.4.2   Protection

Protection can be *link-based*, *segment-based* or *path-based*. We summarize the pros and cons of these approaches below.

In *link-based protection*, for each interior element along the primary path, a protection route is found by omitting that one element from the network topology and recalculating the end-to-end path. Thus for each protected path there is a set of n protection paths where n is the number of interior elements on the primary path. These paths need not be (and usually are not) interior-disjoint from the primary path or from one another. For a single failure, link-based protection may give an efficient alternate route; however, the approach faces combinatorial explosion when protecting against multiple simultaneous failures.

In *segment-based protection*, like in link-based protection, a protected primary path is provided with a set of n protection paths, one for each interior link or node. A given protection path is associated with one of these interior elements; it is not based on the end-to-end service requested but simply defines a route around that element. A classic example of segment-based restoration can be found in SONET Bi-directional Line Switched Rings (BLSR), where any one element can fail and the path is rerouted the other way round the ring. Because segment-based restoration paths are independent of any particular primary path, they may be defined per failed element instead of per path. However, they can also be highly non-optimal from the perspective of a specific service request, and are ill-suited to protecting against multiple simultaneous failures.

Path-based protection defines one or more protection paths for each protected primary path. A primary with one protection path is said to be singly protected; a primary with two protection paths is doubly protected; and similarly for higher numbers. Each protection path for a primary is interior-disjoint with the primary path and interior-disjoint with each of the primary path's other protection paths (if any). Practical algorithms exist for jointly optimizing a primary path and its protection path(s).

In the current PHAROS implementation, we use the path-based protection strategy. Relative to other protection approaches, path-based protection maximizes bandwidth efficiency, provides fast reaction to partial failures, and is readily extended to protection against multiple simultaneous failures. Further, fault-localization is typically not required to trigger the restoration process. In the past, one of the drawbacks was the number of cross-connections that might need to be made to create a new end-to-end path; however, with schemes based on pre-connected subconnections, invoked in the PHAROS implementation, this is less of an issue. The main drawback is higher signaling load for protection.

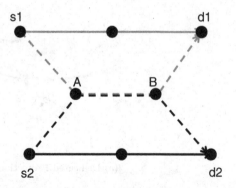

Figure 8.3 Illustration of shared protection.

8.4.2.1 Shared protection

Having selected path-based restoration for the CRA function, there is still a choice of approaches for allocating the network resources required to ensure that each protection path is supported if a failure (or combination of failures) requires its use. Broadly speaking, there are two general protection strategies for path-based restoration: dedicated and shared [18].

In dedicated protection, each protection path has reserved its own network resources for its exclusive use. In shared protection, a subtler strategy significantly reduces the total amount of network resources reserved for protection while providing equal assurance of path restoration after failures. It is based on the observation that for a typical given failure or set of failures, only some primary paths are affected, and only some of their protection paths (in the case of multiple failures) are affected. Thus, a protection resource can be reserved for use by an entire set of protection paths if none of the failures under consideration can simultaneously require use of that resource by more than one path in the set.

PHAROS uses shared (or "joint") protection, which significantly reduces the total amount of network resources reserved for protection while providing equal assurance of path restoration after failures. Shared protection is illustrated by an example in Figure 8.3, where there are two primary paths (the solid gray and black lines), each with its own protection path (the dotted gray and black lines).

No single network failure can interrupt both primary paths, because they are entirely disjoint. So to protect against any single failure, it is sufficient to reserve nodes A and B and the link between them for use by either the gray or the black protection path: a failure that forces use of the dotted gray path will not force use of the dotted black path, and vice versa. For an in-depth treatment of shared protection in practice, the reader is referred to [18].

Shared protection provides substantial savings in bandwidth. Figure 8.4 shows an example network and the capacity used by dedicated and shared protection

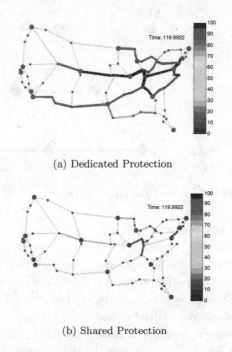

(a) Dedicated Protection

(b) Shared Protection

Figure 8.4 Capacity (in lambdas) used by dedicated (top) and shared (bottom) protection strategies.

strategies respectively. Shared protection uses about 34 percent less overall capacity in this example.

8.4.3   Playbooks

One significant contribution to agility in the PHAROS architecture is a strategy we term playbooks. A *playbook* is a set of pre-calculated alternatives for an action (such as selecting a protection path) that has a tight time budget. The playbook is calculated from the critical path for that action using the CRA function's global knowledge and optimization algorithms. The playbook is stored on each instance that must perform the action; on demand, each instance then makes a fast dynamic selection from among the playbook's alternatives. Playbooks are used to ensure fast, efficient resource use when the time constraints on an action do not allow the computation of paths on demand. In the PHAROS architecture, we use playbooks in two situations: for *Very Fast Service Setup (VFSS)*, and for *Restoration*. We describe the approach used in PHAROS for both of these below.

8.4.3.1   Very Fast Service Setup (VFSS) playbooks

Our current approach is that for each (src, dest, demand rate) combination, we precompute and store two bi-paths:

- First bi-path: The bi-path with the minimum sum of optical edge distances in the working and protection paths. It is computed a priori based only on the topology and as such ignores the network load and protection sharing.
- Second bi-path: Saved copy of the last computed optimal bi-path. The bi-path is optimal in the sense that it minimizes a combined load- and shared-protection-aware cost metric. Since some time has elapsed since the bi-path was initially computed, it may no longer be optimal (or even valid).

The first bi-path is dependent only on the network topology, and needs to be computed only during initialization or after topology changes (a rare event). Note that a link failure is not interpreted as a topology change. For the second bi-path the CRA is constantly running a background process that iterates through the list of valid (src, dest, rate) triplets and computes these optimal bi-paths based on the instantaneous network conditions. Once all (src, dest, rate) combinations have been considered, the process starts once again from the top of the list. Thus, when a new demand arrives, the last saved copy of the corresponding bi-path is just a few seconds old.

In addition, a third bi-path is computed when a new demand arrives. The primary path is computed using Dijkstra's shortest path first (SPF) algorithm where the optical edge costs are related to the current network load. Once the primary path is computed, its links and nodes are removed from the topology, the costs of protection links conditional on the primary path are determined, and then the protection path is computed by running the Dijkstra algorithm again. Since the removal of the primary path may partition the network, there is no guarantee that this bi-path computation will succeed.

These three bi-paths are incorporated into a playbook for that (src, dest, demand rate) combination and cached in the primary CRA (pCRA) instance for the source node. Because VFSS playbooks reside uniquely in the source node's pCRA, there is no possibility of inconsistency. Finally, when an instance receives a demand for that (src, dest, demand rate) combination, it computes the costs of these three bi-paths, taking into account the current network resource availability, and selects the cheapest valid bi-path.

8.4.3.2 Restoration playbooks

A particular failure, such as a fiber cut, may affect thousands of individual demands. Computing alternative paths for all of these demands (for path-based restoration) within the restoration budget is not feasible. Furthermore, unless the resources in the protection path are preallocated, there is no guarantee that a particular demand will successfully find an alternate path after a failure. Thus, path-based protection requires the protection path to be computed along with the primary path, and the resources in the protection path to be reserved.

For each existing demand, there is a playbook entry specifying the path (or paths, for doubly and triply protection demands) to use in case the primary path fails. Each entry specifies the path and regeneration and grooming strategies,

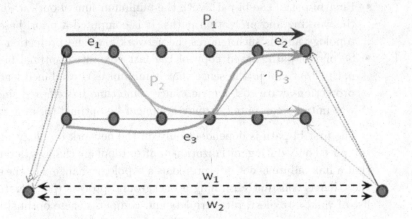

Figure 8.5 PHAROS restoration playbooks allow efficient on-demand selection of restoration paths.

and identifies the pool of resources (such as wavelengths) to choose from upon failure. The playbook does not specify the resource to use, as such assignment can be made (efficiently under shared protection) only after a failure occurs, as illustrated in Figure 8.5.

Wavelengths W_1 and W_2 are enough to protect P_1, P_2, and P_3 against any single failure. However, it is not possible to uniquely assign a wavelength to each demand before the failure occurs. For example, suppose we were to assign W_1 to P_1. Since P_1 and P_2 are both affected by link e1 failure, then P_2 should be assigned W_2. Similarly, since P_1 and P_3 are both affected by link e_2 failure, P_3 should also be assigned W_2. However, P_2 and P_3 should not be assigned the same wavelength, since this will result in blocking if link e_3 fails.

8.4.4   Sub-lambda grooming

Finally, many demands do not fill a full wavelength. If one such demand is uniquely assigned to a full wavelength, without sharing it with other demands, it will result in wasted bandwidth and long-reach transponders. To alleviate this problem, demands can be aggregated into larger flows at the source node. They can also be combined with other nodes' demands at intermediate nodes (a process we refer to as sub-lambda grooming, or SLG) so that wavelength utilization at the core is close to 100%. Once demands have been sub-lambda-groomed, they can be optically bypassed.

Deciding where and when to sub-wavelength-groom demands is a difficult optimization problem. It must take into account different tradeoffs among capacity available, the cost (both capital and operational) of the SLG ports and transponders, and the fact that constantly adding or removing demands will unavoidably result in fragmentation inside a wavelength. What may appear to be a good grooming decision now may hurt performance in the future. Grooming decisions, then, must balance medium- to long-term resource tradeoffs and be based on medium-term traffic patterns.

Within our topology-abstraction-based architecture, grooming is a generalized operation where each level packs its smaller bins into larger bins at the level immediately below. Currently, we have a three-level system where we aggregate and groom sub-lambda demands into full wavelengths, and wavelengths onto fibers. However, aggregation and grooming of smaller bins into larger bins constitutes a fundamental operation that repeats itself at multiple layers.

8.5   Signaling system

The PHAROS signaling architecture is designed to support operations in the control as well as management planes. Its function is the delivery of data between the elements of the architecture in a timely, resilient, and secure fashion. The main requirements for the signaling architecture are:

- *Performance*: the architecture must support the stringent timing requirements for connection setup and failure restoration.
- *Resiliency*: the architecture must be resilient to simultaneous failures of several elements and still be able to perform the most critical functions.
- *Security*: the architecture must support flexible security arrangements among architectural elements to allow for proper authentication, non-repudiation and encryption of messages between them.
- *Extensibility*: the architecture must be extensible to be able to accommodate new features and support the evolution of the PHAROS architecture.

The PHAROS signaling and control network (SCN) is the implementation of the PHAROS signaling architecture. It allows NECs to communicate to potential CRA/NMS, with signaling links segregated from the data plane to minimize the risk of resource exhaustion and interference attacks. The PHAROS architecture supports an SCN topology that is divergent from the fiber-span topology, and does not require that the network element controllers and network elements be co-located. For next-generation core optical networks providing deterministic and minimal delay in signaling for service setup and fault recovery, it is recommended that the SCN be mesh-isomorphic to the fiber-span topology, and the network element controllers be collocated with the network elements as shown in Figure 8.6. This configuration minimizes the signaling delay for service setup and fault recovery.

Based on bandwidth sizing estimates that take into account messaging requirements for connection setup, failure signaling and resource assignment, a 1 Gb/s channel is sufficient to maintain stringent timing for setup and restoration under heavy load and/or recovery from multiple fault scenarios. Two performance goals drive the channel size requirements for the PHAROS SCN: very fast service

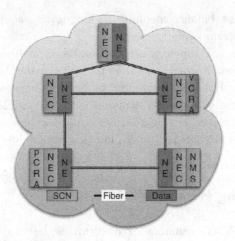

Figure 8.6 The signaling and control network (SCN) connects network elements (NE) and their associated network element controllers (NEC), cross-layer resource allocator (CRA) and network management system (NMS).

setup and 50-ms-class restoration from simultaneous failures. The sizing estimates assume worst case signaling load for a 100-Tb/s-capacity 100-node global fiber network with service granularity ranging from 10 Mb/s to 800 Gb/s. Fibers connecting nodes were presumed to carry 100 100-Gb/s wavelengths.

The majority of the signaling traffic (with some exceptions) travels through the links that constitute the SCN topology. Thus the signaling architecture accommodates both the control and the management planes. Each link in the SCN has sufficient bandwidth to support the peak requirements of individual constituent components. This is done to reduce queueing in the signaling plane, thus expediting the transmission of time-critical messages. Additionally, to ensure that the time-critical messages encounter little to no queueing delay, each link is logically separated into a *Critical Message Channel (CMC)*, and a *Routine Message Channel (RMC)*. All time-critical traffic, such as connection setup messages and failure messages, travels on the CMC, while the rest (including the management traffic) use the RMC.

As in traditional implementations, the SCN is assumed to be packet-based (IP) and to possess a routing mechanism independent of the data plane that allows architectural elements to reach one another outside of the data plane mechanisms.

8.5.1   Control plane operation

In our approach to connection setup, the two competing objectives are:

- The need to perform connection setup very rapidly (for Fast Service Setup (FSS) and Very Fast Service Setup (VFSS) service classes).
- The need for global optimization of protection, which requires the entity responsible for path computation, the primary CRA instance (pCRA), to have a global view of the network.

There are two basic approaches to connection setup: *NEC-controlled* and *CRA-controlled*. They vary in complexity of implementation, the tradeoffs being between the connection setup speed and the need for a global view of resource allocation.

In the NEC-controlled approach, the NEC instance at the source node communicates with its assigned pCRA to receive the information about the routes for working and protection paths and then, in a manner similar to RSVP-TE with explicit route option, sends signaling messages along these paths to the affected NEC instances to service this request. (RSVP-TE reserves resources on the forward path and configures the network elements on the reverse path from destination to source.) This approach has the advantage of fitting into the traditional view of network traffic engineering. One issue in the approach is connection-setup latency: the NEs on each path are configured serially, after the response from the pCRA is received, with processing at each NEC incurring additive delays. Adding to the initial delay of requesting path information from the pCRA makes this approach too slow to be applied in the case of very fast and fast connection classes.

In the CRA-controlled approach, the NEC instance in the source node communicates the service setup request and parameters to its assigned pCRA and leaves it up to this CRA instance to compute the optimal path and to instruct individual NEC instances on the computed path to configure their NEs for the new connection. This approach has several advantages over the NEC-controlled approach. First, NEC configuration occurs in parallel, which serves to speed up connection setup. Second, only CRA instances are allowed to issue NE configuration requests to NECs, which is a desirable property from the viewpoint of network security, as it allows PHAROS to leverage strong authentication mechanisms in NEC-to-CRA communications to prevent unauthorized node configurations. The disadvantage of this approach is its scalability, as a real network may contain a large number of NEC instances, and having a single pCRA presents a scalability limit.

Given our requirement of supporting very fast connection setups, the serialization delay incurred by the NEC-controlled approach is prohibitive. We therefore use the CRA-controlled approach, but address the disadvantage by exploiting the unitary resource management strategy (see Section 8.4). In other words, by dividing the space of all possible service requests into disjoint scopes, a hierarchy

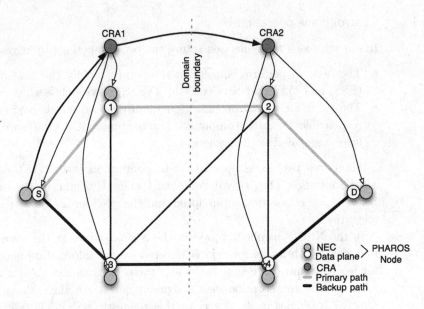

Figure 8.7 The PHAROS *federated* approach for signaling across domains.

of pCRA instances can divide the load while providing more local pCRA access for localized requests.

In an approach that we term the *federated* approach, the initial NEC contacts the pCRA in its scope with a service-setup request as illustrated in Figure 8.7. The pCRA maps the service's path onto domains based on its network state and provisions the service path across its own domain, while at the same time forwarding the request to the appropriate pCRAs in the neighboring domains. This approach deals with the inter-domain state consistency problem by leveraging the fact that a pCRA is likely to have up-to-date information about its own domain and somewhat stale information about other domains. This approach also accommodates security requirements by restricting the number of CRA instances that may configure a given NEC to only those within its own domain. It also retains the parallelizing properties, thus speeding up connection setup.

8.5.2   Failure notification

Traditionally, in MPLS and GMPLS networks, failure notifications are sent in point-to-point fashion to the node responsible for enabling the protection mechanism. This approach works when the number of connections traversing a single fiber is perhaps in tens or hundreds. In PHAROS, assuming the worst-case combination of fine-granularity connections (10 Mbps) and large capacity of a single fiber (10 Tbs), the total number of connections traversing a single fiber may number in tens or hundreds of thousands (a million connections in this case). It is infeasible from the viewpoint of the signaling-plane bandwidth to be able to signal individually to all restoration points in case of a failure, because the number of points and connections whose failure needs to be signaled may be very large. Additionally, point-to-point signaling is not resilient to failures, meaning that, due to other failures in the network, point-to-point failure messages rely on the SCN routing convergence to reach intended recipients and to trigger protection, which may be a lengthy process.

The solution we adopted in PHAROS relies on two simultaneous approaches:

- Signaling on aggregates that can indicate failure of a large number of connections at once.
- Using intelligent flooding as a mechanism to disseminate failure information.

The first approach significantly cuts down on the amount of bandwidth needed to signal a failure of many connections resulting from a fiber cut, but it requires that the nodes receiving failure notifications are able to map the failed aggregates to specific connections requiring protection actions.

The second approach, in addition to reducing bandwidth requirements, also has the desirable property that a signaling message always finds the shortest path to any node in the network even in the presence of other failures, without requiring the signaling-plane routing to converge after a failure.

Combined, these two approaches create a PHAROS solution to the failure-handling problem that is resilient and scalable and addresses the stringent restoration-timing requirements.

8.6   Core node implementation

In this section we discuss a core node implementation that is designed to optimize the capabilities of the PHAROS architecture. We note that the PHAROS architecture does not depend upon the core node being implemented this particular way – as mentioned earlier, it is technology-agnostic.

The PHAROS core node design focuses on maximizing flexibility and minimizing the complexity of intra-node ports required to provide the complete range of PHAROS services and reducing the capital and operational costs per unit of bits. The primary objectives identified to satisfy this vision include: (1) arrange subscriber traffic onto wavelength and sub-wavelength paths to enable switching at the most economic layer, (2) enable shared protection, and (3) enable transponders to be repurposed to service both IP and wavelength services and also service transit optical–electrical–optical (OEO) regeneration functions. When combined with a control plane designed for optimum resource allocation, the PHAROS optical node is highly adaptable to incoming service requests. The PHAROS node architecture defines the principal hardware systems extending from the

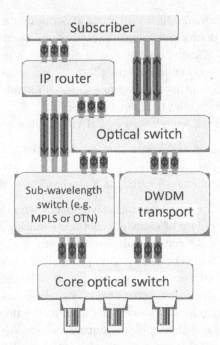

Figure 8.8 PHAROS core node implementation showing various optical network elements.

fiber connections with the subscriber facility to the fiber connections in the physical plant of the core network, as illustrated in Figure 8.8.

The PHAROS node is composed of the following elements:

- Subscriber service layer connections to bring client services into the core node.
- Edge router (packet switch) to support best-effort IP services.
- Fast optical switch to allow sharing of sub-wavelength switch and transport ports.
- Sub-lambda grooming switch and DWDM transport platform to support full and sub-wavelength switched (via MPLS, OTN or SONET) and packet services with fast setup, tightly bounded jitter specifications. This equipment also provides OEO regeneration.
- Core optical switch to manage optical bypass, optical add/drop, and routing between optical fibers.

Note that these elements may or may not be instantiated in the same hardware platform. The PHAROS architecture emphasizes configuration, and can be applied to a variety of different network element configurations.

The core node implementation results in reduced excess network capacity reserved for protection via protection sharing between IP, TDM, and wavelength services that arise at the subscriber ports. The desire to support guaranteed QoS

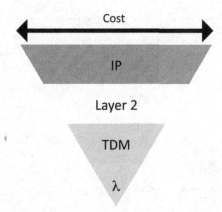

Figure 8.9 Qualitative representation of "hop cost" for services on the network. IP hops are the highest cost, optical (lambda) the lowest. Cross-layer resource management chooses the network path with the minimum total "hop cost."

and high transport efficiency is supported either via TDM switching to realize the hard guarantees for latency or MPLS, for higher transport efficiency, depending on the needs of the particular carrier. Across the network, reduced equipment and port costs are realized by minimizing average hops at high-cost layers via dynamic cross-layer resource allocation. Layer cost is represented pictorially in Figure 8.9. Most high-performance packet-based router/switches include hidden convergence layers in the switch, which adds more buffering and switch PHY costs. TDM switches (SONET, OTN) operate directly at their convergence layer, which is the main reason they are much simpler/less costly. The minimum cost node hop is at the optical layer. The use of colorless and directionless all-optical switching, though not required since OEO processes can be used, can reduce the number of OEO ports by as much as 30 percent in the global network configuration. In colorless switching, any wavelength may be assigned to any fiber port, removing the restriction common in today's reconfigurable optical add/drop multiplexers where a given wavelength is connected to a particular fiber port. Directionless switching means the ability to cross-connect any incoming port to any outgoing port in a multi-degree configuration.

8.7   Performance analysis

We have created a high fidelity OPNET simulation of the PHAROS system. Figure 8.10 compares the performance of three protection approaches: (1) dedicated protection in which each primary path receives its own protection path; (2) shared protection, where a set of protection paths may share a resource as explained in Section 8.4; (3) opportunistic shared protection, a sophisticated

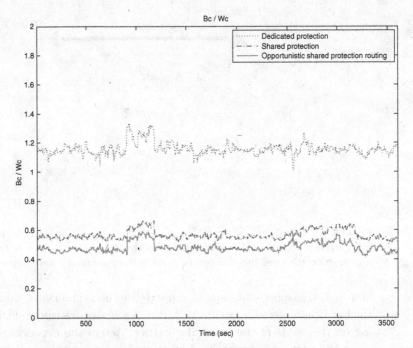

Figure 8.10 B/W comparison of different protection strategies

version of (2) where the protection paths are chosen to maximize shared protection opportunities.

Requests for bandwidth are generated over time. For each approach, we plot the B/W metric as a function of time; B/W is defined as the Backup (Protection) over Working capacity (B and W are in units of km-wavelength), which is a rough measure of the relative cost incurred in protection. Thus, the lower the B/W, the better.

Results shown here are for a 100-node optical network model with 75 nodes in the Continental USA (CONUS), 15 in Europe and 10 in Asia. The line rate is 40 Gb/s, and the aggregate traffic is 20 Tb/s of which 75% is IP traffic and 25% is wavelength services; 90% of the source–destination pairs are within the USA. The bit-averaged distance for the intra-CONUS traffic is about 1808 km. The B/W numbers shown in Figure 8.10 are for CONUS-contained resources only.

We see that the PHAROS shared protection strategies significantly outperform dedicated protection. Specifically, shared protection has about 50% lower B/W than dedicated, and opportunistic improves this further by about 10%.

8.8   Concluding remarks

The emergence of data as the dominant traffic and the resultant unpredictabilty and variability in traffic patterns has imposed several challenges to the design and implementation of the core network – from agile, autonomous resource management, to convergence and L1–L2 integration to the signaling system.

In this chapter we have described the architecture of a future core network control and management system along with a node implementation that enables future scalable and agile optical networks developed as part of the DARPA/STO CORONET program. This work provides control and management solutions that support services across core network dimension with 50-ms-class setup time, and also to respond to multiple network failures in this time frame. It provides a method of cross-layer resource allocation that delivers efficient allocation of bandwidth, both working and protection, to services at all layers in the network, including IP and wavelength services. Preliminary evaluations show significant advantages in using PHAROS.

The architecture described in this chapter enables core network scale beyond 10 times that of today's networks by optimizing path selection to maximize optical bypass, and minimize the number of router hops in the network. As a result, a higher capacity of network services can be supported with less network equipment.

References

[1] Simmons, J. On determining the optimal optical reach for a long-haul network *Journal of Lightwave Technology* 23(3), March 2005.

[2] Simmons, J. Cost vs. capacity tradeoff with shared mesh protection in optical-bypass-enabled backbone networks *OFC/NFOEC07*, Anaheim, CA, NThC2 March 2007.

[3] Dutta, R. and Rouskas, G. N. Traffic grooming in WDM networks: past and future *Network, IEEE*, 16(6) 46–56, November/December 2002.

[4] Iyer, P., Dutta, R., and Savage, C. D. On the complexity of path traffic grooming *Broadband Networks, 2005 2nd International Conference*, pp. 1231–1237 vol. 2, 3–7 October 2005.

[5] Zhou, L., Agrawal, P., Saradhi, C., and Fook, V. F. S. Effect of routing convergence time on lightpath establishment in GMPLS-controlled WDM optical networks *Communications, 2005. ICC 2005. 2005 IEEE International Conference on Communications*, pp. 1692–1696 vol. 3, 16–20 May 2005.

[6] Saleh, A. and Simmons, J. Architectural principles of optical regional and metropolitan access networks, *Journal of Lightwave Technology* 17(12), December 1999.

[7] Simmons, J. and Saleh, A. The value of optical bypass in reducing router size in gigabit networks *Proc. IEEE ICC 99*, Vancouver, 1999.

[8] Saleh, A. and Simmons, J. Evolution toward the next-generation core optical network *Journal of Lightwave Technology* 24(9), 3303, September 2006.

[9] Bragg A., Baldine, I., and Stevenson, D. Cost modeling for dynamically provisioned, optically switched networks *Proceedings SCS Spring Simulation Multiconference*, San Diego, April 2005.

[10] Brzezinski, A and Modiano, E. Dynamic reconfiguration and routing algorithms for IP-over-WDM networks with stochastic traffic *Journal of Lightwave Technology* 23(10), 3188–3205, Oct. 2005.

[11] Strand, J. and Chiu, A. Realizing the advantages of optical reconfigurability and restoration with integrated optical cross-connects *Journal of Lightwave Technology*, 21(11), 2871, November 2003.

[12] Xin, C., Ye, Y., Dixit, D., and Qiao, C. A joint working and protection path selection approach in WDM optical networks *Global Telecommunications Conference, 2001. GLOBECOM '01. IEEE*, pp. 2165–2168 vol. 4, 2001.

[13] Kodialam, M. and Lakshman, T. V. Dynamic routing of bandwidth guaranteed tunnels with restoration *INFOCOM 2000. Nineteenth Annual Joint Conference of the IEEE Computer and Communications Societies. Proceedings.* pp. 902–911 vol. 2, 2000.

[14] Ou, C., Zhang, J., Zang, H., Sahasrabuddhe, L. H., and Mukherjee, B. New and improved approaches for shared-path protection in WDM mesh networks *Journal of Lightwave Technology*, pp. 1223–1232, May 2004.

[15] Ellinas, G., Hailemariam, A. G., and Stern, T. E. Protection cycles in mesh WDM networks *IEEE Journal on Selected Areas in Communications*, 18(10) pp. 1924–1937, October 2000.

[16] Kodian, A., Sack, A., and Grover, W. D. p-Cycle network design with hop limits and circumference limits *Broadband Networks, 2004. Proceedings of the First International Conference on Broadband Networks*, pp. 244–253, 25–29 October 2004.

[17] Gripp, J., Duelk, M., Simsarian, M. *et al.* Optical switch fabrics for ultra-high-capacity IP routers *Journal of Lightwave Technology*, 21(11), 2839, (2003).

[18] Simmons, J. M. Optical network design and planning in *Optical Networks*, B. Mukherjee, Series editor, Springer 2008.

9 Customizable in-network services

Tilman Wolf

University of Massachusetts Amherst, USA

One of the key characteristics of the next-generation Internet architecture is its ability to adapt to novel protocols and communication paradigms. This adaptability can be achieved through custom processing functionality inside the network. In this chapter, we discuss the design of a network service architecture that can provide custom in-network processing.

9.1 Background

Support for innovation is an essential aspect of the next-generation Internet architecture. With the growing diversity of systems connected to the Internet (e.g., cell phones, sensors, etc.) and the adoption of new communication paradigms (e.g., content distribution, peer-to-peer, etc.), it is essential that not only existing data communication protocols are supported but that emerging protocols can be deployed, too.

9.1.1 Internet architecture

The existing Internet architecture is based on the layered protocol stack, where application and transport layer protocols processing occurs on end-systems and physical, link, and network layer processing occurs inside the network. This design has been very successful in limiting the complexity of operations that need to be performed by network routers. In turn, modern routers can support link speeds to tens of Gigabits per second and aggregate bandwidths of Terabits per second.

However, the existing Internet architecture also poses limitations on deploying functionality that does not adhere to the layered protocol stack model. In particular, functionality that crosses protocol layers cannot be accommodated without violating the principles of the Internet architecture. But in practice, many such extensions to existing protocols are necessary. Examples include network address

Next-Generation Internet Architectures and Protocols, ed. Byrav Ramamurthy, George Rouskas, and Krishna M. Sivalingam. Published by Cambridge University Press. © Cambridge University Press 2011.

translation (where transport layer port numbers are modified by a network layer device), intrusion detection (where packets are dropped in a network layer device based on data in the packet payload), etc.

To avoid this tension in the next-generation Internet, it is necessary to include deployment of new functionality as an essential aspect of the network architecture.

9.1.2 Next-generation Internet

The main requirement for a next-generation Internet is to provide data communication among existing and emerging networked devices. In this context, existing protocols as well as new communication paradigms need to be supported. Since it is unknown what kind of devices and communication abstractions may be developed in the future, it is essential that a next-generation network architecture provide some level of extensibility.

When considering extensibility, it is important to focus on the data plane of networks (i.e., the data path in routers). The control plane implements control operations that are necessary to manage network state, set up connections, and handle errors. But the data plane is where traffic is handled in the network. To deploy new protocol functionality into the network, it is necessary to modify the way traffic is handled in the data plane.

Extensions in the data plane have been explored in related research and differ in generality and complexity. Some extensions simply allow selection from a set of different functions. Others permit general-purpose programming of new data path operations. What is common among all solutions is the need for custom processing features in the data path of the routers that implement these extensions.

9.1.3 Data path programmability

The implementation of data communication protocols is achieved by performing processing steps on network traffic as it traverses a network node. The specific implementation of this processing on a particular system or device can vary from ASIC-based hardware implementation to programmable logic and software on general-purpose processors. In the existing Internet, ASIC-based implementations are common for high-performance routers. This approach is possible since the protocol operations that need to be implemented do not change over time. (RFC 1812, which defines the requirements for routers that implement IP version 4, has been around since 1995.)

In next-generation networks, where new functionality needs to be introduced after routers have already been deployed, it is necessary to include software-programmable devices in the data path. By changing the software that performs protocol processing, new protocol features can be deployed. Thus, programmability is no longer limited to end-systems, but gets pushed into the data path of networks.

9.1.4   Technical challenges

Programmability in the data path of routers does not only affect the way traffic is processed, but also places new demands on the control infrastructure and thus on the network architecture. The available programmability needs to be managed and controlled during the operation of the network. There are a number of technical challenges that arise in this context.

- Programmable router systems design: programmable packet processing platforms are necessary to implement custom packet processing. The design and implementation of such systems require high-performance processing platforms that support high-speed I/O and efficient protocol processors to sustain high-bandwidth networking. Secure execution of code, system-level runtime resource management, and suitable programming interfaces need to be developed.
- Control of custom functions: the functionality that is implemented on routers needs to be controlled, as different connections may require different functions. This control may require traffic classification, custom routing, and network-level resource management.
- Deployment of new functions: custom functions that are developed need to be deployed onto router systems. Code development environments need to be provided. The deployment process can range from manual installation to per-flow code distribution. Trust and security issues need to be resolved as multiple parties participate in code creation, distribution, and execution.

Some of these problems have been addressed in prior and related research.

9.1.5   In-network processing solutions

Several solutions to providing in-network processing infrastructure and control have been proposed and developed. Most notably, active networks provide per-connection and even per-packet configurability by carrying custom processing code in each packet [1]. Several active network platforms were developed [2, 3] differing in the level of programmability, the execution environment for active code, and hardware platform on which they were built. Per-packet programmability as proposed in active networks is very difficult to control. In practical networks, such openness is difficult to align with service providers' need for robust and predictable network behavior and performance. Also, while active networks provide the most complete programming abstraction (i.e., general-purpose programmability), the burden of developing suitable code for particular connection is pushed to the application developer.

A less general, but more manageable way of exposing processing capabilities in the network is with programmable routers. While these systems also provide general-purpose programmability, their control interface differs considerably: system administrators (i.e., the network service provider) may install any set of packet processing functions, but users are limited to selecting from this set (rather than providing their own functionality) [4, 5].

In the context of next-generation network architecture, programmability in the data (and control) plane appears in network virtualization [6]. To allow the coexistence of multiple networks with different data path functionality, link and router resources can be virtualized and partitioned among multiple virtual networks. Protocol processing for each virtual slice is implemented on a programmable packet processing system.

The technology used in router systems to provide programmability can range from a single-core general-purpose processor to embedded multi-core network processors [7] and programmable logic devices [8, 9]. Studies of processing workloads on programmable network devices have shown that differences to conventional workstation processing are significant and warrant specialized processing architectures [10, 11]. The main concern with any such router system is the need for scalability to support complex processing at high data rates [12].

One of the key challenges for existing solutions is the question of how to provide suitable abstractions for packet processing as part of the network architecture. On end-systems, per-flow configurability of protocol stacks has been proposed as a key element of next-generation networks [13, 14]. For in-network processing, our work proposes the use of network services as a key element in the network architecture.

9.2   Network services

To provide a balance between generality and manageability, it is important to design the right level of abstractions to access programmability and customization. We discuss how network services provide an abstraction that supports powerful extensibility to the network core while permitting tractable approaches to connection configuration, routing, and runtime resource management.

9.2.1   Concepts

The concept of a "network service" is used to represent fundamental processing operations on network traffic. A network service can represent any type of processing that operates on a traffic stream. Note that the term "service" has been used broadly in the networking domain and often refers to computational features provided on end-system (e.g., on a server). In our context, network service refers to data path operations that are performed on routers in the network. Examples of network services include very fundamental protocol operations as they can be found in TCP (e.g., reliability, flow control) and security protocols (e.g., privacy, integrity, authentication) as well as advanced functionality (e.g., payload transcoding for video distribution on cell phones).

When a connection is established, the sender and receiver can determine a "sequence of services" that is instantiated for this particular communication. The dynamic composition of sequences of services provides a custom network configuration for connections.

We envision that network services are well-known and agreed-upon functions that are standardized across the Internet (or at least across some of the Internet service providers). New network services could be introduced via a process similar to how protocols are standardized by the Internet Engineering Task Force (IETF). Thus, it is expected that the number of network services that a connection can choose from is in the order of tens, possibly hundreds. The network service architecture does not assume that each application introduces its own service (as it was envisioned in active networks), and therefore a very large number of deployed network services is unlikely. Even with a limited number of network services, the number of possible combinations (i.e., the number of possible sequences of services) is very large. For example, just 10 distinct network services and an average of 4 services per connection lead to thousands of possible service sequences. While not all combinations are feasible or desirable, this estimation still shows that a high level of customization can be achieved while limiting the specific data path processing functions to a manageable number.

To further illustrate the concept of network services, consider the following examples:

1. Legacy TCP: conventional Transmission Control Protocol (TCP) functionality can be composed from a set of network services, including: reliability (which implements segmentation, retransmission on packet loss, and reassembly), flow control (which throttles sending rate based on available receive buffer size), and congestion control (which throttles sending rate based on observed packet losses). Network service abstractions support modifications to legacy TCP in a straightforward manner. For example, when a connection wants to use a rate-based congestion control algorithm, it simply instantiates the rate-based congestion control network service (rather than the loss-based congestion control service).

2. Forward Error Correction: a connection that traverses links with high bit-error rates may instantiate a forward error correction (FEC) network service. Similar to reliability and flow control, this functionality consists of a pair of network services (the step that adds FEC and the step that checks and removes FEC). This service could be requested explicitly by the end-systems that initiate the connection or it could be added opportunistically by the network infrastructure when encountering lossy links during routing.

3. Multicast and video transcoding: a more complex connection setup example is video distribution (e.g., IPTV) to a set of heterogeneous clients. The transmitting end-system can specify that a multicast network service be used to

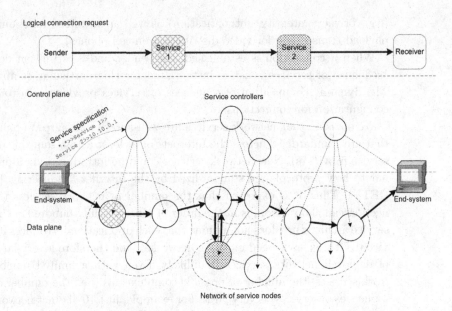

Figure 9.1 Network service architecture.

reach a set of receiving end-systems. In addition, a video transcoding network service can be used to change the video format and encoding. Such a network service is useful when the receiving end-system (e.g., a cell phone) cannot handle the data rate (e.g., due to low-bandwidth wireless links or due to limited processing capacity). In this scenario, network services are used to perform processing from the network layer to the application layer.

Note that the end-systems that set up a connection using network services do not specify where in the network a particular service is performed. It is up to the network to determine the most suitable placement of the network service processing. Leaving the decision on where and how to instantiate service to the network allows the infrastructure to consider the load on the network, policies, etc., when placing services. End-system applications are not (and should not have to be) aware of the state of the infrastructure and thus could not make the best placement and routing decisions. In some cases, constraints can be specified on the placement of network services (e.g., security-related network services should be instantiated within the trusted local network).

9.2.2   System architecture

An outline of the network service architecture is shown in Figure 9.1. There are three major aspects that we discuss in more detail: control plane, data plane, and the interface used by end-systems.

The control plane of the network service architecture determines the fundamental structure and operation. Following the structure of the current Internet, which consists of a set of federated networks (i.e., autonomous systems (AS)), our architecture also groups the network into networks that can be managed autonomously. When exchanging control information (e.g., connection setup, routing information) each AS can make its own local decisions while interacting with other AS globally. In each AS, there is (at least) one node that manages control plane interactions. This "service controller" performs routing and placement computations, and instantiates services for connections that traverse the AS.

In the data plane, "service nodes" implement network service processing on traffic that traverses the network. During connection setup, service controllers determine which nodes perform what service and how traffic is routed between these nodes. Any configuration information that is necessary for performing a network service (e.g., parameters, encryption keys) are provided by the service controller.

The end-system API is used by applications that communicate via the network. Using this interface, communication is set up (similar to how sockets are used in current operating systems) and the desired sequence of services is specified. When initiating a connection setup, the end-system communicates with its local service controller. This service controller propagates the setup request through the network and informs the end-system when all services (and the connections between them) have been set up.

There are several assumptions that are made in this architecture.

- The sequence of services specified by a connection is fixed for the duration of the connection. If a different service sequence is necessary, a new connection needs to be established.
- The underlying infrastructure provides basic addressing, forwarding, etc. There is ongoing research on how to improve these aspects of the next-generation Internet, which is beyond the scope of this chapter. Progress in this domain can be incorporated in the network service architecture we describe here.
- Routes in the network are fixed once a connection is set up. This can be achieved by tunneling or by using a network infrastructure that inherently allows control of per-flow routes (e.g., PoMo [15], OpenFlow [16]).

Given the fundamental concept of network services and the overarching system architecture, there are a number of important technical problems.

- End-system interface and service specification: the interface used by applications on the end-systems using the network service architecture needs to be sufficiently expressive to allow the specification of an arbitrary sequence of services without introducing too much complexity.
- Routing and service placement: during connection setup, the network needs to determine where network service processing should take place and on what route traffic traverses the network. With constraints on service availability,

processing capacity, and link bandwidth, this routing and placement problem is considerably more complex than conventional shortest-path routing.

- Runtime resource management on service nodes: the workload of service nodes is highly dynamic because it is not known a priori what network service processing is used by a connection. Thus, processing resources allocated to particular network services may need to be adjusted dynamically over time. This resource management is particularly challenging on high-performance packet processing platforms that use multi-core processors.

We address solutions to these problems in the following sections. It is important to note that even though the solutions are specific to our network service architecture, similar problems can be solved in other systems that employ in-network processing.

9.3   End-system interface and service specification

When using the network for communication, an end-system application needs to specify which network services should be instantiated. We describe how a "service pipeline" can be used to describe these services and how it can be composed and verified. The service pipeline has been described in our prior work [17]. Our recent work has extended the pipeline concept and integrated it into a socket interface [18]. Automated composition and pipeline verification is described in [19].

9.3.1   Service pipeline

A connection setup in a network with services is conceptually similar to the process in the current Internet. The main difference is that the set of parameters provided to the operating system not only includes the destination and socket type, but also needs to specify the network services. Since we use a sequence of services, we can provide this information in the form of a service pipeline.

The service pipeline is conceptually similar to the pipeline concept in UNIX, where the output of one command can be used as the input of another command by concatenating operations with a '|' symbol. For network services, we use the same concatenation operation (with different syntax) to indicate that the output of one service becomes the input of another. For each service, parameters can be specified. When streams split (e.g., multicast), parentheses can be used to serialize the resulting tree.

Elements of a service specification are:

- Source/sink: source and sink are represented by a sequence of IP address and port number separated by a ":" (e.g., 192.168.1.1:80). the source may leave the IP address and/or port unspecified (i.e., *:*).

- Network service: the service is specified by its name. If configuration parameters are necessary, they are provided as a sequence in parentheses after the name (e.g., `compression(LZ)` specifies a compression service that uses the Lempel-Ziv algorithm).
- Concatenation: the concatenation of source, network service(s), and sink is indicated by a ">>" symbol.

The service specifications for the three examples given in Section 9.2.1 are:

1. Legacy TCP: `*:*>>reliability_tx(local)>>flowcontrol_tx(local)>> congestioncontrol_tx(local)>>congestioncontrol_rx(remote)>> flowcontrol_rx(remote)>>reliability_rx(remote)>>192.168.1.1:80`
   The three key features of TCP (reliability, flow control, and congestion control), which are provided as separate services, need to be instantiated individually. Each consists of a receive and a transmit portion. The `local` and `remote` parameters indicate constraints on the placement of these services.

2. Forward Error Correction: `*:*>>[FEC_tx>>FEC_rc]>>192.168.1.1:80`
   Forward error correction is similar to the services in TCP. The brackets indicate that it is an optional service.

3. Multicast and video transcoding: `*:*>>multicast(192.168.1.1:5000, video_transcode(1080p,H.264)>>192.168.2.17:5000)`
   The multicast service specifies multiple receivers. Along the path to each receiver different services can be instantiated (e.g., video transcoding).

Service pipelines provide a general and extensible method for specifying network services.

9.3.2 Service composition

Clearly, it is possible to specify service combinations that are semantically incorrect and cannot be implemented correctly by the network. This problem leads to two questions: (1) how can the system verify that a service specification is semantically correct and (2) how can the system automatically compose correct specifications (given some connection requirements)? The issue of composition of services has been studied in related work for end-system protocol stacks [20] as well as in our prior work on in-network service composition [19].

To verify if a service specification is valid, the semantic description of a service needs to be extended. For a service to operate correctly, the input traffic needs to meet certain characteristics (e.g., contain necessary headers, contain payload that is encoded in a certain way). These characteristics can be expressed as preconditions for that service. The processing that is performed by a service may change the semantics of the input. Some characteristics may change (e.g., set of headers, type of payload), but others may remain unchanged (e.g., delay-sensitive nature of traffic). The combination of input characteristics and modifications performed by the service determines output characteristics. By propagating these characteristics through the service sequence and by verifying that all preconditions are met for all services, the correctness of a service sequence can be verified. The semantics of a service can be expressed using a variety of languages (e.g., web ontology language (OWL)). The verification operation can be performed by the service controller before setting up a connection.

A more difficult scenario is the automated composition of a service sequence. An application may specify the input characteristics and desired output characteristics of traffic. Based on the formal description of service semantics, a service controller can use AI planning to find a sequence of services that "connects" the input requirements to the output requirements [19]. This feature is particularly important when multiple parties contribute to the service sequence (e.g., an ISP may add monitoring or intrusion detection services to a service sequence). In such a case, the originating end-system cannot predict all possible services and create a complete service sequence. Instead, additional services are included during connection setup.

Once a correct and complete service sequence is available, the services need to be instantiated within the network.

9.4   Routing and service placement

There are a number of different approaches to determining a suitable routing and placement for a given sequence of services. In our prior work, we have explored how to solve this problem given complete information on a centralized node [21] as well as in a distributed setting [22]. We have also compared the relative performance of these approaches [23]. We review some of these results in this section.

9.4.1   Problem statement

The service placement problem is stated as follows (from [23]): the network is represented by a weighted graph, $G = (V, E)$, where nodes V correspond to routers and end-systems and edges E correspond to links. A node v_i is labeled with the set of services that it can perform, $u_i = \{S_k | \text{service } S_k \text{ is available on } v_i\}$, the processing cost $c_{i,k}$ (e.g., processing delay) of each service, and the node's total available processing capacity p_i. An edge $e_{i,j}$ that connects node v_i and v_j is labeled with a weight $d_{i,j}$ that represents the link delay (e.g., communication cost) and a capacity $l_{i,j}$ that represents the available link bandwidth. A connection request is represented as $R = (v_s, v_t, b, (S_{k_1}, \ldots, S_{k_m}))$, where v_s is the source node, v_t is the destination node, b is the requested connection bandwidth (assumed to be constant bit-rate), and $(S_{k_1}, \ldots, S_{k_m})$ is an ordered list of services that are required for this connection. For simplicity, we assume that the processing requirements for a connection are directly proportional to the requested bandwidth b. For service S_k, a complexity metric $z_{i,k}$ defines the amount of computation that is required on node v_i for processing each byte transmitted.

Given a network G and a request R, we need to find a path for the connection such that the source and the destination are connected and all required services can be processed along the path. The path is defined as $P = (E^P, M^P)$ with a sequence of edges, E^P, and services mapped to processing nodes, M^P: $P = ((e_{i_1,i_2}, \ldots, e_{i_{n-1},i_n}), (S_{k_1} \to v_{j_1}, \ldots, S_{k_m} \to v_{j_m}))$, where $v_{i_1} = v_s$, $v_{i_n} = v_t$, $\{v_{j_1}, \ldots, v_{j_m}\} \subset \{v_{i_1}, \ldots, v_{i_n}\}$ and nodes $\{v_{j_1}, \ldots, v_{j_m}\}$ are traversed in sequence along the path. The path P is valid if (1) all edges have sufficient link capacity (i.e., $\forall e_{x,y} \in E^P, l_{x,y} \geq (b \cdot t)$, assuming link $e_{x,y}$ appears t times in E^P), and (2) all service nodes have sufficient processing capacity (i.e., $\forall S_{k_x} \to v_{j_x} \in M^P, p_{j_x} \geq \sum_{y|S_{k_y} \to v_{j_x} \in M^P} b \cdot z_{j_x,k_y})$.

To determine the quality of a path, we define the total cost $C(P)$ of accommodating connection request R as the sum of communication cost and processing cost: $C(P) = \left(\sum_{x=1}^{n-1} d_{i_x,i_{x+1}}\right) + \left(\sum_{\{(j_x,k_x)|S_{k_x} \to v_{j_x} \in M^P\}} c_{j_x,k_x}\right)$. In many cases, it is desirable to find the optimal connection setup. This optimality can be viewed (1) as finding the optimal (i.e., least-cost) allocation of a single connection request or (2) as finding the optimal allocation of multiple connection requests. In the latter case, the optimization metric can be the overall least cost for all connections or the best system utilization, etc. We focus on the former case of a single connection request.

It was shown in [21] that finding a solution to a connection request in a capacity-constrained network can be reduced to the traveling salesman problem, which is known to be NP-complete. Therefore, we limit our discussion to the routing and placement problem without constraints. Using heuristics, the solutions can be extended to consider capacity constraints.

9.4.2   Centralized routing and placement

The centralized routing and placement solution was first described in [21]. The idea is to represent both communication and processing in a single graph and to use conventional shortest-path routing to determine an optimal placement. This solution requires that a single cost metric is used for both communication and processing cost. To account for processing, the network graph is replicated for each of the m processing steps in the connection request (as illustrated in Figure 9.2(a)). Thus, a total of $m + 1$ network graphs ("layers") exist. The top graph, layer 0, represents communication that is performed before the first network service is performed. The bottom graph, layer m, represents communication after all processing steps have been completed. To traverse from one layer to another, vertical edges between layers are added. These edges can only connect the same nodes in neighboring layers. The existence of a (directed) vertical edge indicates that the necessary service processing step to reach the next layer is

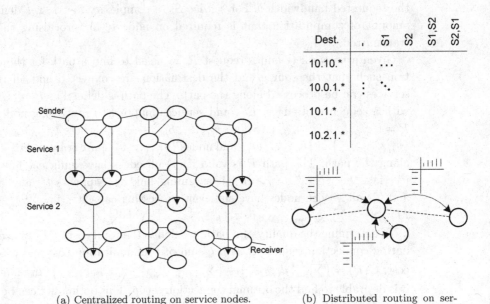

(a) Centralized routing on service nodes.

(b) Distributed routing on service controllers.

Figure 9.2 Routing and placement in network service architecture.

available on that node. The cost of traversing that edge is the cost of processing on that node.

Routing is achieved by finding the least-cost path in the layered graph between the source node on layer 0 and the destination node on layer m. This path is projected back into a single layer with vertical edges indicating placement for network service processing.

The algorithm is guaranteed to find the optimal path for a network without capacity constraints. The computational cost is that of running Dijkstra's shortest path algorithm on the layered graph. Since the layered graph is $m + 1$ times the size of the original network graph, the complexity is $\mathcal{O}(|m||E| + |m||V| + |m||V| \log(|m||V|))$.

9.4.3 Distributed routing and placement

One of the drawbacks of the centralized layered graph solution is the need for complete knowledge of all network links. In an Internet-scale deployment it is unrealistic to assume that such information is available. Thus, we also present a distributed approach, where information can be aggregated and nodes have a limited "view" of the network. This algorithm has been described in [22].

The distributed routing and placement algorithm uses a dynamic programming approach similar to distance vector routing [24]. Let $c_v^{k_1,\ldots,k_m}(t)$ denote the cost of the shortest path from node v to node t where the requested services $S_{k_1},\ldots,S_{k_m}$

are performed along the way. For shortest path computation (i.e., no services), we use the notation $c_v^-(t)$. Thus, a node v can determine the least-cost path by considering to process i $(0 \le i \le m)$ services and forwarding the request to any neighboring node n_v $(n_v \in \{x \in V | e_{v,x} \in E\})$:

$$c_v^{k_1,\dots,k_m}(t) = \min_{0 \le i \le m} \left(\sum_{l=1}^{i} c_v^{k_l}(v) + \min_{n_v} \left(c_v^-(n_v) + c_{n_v}^{k_{i+1},\dots,k_m}(t) \right) \right).$$

The argument i on the right side determines how many of the m services that need to be performed should be processed on node v. Note that if $i = 0$, no service is processed, i.e., $\sum_{l=1}^{i} c_v^{k_l}(v) = 0$. If $i = m$, all the services are processed on node v, i.e., $c_{n_v}^{k_{i+1},\dots,k_m}(t) = c_{n_v}^-(t)$. The argument n_v determines to which neighbor of v the remaining request should be sent.

To acquire the necessary cost information, nodes exchange a "service matrix" with their neighbors (as illustrated in Figure 9.2(b)). This matrix contains costs for all destinations and all possible service combinations. Since the number of service combinations can be very large, a heuristic solution has been developed that only uses cost information for each individual service. This approach is discussed in detail in [22].

9.5   Runtime resource management

The network service architecture presents a highly dynamic environment for the processing system on which services are implemented. Each connection may request a different service sequence, which can lead to variable demand for any particular service. This type of workload is very different from conventional IP forwarding, where each packet requires practically the same processing steps. While operating systems can provide a layer of abstraction between hardware resources and dynamic processing workloads, they are too heavy-weight for embedded packet processors that need to handle traffic at Gigabit per second data rates. Instead, a runtime system that is specialized for dealing with network service tasks can be developed. Of particular concern is to handle processing workloads on multi-core packet processing systems. We discuss the task allocation system developed in our prior work [25].

9.5.1   Workload and system model

The workload of a router system that implements packet forwarding for the current Internet or service processing for next-generation networks can be represented by a task graph. This task graph is a directed acyclic graph of processing steps with directed edges indicating processing dependencies. Packet processing occurs along one path through this graph for any given packet. Different packets may traverse different paths. An example of such a graph representation of packet processing is the Click modular router abstraction [26].

As discussed above, changes in traffic may cause more or less utilization along any particular path in the graph and thus more or less utilization for any particular processing step. To determine the processing requirements at runtime, it is necessary to do runtime profiling and track (at least) the following information:

- Processing requirements for each task.
- Frequency of task usage.

In our runtime system prototype, we represent the processing requirement as a random variable S_i, which reflects the processing time distribution of task t_i. For any given packet, the task service time is s_i. The frequency of usage is represented by the task utilization $u(t_i)$, which denotes the fraction of traffic traversing task t_i.

Based on this profiling information, the runtime system determines how to allocate resources to tasks.

9.5.2   Resource management problem

The formal problem statement for runtime management of multi-core service processors is as follows (from [25]): Assume we are given the task graph of all subtasks in all applications by T task nodes $t_1, \ldots, t_T$ and directed edges $e_{i,j}$ that represent processing dependencies between tasks t_i and t_j. For each task, t_i, its utilization $u(t_i)$ and its service time S_i are given. Also assume that we represent a packet processing system by N processors with M processing resources on each (i.e., each processor can accommodate M tasks and the entire system can accommodate $N \cdot M$ tasks). The goal is to determine a mapping m that assigns each of the T tasks to one of N processors: $m : \{t_1, \ldots, t_T\} \to [1, N]$. This mapping needs to consider the constraint of resource limitations: $\forall j, 1 \leq j \leq N : |\{t_i | m(t_i) = j\}| \leq M$.

The quality of the resource allocation (i.e., mapping) can be measured by different metrics (e.g., system utilization, power consumption, packet processing delay, etc.). Our focus is to obtain a balanced load across processing components, which provides the basis for achieving high system throughput.

9.5.3   Task duplication

One of the challenges in runtime management is the significant differences in processing requirements between different tasks. Some tasks are highly utilized and complex and thus require much more processing resources than tasks that are simple and rarely used. Also, high-end packet processing systems may have more processor cores and threads than there are tasks.

To address this problem, we have developed a technique called "task duplication" that exploits the packet-level parallelism inherent to the networking domain. Task duplication provides a straightforward way to distributing processing tasks onto multiple processing resources. For the discussion, we assume processing is stateless between packets. If stateful processing is performed, the runtime system can ensure that packets of the same flow are sent to the same instance of the processing task.

Task duplication creates additional instances of tasks with high work requirements. The amount of work, w_i, a task performs is the product of the processing requirements for a single packet and the frequency with which the task is invoked: $w_i = u(t_i) \cdot E[S_i]$. This amount of work can be reduced if the number of task instances is increased. If a task is duplicated such that there are d_i instances and traffic is spread evenly among these instances, then the amount of utilization for each instance decreases to $u(t_i)/d_i$. Thus, the effective amount of work per instance is $w_i' = u(t_i)/d_i \cdot E[S_i]$. Therefore, a more balanced workload can be obtained by greedily duplicating the task with the highest amount of work until all $M \cdot N$ resources are filled with tasks. This also allows the use of all resources if there are fewer tasks than resources.

Note that the work equation also shows that the differences in the amount of work per task are not only due to the inherent nature of the task (i.e., the expected service time $E[S_i]$), but also due to the dynamic nature of the network (i.e., the current utilization of the task $u(t_i)$). Thus, the imbalance between tasks cannot be removed by achieving a better (i.e., more balanced) offline partitioning, and there is always need to adapt to current conditions at runtime.

9.5.4 Task mapping

Once the tasks and their duplicates are available, the mapping of tasks to processors needs to be determined. There are numerous different approaches to placing tasks. When using tasks with vast differences in the amount of work that they need to perform, then a mapping algorithm needs to take care in co-locating complex tasks with simple tasks. If too many complex tasks are placed on a single processor, then that system resource becomes a bottleneck and the overall system performance suffers. Solving this type of packing problem is NP-complete [27].

The benefit of having performed task duplication is that most tasks require nearly equal amounts of work. Thus, the mapping algorithm can place any combination of these tasks onto a processor without the need for considering difference in processing work. Instead, secondary metrics (e.g., locality of communication) can be considered to make mapping decisions. We have shown that a depth-first search to maximize communication locality is an effective mapping algorithm. Our prototype runtime system that uses duplication and this mapping strategy shows an improvement in throughput performance over a system with conventional symmetric multiprocessing (SMP) scheduling provided by an operating system [25]. More recent work considers not only processing resource allocation, but also memory management [28]. In particular, the partitioning of data structures among multiple physical memories with different space and performance characteristics is an important issue. Static partitioning used in traditional packet processing systems is not sufficient for the same reasons that static processing allocations cannot adapt to changing networking conditions.

Overall, runtime management of processing resources is an important aspect of packet processing platforms in next-generation networks – especially as the complexity and diversity of such services continues to increase.

9.6   Summary

The functionality provided by the networking infrastructure in the next-generation Internet architecture encompasses not only forwarding, but also more advanced protocol and payload processing. A key challenge is to find suitable abstractions that allow end-systems to utilize such functionality, while maintaining manageability and controllability from the perspective of service providers. We presented an overview of a network service architecture that uses network services as fundamental processing steps. The sequence of services that is instantiated for each connection can be customized to meet the end-system application's needs. We discussed how service specifications can be used to express these custom processing needs and how they can be translated into a constrained mapping problem. Routing in networks that support services is a problem that needs to consider communication and processing costs. We presented two solutions, one centralized and one distributed, to address the routing problem. We also presented how runtime management on packet processing systems can ensure an effective utilization of system resources.

The use of network service abstractions to describe in-network processing service can be used beyond the work presented here. For example, when developing virtualized network infrastructure, network service specifications can be used to describe data path requirements for virtual slices.

In summary, in-network processing services are an integral part of the next-generation Internet infrastructure. The work we presented here can provide one way of making such functionality possible.

References

[1] Tennenhouse, D. L., Wetherall, D. J. Towards an Active Network Architecture. *ACM SIGCOMM Computer Communication Review*. 1996 Apr;26(2): 5–18.

[2] Tennenhouse, D. L., Smith, J. M., Sincoskie, W. D., Wetherall, D. J., Minden, G. J. A Survey of Active Network Research. *IEEE Communications Magazine*. 1997 Jan;35(1):80–86.

[3] Campbell, A. T., De Meer, H. G., Kounavis, M. E., *et al.* A Survey of Programmable Networks. *ACM SIGCOMM Computer Communication Review.* 1999 Apr;29(2):7–23.

[4] Wolf, T. Design and Performance of Scalable High-Performance Programmable Routers. Department of Computer Science, Washington University. St. Louis, MO; 2002.

[5] Ruf, L., Farkas, K., Hug, H., Plattner, B. Network Services on Service Extensible Routers. In: Proc. of Seventh Annual International Working Conference on Active Networking (IWAN 2005). Sophia Antipolis, France; 2005.

[6] Anderson, T., Peterson, L., Shenker, S., Turner, J. Overcoming the Internet Impasse through Virtualization. *Computer.* 2005 Apr;38(4):34–41.

[7] Wolf, T. Challenges and Applications for Network-Processor-Based Programmable Routers. In: Proc. of IEEE Sarnoff Symposium. Princeton, NJ; 2006.

[8] Hadzic, I., Marcus, W. S., Smith, J. M. On-the-fly Programmable Hardware for Networks. In: Proc. of IEEE Globecom 98. Sydney, Australia; 1998.

[9] Taylor, D. E., Turner, J. S., Lockwood, J. W., Horta, E. L. Dynamic Hardware Plugins: Exploiting Reconfigurable Hardware for High-Performance Programmable Routers. *Computer Networks.* 2002 Feb;38(3):295–310.

[10] Crowley, P., Fiuczynski, M. E., Baer, J. L., Bershad, B. N. Workloads for Programmable Network Interfaces. In: IEEE Second Annual Workshop on Workload Characterization. Austin, TX; 1999.

[11] Wolf, T., Franklin, M. A. CommBench – A Telecommunications Benchmark for Network Processors. In: Proc. of IEEE International Symposium on Performance Analysis of Systems and Software (ISPASS). Austin, TX; 2000. pp. 154–162.

[12] Wolf, T., Turner, J. S. Design Issues for High-Performance Active Routers. *IEEE Journal on Selected Areas of Communication.* 2001 Mar;19(3):404–409.

[13] Dutta, R., Rouskas, G. N., Baldine, I., Bragg, A., Stevenson, D. The SILO Architecture for Services Integration, controL, and Optimization for the Future Internet. In: Proc. of IEEE International Conference on Communications (ICC). Glasgow, Scotland; 2007. pp. 1899–1904.

[14] Baldine, I., Vellala, M., Wang, A., *et al.* A Unified Software Architecture to Enable Cross-Layer Design in the Future Internet. In: Proc. of Sixteenth IEEE International Conference on Computer Communications and Networks (ICCCN). Honolulu, HI; 2007.

[15] Calvert, K. L., Griffioen, J., Poutievski, L. Separating Routing and Forwarding: A Clean-Slate Network Layer Design. In: Proc. of Fourth International Conference on Broadband Communications, Networks, and Systems (BROADNETS). Raleigh, NC; 2007. pp. 261–270.

[16] McKeown, N., Anderson, T., Balakrishnan, H., *et al.* OpenFlow: Enabling Innovation in Campus Networks. *SIGCOMM Computer Communication Review.* 2008 Apr;38(2):69–74.

[17] Keller, R., Ramamirtham, J., Wolf, T., Plattner, B. Active Pipes: Program Composition for Programmable Networks. In: Proc. of the 2001 IEEE Conference on Military Communications (MILCOM). McLean, VA; 2001. pp. 962–966.

[18] Shanbhag, S., Wolf, T. Implementation of End-to-End Abstractions in a Network Service Architecture. In: Proc. of Fourth Conference on emerging Networking EXperiments and Technologies (CoNEXT). Madrid, Spain; 2008.

[19] Shanbhag, S., Huang, X., Proddatoori, S., Wolf, T. Automated Service Composition in Next-Generation Networks. In: Proc. of the International Workshop on Next Generation Network Architecture (NGNA) held in conjunction with the IEEE 29th International Conference on Distributed Computing Systems (ICDCS). Montreal, Canada; 2009.

[20] Vellala, M., Wang, A., Rouskas, G. N., et al. A Composition Algorithm for the SILO Cross-Layer Optimization Service Architecture. In: Proc. of the Advanced Networks and Telecommunications Systems Conference (ANTS). Mumbai, India; 2007.

[21] Choi, S. Y., Turner, J. S., Wolf, T. Configuring Sessions in Programmable Networks. In: Proc. of the Twentieth IEEE Conference on Computer Communications (INFOCOM). Anchorage, AK; 2001. pp. 60–66.

[22] Huang, X., Ganapathy, S., Wolf, T. A Scalable Distributed Routing Protocol for Networks with Data-Path Services. In: Proc. of 16th IEEE International Conference on Network Protocols (ICNP). Orlando, FL; 2008.

[23] Huang, X., Ganapathy, S., Wolf, T. Evaluating Algorithms for Composable Service Placement in Computer Networks. In: Proc. of IEEE International Conference on Communications (ICC). Dresden, Germany; 2009.

[24] Bellman, R. On a Routing Problem. *Quarterly of Applied Mathematics.* 1958 Jan;16(1):87–90.

[25] Wu, Q., Wolf, T. On Runtime Management in Multi-Core Packet Processing Systems. In: Proc. of ACM/IEEE Symposium on Architectures for Networking and Communication Systems (ANCS). San Jose, CA; 2008.

[26] Kohler, E., Morris, R., Chen, B., Jannotti, J., Kaashoek, M. F. The Click Modular Router. *ACM Transactions on Computer Systems.* 2000 Aug;18(3):263–297.

[27] Johnson, D. S., Demers, A. J., Ullman, J. D., Garey, M. R., Graham, R. L. Worst-Case Performance Bounds for Simple One-Dimensional Packing Algorithms. *SIAM Journal on Computing.* 1974 Dec;3(4):299–325.

[28] Wu, Q., Wolf, T. Runtime Resource Allocation in Multi-Core Packet Processing Systems. In: Proc. of IEEE Workshop on High Performance Switching and Routing (HPSR). Paris, France; 2009.

10 Architectural support for continuing Internet evolution and innovation

Rudra Dutta[†] and Ilia Baldine[‡]

[†]North Carolina State University, USA [‡]Renaissance Computing Institute, USA

Starting in August of 2006 our collaborative team of researchers from North Carolina State University and the Renaissance Computing Institute, UNC-CH, have been working on a Future InterNet Design (NSF FIND) project to envision and describe an architecture that we call the Services Integration, controL, and Optimization (SILO). In this chapter, we describe the output of that project. We start by listing some insights about architectural research, some that we started with and some that we gained along the way, and also state the goals we formulated for our architecture. We then describe that actual architecture itself, connecting it with relevant prior and current research work. We show how the promise of enabling change is validated by showing our recent work on supporting virtualization as well as cross-layer research in optics using SILO. We end with an early case study on the usefulness of SILO in lowering the barrier to contribution and innovation in network protocols.

10.1 Toward a new Internet architecture

Back in 1972 Robert Metcalfe was famously able to capture the essence of networking with a phrase "Networking is inter-process communication," however, describing the *architecture* that enables this communication to take place is by no means easy. The architecture of something as complex as the modern Internet encompasses a large number of principles, concepts and assumptions, which necessarily bear periodic revisiting and reevaluation in order to assess how well they have withstood the test of time. Such attempts have been made periodically in the past, but really started coming into force in the early 2000s, with programs like DARPA NewArch [24], NSF FIND [11], EU FIRE [12] and China's CNGI all addressing the question of the "new" Internet architecture. The degree to which the Internet continues to permeate modern life with hundreds of new uses and applications, adapted to various networking technologies (from optical, to mobile

Next-Generation Internet Architectures and Protocols, ed. Byrav Ramamurthy, George Rouskas, and Krishna M. Sivalingam. Published by Cambridge University Press. © Cambridge University Press 2011.

wireless, to satellite), raises concerns with the longevity of Internet architecture. The original simple file transfer protocols and UUNET gave way to e-mail and WWW, which by now are becoming eclipsed by streaming media, compute grids and clouds, instant messaging and peer-to-peer applications. Every step in this evolution raises the prospect of reevaluation of the fundamental principles and assumptions underlying the Internet architecture. So far this architecture has managed to survive and adapt to the changing requirements and technologies while providing immense opportunities for innovation and growth. On the one hand, such adaptability seems to confirm that some of the original principles have truly been prescient to allow the architecture to survive for over 30 years. On the other, it begs the question if the survival of the architecture is in fact being ensured by the reluctance to question those principles, cemented by shoehorning novel applications and technologies into the existing architecture without giving thought to its suitability.

Such contradiction will not be easily resolved, nor should it be. A dramatic shift to a new architecture should only be possible for the most compelling of reasons, and so, the existence of this contradiction creates the ultimate "tussle" [7] for the networking researcher community. This tussle pits the investment in time, technologies and capital made in the existing architecture against the possibilities which open up by adapting the new architecture in allowing for creation of novel and improved services over the Internet as well as opening new areas of research and discovery. It also allows us to continually refine the definition of the Internet architecture and separate and reexamine the various aspects of it. A sampling of the projects funded through the NSF FIND program, targeted at re-examining the architecture of the Internet, illustrates the point: there are projects concerned with naming [17, 14]), routing [4, 15], protocol architectures [8], which examine these and other aspects from perspectives of security, management [21], environmental impact [2] and economics [14]. Another dimension is presented by the range of technologies allowing devices to communicate: wireless, cellular, optical [5] and adaptations of the Internet architecture to them.

This diversity of points of view makes it difficult to see clearly the fundamental elements of the architecture and their influence over each other. Most importantly for the researcher interested in architecture, this makes it nearly impossible to answer concisely the question of what the Internet architecture actually is, or even what concerns are encompassed by the term "Internet architecture." What things should be considered part of the architecture of a complex system, and what should be considered specific design decisions, comparatively more mutable? This further fuels the "to change or not to change" tussle we alluded to above.

One way to make progress in the tussle appears to be in creating modifications in the current architecture, which enable new functionality or services not possible today, while limiting the impact on the rest of the architecture, in essence evolving the architecture while preserving backward compatibility. This approach has the additional merit of taking the concerns expressed in some recent papers regarding the potential of clean-slate approaches to be far divorced from reality, with no reasonable chance of translating to deployment; such concerns have been epitomized by the phrase "Clean-slate is not blue-sky."

In our project named SILO (**S**ervices **I**ntegration, contro**L** and **O**ptimization) we started, in a way, by following this approach. We did not attempt to rethink the Internet as a whole. Instead, we identified one particular aspect of the Internet architecture that, in our opinion, created a significant barrier to its future development. We proposed a way to modify this aspect of the architecture in a way that is least impactful on the rest of the architecture and demonstrated the use of this new architecture via a prototype implementation and case studies.

Somewhat to our surprise, however, what emerged from our research, was a new understanding regarding the problem at hand. The important problem is *not* to obtain a particular design or arrangement of specific features, but rather, to obtain a *meta-design* that explicitly allows for future change. With a system like the Internet, the goal is not to design the "next" system, or even the "best next" system, but rather a system that can sustain continuing change and innovation.

This principle, which we call *designing for change*, became fundamental to our project. In the process, we have come to develop our own answer to the question of what architecture actually is: it is precisely the characteristics of the system that does not change itself, but provides a framework within which the system design can change and evolve. The current architecture houses an effective design, but is not itself effective in enabling evolution. Our challenge has been to articulate the necessary minimum characteristics of an architecture that will be successful in doing so.

10.2   The problems with the current architecture

As witnessed by the breadth of scope of the various FIND-related projects, ideas on how to improve the current Internet cover a wide range of approaches. These ideas are frequently driven by the difficulties in attempting to integrate some new functionality into the Internet architecture. In the SILO project we began with a single basic observation: that protocol research has stagnated despite the clear need to improve data transfers over the new high-speed optical and wireless technologies and has been reduced to designing variants of TCP. This stagnation points to a weak point in the original Internet architecture, that somehow has disallowed the evolution and development of this aspect of the architecture. The cause of this stagnation, in our opinion, lies in (a) the difficult barrier to entry in implementing new data transfer protocols in the TCP/IP stack, except for user-space, (b) perhaps more importantly, the lack of clear separation between policies and mechanisms in TCP/IP design (e.g., window-based flow control vs. the various ways in which the window size can respond to changes in the network environment) preventing the reuse of various components, and (c) the lack of a predefined agreed-upon way for protocols at different layers to share information with each other for the purpose of optimizing their behavior for different optimization criteria (of the user, the system or the network).

Such lack of flexibility, for example, prevented applications that would ideally prefer to use some parts of the functionality of the TCP/IP stack, but not others, in transmitting data. For instance, being able to request a specific mode of flow control (or totally remove it), while still retaining in-order delivery of TCP may be desirable. However, the current implementations make no allowance for such flexibility.

The lack of explicit and well-defined cross-layer interaction mechanisms resulted in more subtle problems: these interactions are implemented anyway, but in an ad-hoc fashion, resulting in a monolithic implementation where TCP and IP codes are intermingled to achieve higher efficiencies. As a result, clarity and reusability are sacrificed, with the unintended consequence of making each further unit of development and research more difficult. In a way, this is a self-reinforcing process, each modification making further modificiations of the whole structure more difficult, ensuring that in the long run TCP and its modifications remain the dominant mode of data transport. When it comes to adding new cross-layer interactions, particularly with the physical layer, the problem is even more pronounced, as is indicated by the fact that no standard cross-layer solution has been widely adopted, for example, to assist TCP over wireless by taking advantage of physical layer conditions, despite a clear need.

Finally, the proliferation of half-layer solutions, like MPLS or IPSec, pointed at another aspect of this problem: that the protocols layers as we know them (TCP/IP or OSI stacks) were no longer relevant and were merely markers for some vague functional boundaries within the architecture. These half-layer solutions clearly addressed important needs, yet the Internet architecture had no way of describing their place within a data flow.

In essence, the TCP/IP stack has become ossified, preventing further development and evolution of protocols within its framework. Applications written today that require data services not accommodated by the TCP/UDP dichotomy are left to take one of several paths: (a) implement their own UDP-based data transfer mechanisms without the ability to reuse the elements of the existing architecture or to take advantage of kernel-space optimizations in buffer management, (b) adapting an existing TCP implementation to new situations, e.g., new media like wireless, or large bandwidth–delay product in optical networks, and (c) abandoning the old and "rolling their own" implementation, as has occurred in sensor networks, where TCP/IP has been supplanted by a simpler implementation suitable for the low-cost/low-power sensor hardware.

These approaches point to a significant risk of fracturing the future protocol development into their applicable domains (wireless, optical, sensor, mobile). In turn, these networks are then forced to communicate with each other or "the (canonical) Internet" via proxies or gateways. In a way, this is a "balkanization"

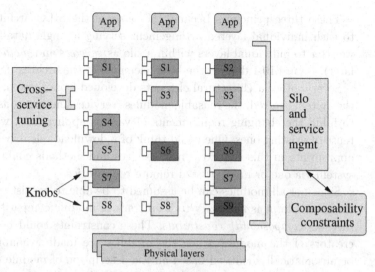

Figure 10.1 Generalization of layering in SILO.

of the network as apprehended in [18]. From our perspective, such an outcome is undesirable and presents the fundamental challenge to the concept of IP as a simple convergence point (often referred to as IP being the "narrow waist" of the "hourglass" protocol stack), which stands as one of the fundamental assumptions of the current Internet architecture.

Based on the identified shortcomings of the current Internet architecture, it became clear that what is needed is a new architectural framework that will address these deficiencies and allow for a continuing evolution of protocols and their adaptation to new uses and media types.

10.3   SILO architecture: design for change

As a starting point, we adopted a view that layering of protocol modules within a dataflow was a desirable feature that has withstood the test of time, as it made data encapsulation easy, and simplified buffer management. The layer *boundaries*, on the other hand, do not have to be in specific places; to our minds, this caused entrenchment of existing protocols, and is one of the causes of the identified ossification of the Internet architecture. Based on this initial assumption, the desirable characteristics of the new architecture started to emerge: that (a) each data flow should have its own arrangement of layered modules, such that the application or the system could create such arrangements based on application needs and underlying physical layer attributes; (b) the constituent modules should be small and reusable to assist in the evolution by providing ready-made partial solutions; and (c) that the modules should be able to communicate with each other over a well-defined set of mechanisms.

These three principles became the basis of the SILO architecture. We refer to each individual layered arrangement serving a single dataflow as a *silo* and we refer to individual layers within a silo as *services* and *methods* (more on this later). Figure 10.1 depicts the basic elements of the architecture.

Several other architectural elements developed from these basic principles. As the system evolved, the reusable modules (services and methods) could be added to fulfill the changing requirements of various applications, while allowing the reuse of existing ones. One could think of a downloadable driver model as being appropriate in this context – new services and methods could be added to the system via one or more trusted remote repositories.

Since not all modules can be assumed to be able to coexist with each other in the same silo, it is necessary to keep track of module compatibility. We refer to these as *composability constraints*. These constraints could be specified by the creators of the modules when the modules are made available, or they could be automatically deduced based on the description of module functionality. We envision that knowledge distilled from deployment experience of network operators, collectively, can also be stored here. The number of such constraints can be expected to be large and grow with time. This pointed out to us the need for automated silo composition, which can be accomplished by one or more algorithms based on the application specification. This automated construction of silos became a crucial part of the architecture.

From the perspective of cross-layer interactions, it also became desirable to not simply allow modules to communicate with each other outside the dataflow, but to allow for an external entity to access module states for purposes of optimizing the behavior of individual silos and/or the system as a whole. We referred to this function as *cross-service tuning*, which was accomplished by querying individual modules via *gauges* and modifying their state via *knobs*. Both gauges and knobs had to be well defined and exposed as part of the module interface. The important aspect of this approach is that the optimization algorithm could be pluggable, just like the modules within a silo, allowing for easy retargeting of optimization objectives by a substitution of the optimization algorithm. This addresses the previously identified deficiency of the current architecture, where policies and methods in protocol implementations were frequently mixed together, not allowing for evolution of one without the other.

The service/method dichotomy introduced earlier becomes important from the point of view of system scalability. Borrowing from object-oriented programming concepts, what we call services are generic functions like *encryption* or *header checksum* or *flow control*, while methods are specific implementations of services. Thus, in some sense, methods are polymorphic on services. This relationship allows for aggregation of some composability constraints based on generic service definitions, which necessarily propagate to the methods implementing this service, thus making the job of the developer, as well as of the composition algorithm, substantially easier.

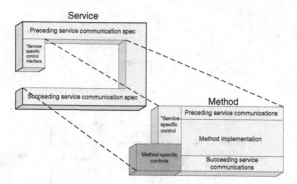

Figure 10.2 Services vs. methods.

Each service is described from the point of view of its functionality, its generic interfaces (to the services immediately above and below it in a silo), as well as the knobs and gauges it exposes. These, as well as composability constraints are inherited by methods implementing this service. The methods implementing services must conform to this interface definition; however, they may be allowed to expose method-specific knobs and gauges, as seen in Figure 10.2.

Another way to look at the SILO architecture is from the point of view of functional blocks. This architectural view also served as the basis of the proto- type implementation of this architecture. At the heart of the system is the *Silo Management Agent* (SMA), which is responsible for maintaining the state of indi- vidual dataflows and associated silos. The application communicates with this entity via a standard API passing both data, as well as silo meta-information, like descriptions of desired services. The SMA is assisted by a *Silo Composition Agent* (SCA) which contains algorithms responsible for assembling silos based on application requests and known composability constraints between services and methods. All service descriptions, method implementations, constraints and interface definitions are stored in a *Universe of Services Storage* (USS). Both SMA and SCA consult this module in the course of their operations. Finally there is a separate *Tuning Strategies Storage*, which houses various algorithms capa- ble of optimizing the behavior of individual silos (or their collections) for specific objectives. This optimization is achieved by monitoring gauges and manipulating knobs that methods inside instantiated silos expose. This architecture is shown in Figure 10.3.

The normal sequence of operations for this architecture looks something like this: (a) an application requests a new silo from the SMA specifying, possibly in some vague form, its communications preferences, (b) SMA passes the request to the SCA, which invokes one of the composition algorithms and, when successful, passes back to the SMA a silo *recipe*, which explicitly describes the ordered list of services which will make up the new silo, (c) SMA instantiates a new silo by loading the methods described in the recipe and instantiating a state for the new data flow; it passes a silo handle back to the application, (d) the application

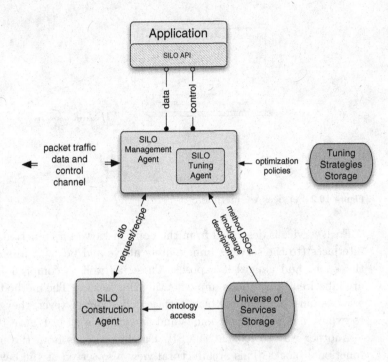

Figure 10.3 SILO functional architecture.

begins communicating while an appropriate tuning/optimization algorithm is applied to the silo via the tuning agent.

It is quite clear that, while this architecture offers a great deal of flexibility in arranging communication services, this flexibility comes at some cost. One important problem that needs to be addressed is that of an agreement on the silo structure between two communicating systems, noting that the silos need not be identical to accomplish communications tasks (monitoring or accounting services are a trivial example of services that require no strict counterpart in the far-end silo). The solution to this problem comes in one of several flavors. One approach may be an out-of-band channel, which allows two SMAs to communicate and create an agreement between them prior to applications commencing their communications. This approach may be suitable for peer-to-peer models of communications. Another, more suitable for client-server models, allows for a just-in-time analysis of compatibility between two silos by embedding a fingerprint of the client silo structure in the first packet that is sent out. Based on the information in that packet, the SMA can determine if the communication between a client and an already instantiated server is possible. In our work, this remains a problem still open for further investigation.

The last important question that remains to be addressed is why this architecture is better suited for evolution than the current one. As was mentioned in

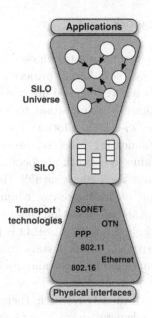

Figure 10.4 The SILO hourglass.

the previous section, our mantra for this project has been "design for change," and we believe we have succeeded in that. The architecture we have described does not mandate that any specific services be defined or methods implemented. It does not dictate that the services be arranged in a specific fashion and leaves a great deal of freedom to the implementors of services and methods. What it does define is a generic structure in which these services can coexist to help applications fulfill their communications needs, which can vary depending on the type of application, the system it is running on, and the underlying networking technologies available to it. Thus, as the application needs evolve along with the networking technologies, new communications paradigms can be implemented by adding new modules into the system. At the same time, all previously developed modules remain available, ensuring the smooth evolution.

The described architecture is a *meta-design* which allows its elements (the services and methods, the composition and tuning algorithms) to evolve independently, as application needs change and networking technologies evolve. Where, in the current architecture, the IP protocol forms the narrow waist of the hourglass (i.e., the fundamental invariant), in the SILO architecture the convergence point is the conformance to the meta-design, not a protocol (which is part of the design itself). Rather than a protocol which all else must be built on and under, SILO offers the silo abstraction as an invariant, the narrow waist in the hourglass of this particular meta-design (Figure 10.4).

10.4   Prior related work

One of the earliest attempts to impose orderly design rules on networking protocols is definitely the x-kernel project [13]. While SILO is similar to the x-kernel in introducing well-defined interfaces for protocol modules and organizing module interactions, it is important to recognize several major differences: (a) x-kernel was an OS-centric effort in implementing existing network protocols as sets of communicating processes inside the novel kernel, while SILO is an attempt to introduce a network protocol meta-design that is independent of any assumptions about the underlying OS; (b) x-kernel made an early attempt at streamlining some of the cross-layer communications mechanisms; SILO makes cross-layer tuning and optimization enabled by such mechanisms an explicit focus of the framework; and finally (c) SILO is focused on the problem of automated dynamic composition of protocol stacks based on individual application requests and module composability constraints, while the x-kernel protocols are prearranged statically at boot time.

Among recent clean-slate research, there are two projects whose scope extends to include the whole network stack and hence are most closely related to our own project.

The first is work on the role-based architecture (RBA) [6], carried out as part of the NewArch project [24]. Role-based architecture represents a non-layered approach to the design of network protocols, and organizes communication in functional units referred to as "roles." Roles are not hierarchically organized, and thus may interact in many different ways; as a result, the metadata in the packet header corresponding to different roles form a "heap," not a "stack" as in conventional layering, and may be accessed and modified in any order. The main motivation for RBA was to address the frequent layer violations that occur in the current Internet architecture, the unexpected feature interactions that emerge as a result [6], and to accommodate "middle boxes."

The second is the recursive network architecture (RNA) [26, 20] project, also funded by FIND. Recursive network architecture introduces the concept of a "meta-protocol" which serves as a generic protocol layer.

The meta-protocol includes a number of fundamental services, as well as configurable capabilities, and serves as a building block for creating protocol layers. Specifically, each layer of a stack is an instantiation of the same meta-protocol; however, the meta-protocol instance at a particular layer is configured based on the properties of the layers below it. The use of a single tunable meta-protocol module in RNA makes it possible to support dynamic service composition, and facilitates coordination among the layers of the stack; both are design goals of our own SILO architecture, which takes a different approach in realizing these capabilities.

10.5   Prototype and case studies

As part of the outcome of our current FIND project, we have produced a working prototype implementation, which serves as proof-of-concept demonstration of the feasibility of the SILO framework. This prototype, which is publicly available from the project website [23], is implemented in portable C++ and Python as a collection of user-space processes running in a recent version of Linux (although the prototype carries no explicit dependencies on the Linux kernel). Individual services as well as tuning algorithms are implemented as dynamically loadable libraries (DLLs or DSOs). The general structure of the prototype follows Figure 10.3.

One of the important issues we encountered when addressing the problem of dynamically composable silos was the problem of representation of the relationships (composability constraints) between different services and modules. This is, essentially, a problem of knowledge representation. These constraints take the form of statements similar to "Service A requires Service B" or "Service A cannot coexist with Service B," which can be modulated by additional specifications like "above" or "below" or "immediately above" or "immediately below." Additionally, we also needed to deal with the problem of specifying application preferences or requests for silos, which can be described as *application-specific* composability constraints. To address this problem we turned to *ontologies*, specifically, ontology tools developed by the Semantic Web community.

We adopted RDF (Resource Description Framework) as the basis for ontology representation in the SILO framework. Relying on RDF-XML syntax we were able to create a schema defining various possible relationships between the elements of the SILO architecture: services and methods. These relationships include the aforementioned composability constraints, which can be combined into complex expressions using conjunction, disjunction, and negation. Using this schema we have defined a sample ontology for the services we implemented in the current prototype. The application constraints/requests are expressed using the same schema. This uniform approach to describing both application requests as well as the SILO ontology is very advantageous in that a request, issued by the application, and expressed in RDF-XML, can be merged into the SILO ontology to create a new ontology with two sets of constraints – the original SILO constraints, and those expressed by the application, on which the composition algorithm then operates. Using existing Semantic Web tools we have implemented several composition algorithms that operate on these ontologies and create silo recipes, from which silos can be instantiated.

Our RDF schema also allows us to express other knowledge, such as the functions of services (an example of a service function could be "Congestion Control" or "Error Correction" or "Reliable Delivery"), as well as their data effects (examples include cloning of a buffer, splitting or combining of buffers, transformation, and finally null, which implies no data effect). These are intended to aid composition algorithms in deciding the set of services that need to be included in a silo, when an application is unable to provide precise specifications in the request. Using this additional information in the composition algorithm is an active area of our research.

10.6   Future work: SDO, stability, virtualization, silo-plexes

As was mentioned throughout this chapter, the SILO architecture enables a number of potential areas for research. In this section we try to address some of the more interesting and forward-looking ones.

10.6.1   Virtualization

Network virtualization efforts have attracted growing interest recently. Virtualization allows the same resource to be shared by different users, with independent and possibly different views. In network virtualization, a substrate network is shared by a virtualization system or agent which provides interfaces to different clients.

Testbeds such as PlanetLab have demonstrated network virtualization, and other efforts such as Emulab have allowed investigation of a virtualized network through an emulated environment. The Global Environment for Networking Innovation (GENI) has identified virtualization as a basic design strategy for the envisioned GENI facility to enable experimentation support for diverse research projects. More importantly, it has been conjectured that virtualization itself could become an essential part of the future Internet architecture. The FIND portfolio also contains projects on virtualization [27, 3]. Nevertheless, network virtualization is comparatively less mature than OS virtualization, being a significantly more recent field. We can expect there will be substantial ongoing work in increasing the isolation, generality and applicability of this area in the short- to mid-term. As such, it is an important area for any new architecture to consider.

Accordingly, we consider how virtualization can be realized in the SILO framework in order to achieve greater reusability. We go on to conjecture that such a realization might allow the concept of network virtualization to be generalized.

10.6.1.1   Virtualization as service

So far, network virtualization has been strongly coupled to the platform and hardware of the substrate. Logically, however, network virtualization consists of many coordinated individual virtualization capabilities, distributed over networking elements, that share the common functionality of maintaining resource partitions and enforcing them. In keeping with the SILO vision, we can view these functions as separate and composable services.

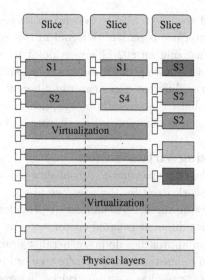

Figure 10.5 Successive virtualization.

The most basic of these services is that of *splitting* and *merging* flows; these services must obviously be paired. This is no more than the ability to mux/demux multiple contexts. Note that this service is a highly reusable one, and can be expected to be useful in diverse scenarios whenever there is aggregation/disaggregation of flows, such as mapping to/from physical interfaces, or at intermediate nodes for equivalence classes on a priority or other basis, or label stacking.

In the networking context, virtualization is usually interpreted as implying two capabilities beyond simple sharing. The first is *isolation*: each user should be unaware and unaffected by the presence of other users, and should feel that it operates on a dedicated physical network. This is sometimes also called "slicing." This can be broken down into two services: (i) slice maintenance, which keeps track of the various slices and the resources used by them, and (ii) access control, which monitors and regulates the resource usage of each slice, and decides whether any new slices requested can be allowed to be created or not; for example, rate control such as leaky bucket would be an access control function.

The second capability is *diversity*: each user should be able to use the substrate in any manner in which it can be used, rather than being restricted to use a single type of service (even if strictly time-shared). This is akin to the ability to run different operating systems on different virtual machines. In SILO, this capability is natively supported, through the composable nature of the stack. Not only do different silos naturally contain different sets of services, but the composability constraints provide a way to indicate what set of upper services may be chosen by different slices when building on a particular virtualized substrate.

The definition of a standard set of services for virtualization means that every realization of this service (for different substrates) would implement the functional interfaces specified by the service itself, thus any user of the virtualization agent would always be able to depend on these standard interfaces. Articulating this basic interface is part of our goal in this regard. For example, consider the case of virtualizing an 802.11 access point with the use of multiple SSIDs; the interface must allow specification of shares, similar to the example above. However, since the different slices can use different varieties of 802.11 with different speeds, the sharing must really be specified in terms of time shares of the wireless medium, which is the appropriate sharing basis in this context.

10.6.1.2 Generalizing virtualization

Following the principle that a virtual slice of a network should be perceived just like the network itself by the user, we are led to the scenario that a slice of a network may be further virtualized. A provider who obtains a virtual slice and then supports different isolated customers may desire this scenario. The current virtualization approaches do not generalize gracefully to this possibility, because they depend on customized interfaces to a unique underlying hardware. If virtualization is expressed as a set of services, however, it should be possible to design the services so that such generalization is possible simply by reusing the services (see Figure 10.5). There would obviously be challenges and issues. One obvious question is whether the multiple levels of virtualization should be mediated by a single SMA or whether the SMA itself should run within a virtualization, and thus multiple copies of SMA should run on the multiple stacks. Either approach is possible to proceed with, but we believe the former is the correct choice. In OS virtualization, the virtualization agent is itself a program and requires some level of abstraction to run, though it maps virtual machine requests to the physical machine down to a high level of detail. Successive levels of virtualization with agents at all levels being supported by the same lower-level kernel are difficult to conceive. However, networking virtualization agents do not seek to virtualize the OS which supports them. As such, the kernel support they require can be obtained through a unique SMA.

It may appear from this discussion that in fact with per-flow silo states, there is no need to virtualize, and in fact it is possible to extend all the slices to the very bottom (dotted lines in Figure 10.5). However, the advantage lies precisely in state maintenance; a service which is not called upon to distinguish between multiple higher-level users can afford to keep state only for a single silo, and the virtualization service encapsulates the state keeping for the various users.

10.6.1.3 Cross-virtualization optimization

Finally, it is possible to conceive of cross-layer interaction across virtualization boundaries, both in terms of composability constraints, and tuning. The service S1 in Figure 10.5 may require the illusion of a constant bit-rate channel below it, and the virtualization below it may be providing it with this by isolation. If, however, there is some service still further down that does not obey this restriction (some form of statistical multiplexing, for example), then the illusion will fail. It must be possible to express this dependence of S1 as a constraint, which must relate services across a virtualization boundary. It appears harder to motivate the need to tune performance across boundaries, or even (as the SMA could potentially allow) across different slices. Although we have come up with some use cases, they are not persuasive. However, we recall that the same was true of the layering abstraction itself, and it is only recently that cross-layer interactions have come to be perceived as essential. We feel that cross-virtualization optimization is also an issue worth investigation, even if the motivation cannot be clearly seen now.

10.6.2   SDO: "software defined optics"

In today's networks, the physical layer is typically considered as a black box: sequences of bits are delivered to it for transmission, without the higher layers being aware of exactly how the transmission is accomplished.

This separation of concerns imposed by the layering principle has allowed the development of upper-layer protocols that are independent of the physical channel characteristics, but it has now become too restrictive as it prevents other protocols or applications from taking advantage of additional functionalities that are increasingly available at the physical layer.

Specifically, in the optical domain, we are witnessing the emergence of what we call *software defined optics (SDO)*, i.e., optical layer devices that are:

1. *intelligent and self-aware*, that is, they can sense or measure their own characteristics and performance, and
2. *programmable*, that is, their behavior can be altered through software control.

Our use of the term SDO is a deliberate attempt to draw a parallel to another recent exciting technology, software defined radios (SDR), devices for which nearly all the radio waveform properties and applications are defined in software [25, 10, 16, 1].

The software logic defining more and more of these SDO devices requires cross-layer interactions, hence the current strictly layered architecture cannot capture the full potential of the optical layer. For instance, the optical substrate increasingly employs various optical monitors and sensors, as well as pools of amplifiers and other impairment compensation devices.

The monitoring and sensing devices are capable of measuring loss, polarization mode dispersion (PMD), or other signal impairments; based on this information, it should then be possible to use the appropriate impairment compensation to deliver the required signal quality to the application.

But such a solution cannot be accomplished within the current architecture, and has to be engineered *outside of it* separately for each application and impairment type; clearly, this is not an efficient or scalable approach.

Reconfigurable optical add-drop multiplexers (ROADMs) and optical splitters with tunable fanout (for optical multicast) are two more examples of currently available SDO devices whose behavior can be programmed according to the wishes of higher-layer protocols. Looking several years into the future, one can anticipate the development of other sophisticated devices such as programmable mux-demux devices (e.g., that allow the waveband size to adjust dynamically), or even hardware structures in which the slot size can be adjustable

In the SILO architecture, all these new and diverse functionalities within (what is currently referred to as) the physical layer will typically be implemented as separate services, each with its own control interfaces (knobs) that would allow higher-level services and applications direct access to, and control of, the behavior of the optical substrate.

Hence, the SILO architecture has the ability to facilitate a diverse collection of critically important cross-layer functions, including traffic grooming [9], impairment-aware routing [22, 28], and multi-layer network survivability [19] that have been studied extensively, as well as others that may emerge in the future.

We also note that there is considerable interest within the GENI community to extend the programmability and virtualization functionality that is core to the GENI facility, all the way down to the optical layer so as to enable meaningful and transforming optical networking research. Currently, however, a clear road map on how to achieve such a "GENI-ized" optical layer has not been articulated, mainly due to the lack of interfaces that would provide GENI operations access to the functionality of the optical layer devices.

We believe that the SILO architecture would be an ideal vehicle for enabling optical-layer-aware networking within GENI, as well as enabling cross-layer research through explicit control interfaces (e.g., such as SILO knobs). Therefore, we are in the process of outlining specific strategies for incorporating the SILO concepts within the GENI architecture whenever appropriate.

10.6.3   Other open problems

In this section we identify and briefly describe additional open problems associated with the SILO architecture we plan to study:

Agreement on silo structure with remote end: as was mentioned in Section 10.3, the flexibility offered by the SILO architecture comes at a price: the need for an agreement between communicating applications about the structure of silos on both ends. We have already identified several solutions to this problem, however it is an interesting enough problem to continue keeping it open. Some desirable characteristics of an ideal solution are: low overhead of the agreement protocol, high degree of success in establishing agreement, and security of the agreement process.

Stability and fairness: the stability of today's Internet is in part guaranteed by the fact that the same carefully designed algorithms govern the behaviors of TCP flows to achieve fairness between flows. As demonstrated in the literature, this stability is fragile and can be taken advantage of by non-

compliant TCP implementations to achieve higher throughput rates compared to unmodified versions. SILO allows a plug-and-play approach to substituting optimization policies into protocol stacks, thus ensuring stability and
fairness of the system within some predefined envelope of behavior becomes
paramount.

SILO in the core and associated scalability problems: all examples in
this chapter concentrated on the edge of the network where applications construct silos to communicate with one another. Nothing so far has been said
about the structure of the networking stacks in the core. It is clear that the
SILO concept can be extended to the core, by providing value addition mod-
ules/services to individual flows or groups of flows, as long as the problem of
scalability of this approach is addressed.

Silo composition based on fuzzy application requests: as indicated in
Section 10.3, the problem of composition of silos based on application requests
remains open. One of the important areas to be studied is the ability to
construct silos based on vague specifications from the application which
may provide minimal information about its needs like "reliable delivery with
encryption." This type of fuzzy or inexact specification requires an extended
ontology of services in which some reasoning can take place. The solutions will
be multiple and the system must pick the one that by some criteria optimizes
overall system behavior, or perhaps addresses some other optimization goal.

10.7   Case study

Does SILO work? Is there any evidence to show that it lowers the barrier to
continuing innovation, its stated goal? Of course, the answer to such a question
would take a long and diverse experimental effort, and to be convincing, would
have to come at least partly from actual developer communities after at least
partial deployment.

However, we have been able to conduct a small case study which gives us hope.
In the Fall of 2008, we made a simplified version of the SILO codebase available
to graduate students taking the introductory computer networks course at North
Carolina State University. Students are encouraged to take this course as a pre-
requisite to advanced graduate courses on networking topics, and most students
taking the course have no prior networking courses, or a single undergraduate
course on general networking topics. Students are required to undertake a small
individual project as one of the deliverables, which is typically a reading of a
focused topic in the literature and synthesizing in a report. In this instance,
students were told that they could try their hand at programming a small networking protocol as a SILO service as an alternative project. Nine out of the
around fifty students in the class chose to do so. All but one of these students
had not coded any networking software previously. To our satisfaction, all nine

produced code to perform non-trivial services, and the code not only worked, but it was possible to compose the services into a stack and interoperate them, although there was no communication or effort among the students to preserve interoperability during the semester. In one case, the code required reworking by the teaching assistant, because the student concerned had (against instructions) modified the SILO codebase distribution. The services coded by the students were implementations of ARQ, error control, adaptive compression, rate control, and bit stuffing. Testing services such as bit-error simulators were also coded, and two students attempted to investigate source routing and label switching, going into the territory of SILO services over multiple hops, which are as yet comparatively unformed and malleable in our architectural vision.

While this is only the veriest beginnings of trying to validate SILO, we feel that it at least shows that the barrier to entry into programming networking services has been lowered, in that the path from conceptual understanding of a networking protocol function to attaining the ability to produce useful code for the same is dramatically shorter. In future similar case studies, we hope to study the reaction to such beginning programmers to the tuning agent and ontology capabilities. And as always, we continue to invite the community to download the SILO code from our project website, try using it, and send us news about their positive and negative experiences.

Acknowledgements

This work was supported in part by the NSF under grants CNS-0626103 and CNS-0732330.

References

[1] Software defined radio forum, focusing on open architecture reconfigurable radio technologies. www.sdrforum.org.

[2] M. Allman, V. Paxson, K. Christensen, and B. Nordman. Architectural support for selectively-connected end systems: Enabling an energy-efficient future internet.

[3] T. Anderson, L. Peterson, S. Shenker, and J. Turner. Overcoming the Internet impasse through virtualization. *IEEE Computer*, 38(4):34–41, April 2005.

[4] B. Bhattacharjee, K. Calvert, J. Griffioen, N. Spring, and J. Sterbenz. Postmodern internetwork architecture.

[5] D. Blumenthal, J. Bowers, N. McKewon, and B. Mukherjee. Dynamic optical circuit switched (docs) networks for future large scale dynamic networking environments.

[6] R. Braden, T. Faber, and M. Handley. From protocol stack to protocol heap – role-based architecture. *ACM Computer Communication Review*, 33(1):17–22, January 2003.

[7] D. D. Clark, J. Wroclawski, K. Sollins, and R. Braden. Tussle in cyberspace: Defining tomorrow's internet. In *Proceedings of the 2002 ACM SIGCOMM Conference*, pages 347–356, Pittsburgh, PA, August 2002.

[8] D. Duchamp. Session layer management of network intermediaries.

[9] R. Dutta, A. E. Kamal, and G. N. Rouskas, editors. *Traffic Grooming in Optical Networks: Foundations, Techniques, and Frontiers*. Springer, 2008.

[10] W. Tuttlebee (ed.). *Software Defined Radio*. John Wiley, 2002.

[11] D. Fisher. US National Science Foundation and the future internet design. *ACM Computer Communication Review*, 37(3):85–87, July 2007.

[12] A. Gavras, A. Karila, S. Fdida, M. May, and M. Potts. Future internet research and experimentation: The FIRE intitiative. *ACM Computer Communication Review*, 37(3):89–92, July 2007.

[13] N. Hutchinson and L. Peterson. The x-kernel: An architecture for implementing network protoccols. *IEEE Transactions on Software Engineering*, 17(1):64–76, 1991.

[14] R. Kahn, C. Abdallah, H. Jerez, G. Heileman, and W. W. Shu. Transient network architecture.

[15] D. Krioukov, K. C. Claffy, and K. Fall. Greedy routing on hidden metric spaces as a foundation of scalable routing architectures without topology updates.

[16] J. Mitola. The software radio architecture. *IEEE Communications Magazine*, 33(5):26–38, May 1995.

[17] R. Morris and F. Kaashoek. User information architecture.

[18] Computer Business Review Online. ITU head foresees internet balkanization, November 2005.

[19] M. Pickavet, P. Demeester, D. Colle, D. Staessensand, B. Puype, L. Depré, and I. Lievens. Recovery in multilayer optical networks. *Journal of Lightwave technology*, 24(1):122–134, January 2006.

[20] The RNA Project. RNA: recursive network architecture. www.isi.edu/rna/.

[21] K. Sollins and J. Wroclawski. Model-based diagnosis in the knowledge plane.

[22] J. Strand, A. L. Chiu, and R. Tkach. Issues for routing in the optical layer. *IEEE Communications*, 39:81–87, February 2001.

[23] The SILO Project Team. The SILO NSF FIND project website. www.net-silos.net/, 2008.

[24] D. D. Clark *et al.* NewArch project: Future-generation internet architecture. www.isi.edu/newarch/.

[25] M. Dillinger *et al. Software Defined Radio: Architectures, Systems and Functions*. John Wiley, 2003.

[26] J. Touch and V. Pingali. The RNA metaprotocol. In *Proceedings of the 2008 IEEE ICCCN Conference*, St. Thomas, USVI, August 2008.

[27] J. Turner, P. Crowley, S. Gorinsky, and J. Lockwood. An architecture for a diversified internet. www.nets-find.net/projects.php.

[28] Y. Xin and G. N. Rouskas. Multicast routing under optical layer constraints. In *Proceedings of IEEE INFOCOM 2004*, pages 2731–2742, March 2004.

Part III

Protocols and practice

11 Separating routing policy from mechanism in the network layer

James Griffioen, Kenneth L. Calvert, Onur Ascigil, and Song Yuan
University of Kentucky, USA

11.1 Introduction

Despite the world-changing success of the Internet, shortcomings in its routing and forwarding system (i.e., the network layer) have become increasingly apparent. One symptom is an escalating "arms race" between users and providers: providers understandably want to control use of their infrastructure; users understandably want to maximize the utility of the best-effort connectivity that providers offer. The result is a growing accretion of hacks, layering violations and redundant overlay infrastructures, each intended to help one side or the other achieve its policies and service goals.

Consider the growing number of overlay networks being deployed by users. Many of these overlays are designed specifically to support network layer services that cannot be supported (well) by the current network layer. Examples include resilient overlays that route packets over multiple paths to withstand link failures [4], distributed hash table overlays that route packets to locations represented by the hash of some value [19, 16, 24], multicast and content distribution overlays that give users greater control of group membership and distribution trees [10, 14], and other overlay services. In many of these examples, there is a "tussle" between users and providers over how packets will be routed and processed. By creating an overlay network, users are able to, in a sense, impose their own routing policies – possibly violating those of the provider – by implementing a "stealth" relay service.

The lack of support for flexible business relationships and policies is another problem area for the current network layer. The provider–customer relationship is largely limited to the first-hop (local) provider. Customers only form business relationships with (i.e., pay) their local provider, and then rely on the provider to get their packets to the destination. This limits customers to the paths selected (purchased) by the local provider, and makes it difficult to obtain paths with specific properties (e.g., QoS), since almost all end-to-end paths involve multiple

Next-Generation Internet Architectures and Protocols, ed. Byrav Ramamurthy, George Rouskas, and Krishna M. Sivalingam. Published by Cambridge University Press. © Cambridge University Press 2011.

providers. Ideally, a customer would be allowed to negotiate and purchase service from any provider along the path(s) to the destination, thereby controlling its own business policies and economic interests rather than being subject to the business policies of the local provider. As others have observed, the current network layer (IP) is not designed so that users and providers can negotiate solutions to their "tussles" [11]; thus, the "arms race" continues.

While some of these problems might be addressed piecemeal in the context of the present architecture, their root causes are so embedded in the current network layer that solving all of them may not be feasible within the constraints imposed by the current protocol specifications; a fresh start is required. Even solutions like IPv6 do not get to the fundamental problems mentioned above. In other words, it is time to start over from scratch. While the original Internet's primary goal was to support end-to-end connectivity from any host to any host, today's networks must go farther, providing support for features such as protection, security, authentication, authorization, quality of service, flexible billing/payment, and simplified network configuration and management.

If we were to start over and design a new network layer, what would it look like? We present our answer – a "clean-slate" approach to the network layer that enables users and providers to jointly control routing and forwarding policies, thus allowing tussles to play out in ways that do not interfere with the deployment of new services. Our design separates routing from forwarding, addressing, and topology discovery; uses a flat, topology-independent identifier space; and achieves scalability via hierarchy based on the hiding of topology information.

In the next section we present design goals for our new PFRI network layer architecture. Section 11.3 then describes how a simple PFRI network is organized and operates. Section 11.4 then addresses the issue of scalability, showing how PFRI can be scaled to support large networks. Section 11.5 discusses issues that arose while designing the architecture, and Section 11.6 describes an initial prototype implementation. Finally, Section 11.7 discusses related approaches and compares them with PFRI.

11.2   PoMo design goals

The *Postmodern Internet Architecture (PoMo)* project [6] is a collaborative research project between the University of Kentucky, the University of Maryland, and the University of Kansas that is exploring a clean-slate redesign of the network layer to address the issues raised in Section 11.1. In what follows, we focus on the forwarding and routing features of the PoMo architecture, which we call the *Postmodern Forwarding and Routing Infrastructure (PFRI)*. The authors bear primary responsibility for this design.

PFRI's overarching design goal is the separation of routing *policy* from forwarding *mechanisms*, thereby allowing the "tussles" between users and providers to occur outside the forwarding plane. In the current Internet architecture, routing, forwarding, and addressing are tightly interwoven, and we believe this is the source of many of its limitations. For example, each forwarding element in the Internet takes part in the routing protocol and makes its own routing decisions, based on the hierarchical addresses carried in packets. The end-to-end path followed by a packet has a particular property (e.g., QoS) only if *every forwarding element along the path* chooses routes according to a policy consistent with that property. Moreover, trying to change either the routing or the addressing scheme of the current network layer requires changing every forwarding element in the network.

A key goal of PFRI is to disentangle the roles of routing, forwarding, and addressing, to allow users and network designers to deploy and use alternative addressing and routing policies. The role of the forwarding plane is simply to act on (enforce) those policies. To achieve this separation, we identified several design goals for PFRI.

- Users and providers should be empowered to arrive at a collective decision about how packets should be handled by the network layer. More precisely, the (routing) policies should be decided independent of the forwarding infrastructure, separating (routing) policy from (forwarding) mechanisms.
- There must be greater flexibility in customer–provider relationships. Customers should be able to arrange transit services from providers other than the local provider to which they are attached. This provides for competition among transit providers and incentives for deployment of advanced features that users may leverage (e.g., to support end-to-end QoS).
- Users should be able to negotiate the infrastructure's treatment of their packets (including routing) on a per-packet basis. Although one could envision a system that supported control only at flow-level granularity, the architecture should not rule out per-packet control. Instead, it should provide mechanisms that support flow-level control as an optimization for flows that do not need the finer level of control.
- To promote the alignment of incentives and to raise the cost of misusing the network, each packet should carry an explicit indication that it has been checked for policy compliance, as well as an indication of who benefits from its being forwarded.
- Hierarchical identifiers (addresses) that contain embedded structure, meaning, and policies are counter to our other goals. Instead, identifiers should be location-independent, semantics-free identifiers whose only purpose is to uniquely identify network elements.
- Address assignment (and network configuration in general) should be automated to the greatest extent possible.
- The architecture should be scalable both in the forwarding (data) plane and in the routing (control) plane.

PFRI achieves these goals through a variety of features, including:

- User- and provider-controlled *packet-based routing state* (as opposed to provider-controlled, router-based routing state).
- A flat, unstructured address space.
- A *motivation* mechanism that ensures that only policy-compliant packets are forwarded.
- An *accountability* mechanism that enables users to verify that they receive the service purchased.
- An autoconfiguration mechanism that allows the network to configure itself using only information about administrative boundaries.
- A recursive architecture that allows for multiple levels of hierarchy and abstraction.

In the following sections we present the PFRI architecture and show how it achieves these design goals.

11.3   Architecture overview

The PFRI network layer is defined in a recursive manner, starting with simple packet delivery across a flat network structure (i.e., the base case), and then moving to more complex packet delivery across a hierarchical network structure (i.e., the induction step). We will begin our description with the base case which highlights PFRI's main features and operation. After presenting the basic features we turn our attention to scaling the architecture in Section 11.4.

11.3.1   PFRI network structure and addressing

PFRI, like most networks, is structured as a set of nodes and channels. *Channels* provide a bi-directional best-effort packet transit service between nodes. Nodes come in two types: *forwarding nodes (FNs)* are used to relay packets from one channel to another, while *endpoints* act as the source (sender) of packets or the sink (receiver) of packets. For purposes of this discussion, we classify channels as either *infrastructure channels*, which provide connectivity between FNs, or *end-channels*, which connect an FN to an endpoint. The basic network components and structure are shown in Figure 11.1.

Unlike most networks, PFRI assigns addresses to channels, not nodes. As we will see in Section 11.4 assigning addresses to channels rather than nodes allows PFRI to define the network structure recursively (i.e., hierarchically) without the need to name hierarchical nodes. Each channel is assigned a unique, flat, unstructured *channel ID (CID)*. CIDs assigned to end-channels are called *Endpoint IDs (EIDs)* to highlight the fact that the channel attaches to an endpoint.

When assigning CIDs to channels, it is sufficient that each CID be globally unique. Consequently, CIDs can be assigned (pseudo)randomly – without

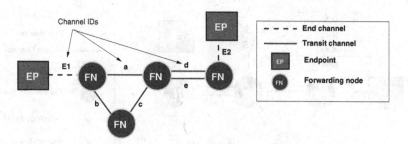

Figure 11.1 An example base-level PFRI network structure.

requiring a centralized naming authority – if the namespace from which they are selected is sufficiently large. However, PFRI takes it a step further and creates an unforgeable binding between a channel and the two nodes the channel connects. In particular, PFRI computes a channel's CID by cryptographically hashing the public keys of the two endpoints together. It uses this binding between channel and the nodes it connects to prove that packets traversed the specified channels. The downside of this approach is that CIDs must be large, and thus they consume more space in packet headers. Although this concern should not be overlooked, we prefer not to base our design on yesterday's resource constraints, when header bits were individually justified and packet sizes were small. In cases where header size is an issue, various techniques such as compression or tag switching can be used to reduce packet header sizes.

11.3.2 PFRI forwarding

Unlike current Internet routers that each make independent routing decisions and must converge in order for routing to stabilize, FNs do not make routing (policy) decisions, participate in routing protocols, or maintain state needed to make routing decisions. Instead, routing policy decisions are made by the source of a packet and (possibly) by routing services along the path to the destination. These decisions are then carried in the PFRI packet header to inform FNs along the path. The selected path is represented in the packet by a sequence of CIDs. Although the forwarding infrastructure does not select paths, it assists sources in two ways.

First, information about the network topology is discovered and maintained by a network-level *topology service (TS)*. All FNs send link-state announcements (LSAs) to the topology server, which assembles the LSAs into a complete graph of the topology. Each announcement carries the FN's adjacent CIDs, channel properties, transit properties, and motivation information (motivation is discussed in the next section). Once the TS knows the topology, the TS can be queried for information about the topology, returning subsets of the topology (e.g., paths). To reduce the size of the (transmitted and stored) LSAs and to avoid the volatility of endpoints going up and down, FNs only include infrastructure channels

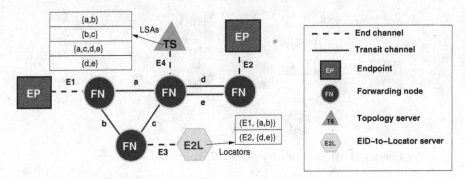

Figure 11.2 PFRI network structure with TS and E2L services.

in each LSA, omitting all end-channels. (Scalability in general is discussed in Section 11.4 and Section 11.5.)

Second, the locations of endpoints in the topology are maintained by an *EID-to-Locator (E2L)* service. The E2L service does not know the location of all endpoints. Only endpoints that want to be "found" will register with the E2L service. All other endpoints will remain hidden. It should be noted that "hidden" does not imply unreachable. A "hidden" endpoint may initiate a packet flow to a "findable" endpoint, which can respond to the "hidden" endpoint using the reverse of the path carried in the packet header. When an endpoint registers with the E2L service, it sends a tuple called a *locator* to the E2L server. A locator maps an EID to the set of access channels (aCIDs) used to reach it. More precisely, a locator is a tuple consisting of $\langle EID, (aCID_0, aCID_1, \ldots, aCID_n)\rangle$. For example, in Figure 11.2, the access channels for $E1$ are a and b. Given a destination EID's locator and topology information from the TS, senders can form the complete path to the destination. The paths needed to reach the TS and the E2L servers are discovered (by all nodes) during the network's autoconfiguration process [5].

Consider the base-level topology shown in Figure 11.2. Initially all FNs send LSA advertisements to the TS. The endpoints that want to be "found" register with the E2L server resulting in the locator tables shown in Figure 11.2. To send a packet from endpoint $E1$ to endpoint $E2$, the source first contacts the EID-to-Locator server to obtain the locator $(E2, \{d, e\})$. Knowing that its own locator is $(E1, \{a, b\})$, the sender contacts the TS to find the path(s) from {a,b} to {d,e} and selects one of the paths. The source then places the path, say $\langle E1{\rightarrow}b{\rightarrow}c{\rightarrow}e{\rightarrow}E2\rangle$, in the packet header and gives it to the PFRI layer to forward. (Note we are omitting – for the moment – a critical step related to motivation that will be discussed in the next section.) Each FN along the path examines the packet header to find the *next channel ID* and forwards the packet out the associated interface.

Note that packets can in principle be sent from a source to a destination as soon as the TS and E2L servers receive enough information from the FNs to discover a path through the network – unlike in the current Internet, where routing protocols must converge (sometimes at multiple levels) before any packet can be reliably sent. The TS collects link-state announcements, but there is no distributed link-state computation that must converge. Likewise, when the topology changes, forwarding can continue unaffected along any path that is not changed. Moreover, flows affected by the change can be quickly rerouted along different paths.

11.3.3   PFRI routing policies

In the previous example, the (routing) policy decisions were made entirely by the sender. To facilitate and support the types of "tussles" outlined in Section 11.1, the PFRI forwarding infrastructure is designed to interact with policy services controlled by the various stakeholders, each supplying some part of the overall policy to determine the path traversed. Precisely how these services are implemented – say, as a single do-it-all service or as multiple distributed services – is not as important as the way in which these services interface with and control the forwarding infrastructure, and so our discussion will focus on the interfaces.

Conceptually, PFRI separates the process of path discovery from the process of path selection. Discovering that a path exists is not sufficient to use the path, however. Although the TS may have complete information about the topology, senders cannot use the TS's paths without first obtaining permission to use the channels along the path. Permission to use a channel is obtained by contacting a *motivation server (MS)*. A motivation server decides which senders should be allowed to use a channel, or more precisely, whether a sender is allowed to use the relay service at an FN to transit a packet between two channels. Motivation servers grant use of a channel by returning a *capability* to the requestor. The capability can be used to create a motivation token, which is carried in the packet and serves as "proof" to the FN that the sender has the right to use the relay service. In other words, a *motivation token* is required at each hop to check for policy compliance. The details of how motivation tokens are generated and checked are beyond the scope of this chapter, but the basic idea is that the motivation server shares secrets with each FN for which it is responsible, and uses those shared secrets to generate capabilities whose motivation tokens can be verified by the FN. In summary, before a sender can use a path discovered via the TS, the sender must first obtain motivation tokens for all channels along the path and include them in the packet header. To ensure that capabilities are not misused, the motivation tokens must be bound to the packet contents (and in some cases to the path segment carried in the packet).

Motivation servers provide a way for users to "purchase" transit capabilities from the providers that own and operate the infrastructure. However, senders may not be interested in selecting all the channels along a path, and instead would prefer to "purchase" a single motivation token to go from the source EID to the destination EID, letting some other entity decide the path. To support

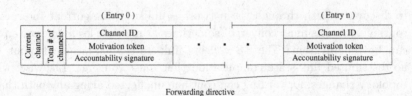

Figure 11.3 The structure of the forwarding directive (FD) in a PFRI packet header.

a model in which providers select some or all of the path, PFRI supports the concept of *partial paths* and *path faults*. A *partial path* is a sequence of channels where some of the channels in the sequence do not share a common FN (i.e., are not directly connected). When an FN encounters a packet whose next channel in the header is not directly attached to the FN, we say a *path fault* occurs. In this case, the FN forwards the packet to a *path fault handler (PFH)* to resolve the path fault. The PFH maps the faulting next channel to a *gap-filling segment (GFS)*, where the gap-filling segment is a sequence of channel IDs representing a (possibly partial) path that will get the packet to the next channel.

The path fault handler itself does not make any policy decisions, but rather acts as an interface between the forwarding plane and the policy services. Policy services preload the PFHs with gap-filling segments (i.e., policy decisions), and thus a PFH only needs to return the gap-filling segment (i.e., policy decision). This enables providers to select paths and influence the routes that packets take. Like senders, providers must also supply motivation for the gap-filling segment so that FNs along the path will relay the packet.

11.3.4   PFRI packet header mechanisms

Having described the basic network structure, addresses, and network layer services, we are now able to describe the PFRI packet header structure. As noted earlier, PFRI uses the packet header to carry policy decisions rather than storing policy state in routers (as is the case in the current Internet).

The PFRI packet header consists of various fixed fields that describe the version, packet length, etc., followed by a variable length *forwarding directive(FD)*. The forwarding directive describes the (partial) path the packet must traverse using an array of entries representing channels. Each entry contains a CID identifying a channel, a motivation token needed to relay the packet to the next channel, and an *accountability* field that can be filled in with a "signature" indicating the processing that the packet actually received (for example, that it actually passed through the indicated link). The first field in an FD is the *current channel* pointer indicating the channel in the FD where processing should be resumed (i.e., the channel over which the packet just arrived). The channel following the current channel is the next channel over which the packet should be forwarded. The FD structure is illustrated in Figure 11.3.

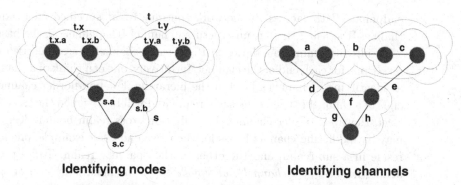

Identifying nodes   **Identifying channels**

Figure 11.4 Because PFRI identifies channels rather than nodes, hierarchy can be added without assigning (or reassigning) identifiers to the realms.

11.4   Scaling the PFRI architecture

The flat network structure that we have assumed until this point highlighted several of PFRI's features and showed how policy can be separated from the forwarding and addressing mechanisms. However, the flat network structure described so far does not scale to the size of the entire Internet, and is not capable of supporting the explosive growth next-generation networks will experience, particularly with the widespread deployment of mobile and sensor devices.

To address the scalability problem, PFRI leverages abstraction and hierarchy to hide information and minimize the scope of the network level services described earlier. While other architectures also use hierarchy to improve scalability, PFRI's approach is different in that it attempts to maintain the same network structure, services, and abstractions at all levels of the hierarchy. Because layers of the hierarchy have the same properties, we can define the overall network architecture recursively using an inductive model. Moreover, PFRI can support any number of levels of hierarchy, as opposed to the two levels commonly associated with the Internet architecture.

Section 11.3 described how routing and forwarding work in the base case: a flat network topology. In the induction step, PFRI allows a set of channels and nodes to be abstracted and represented as a single node, which we call a *realm*. Because PFRI assigns identifiers to channels rather than nodes, the resulting realm does not need to be identified and can remain nameless. In contrast, architectures that assign identities to nodes must come up with a new identity for each aggregated node at the next level – typically assigning some sort of hierarchical identifier to the new entity. Figure 11.4 illustrates the two approaches.

Because realms do not need to be assigned new names, PFRI only needs to specify the *border channels* that define the boundary of a realm in order to define the hierarchy. Once the border channels have been defined, the resulting realm takes on the appearance and behavior of a (nameless) forwarding node or endpoint. In other words, realms become indistinguishable from FNs and endpoints. Paths across the network continue to be specified as a sequence of channel IDs, using the flat unstructured channel IDs defined at the base level.

To identify realm boundaries in PFRI each channel is assigned two *channel-levels*, one for each end of the channel. Intuitively, one can think of a channel-level as identifying a channel's level in the hierarchy. For non-border channels, both ends of a channel reside in the same realm and will be assigned the same channel-level. However, border channels by definition cross realm boundaries, and thus may have differing channel-levels for their two ends. For example, one level may reside in a sub-realm, and the other in the "parent" realm. To be precise, we define the *level of a channel c at a node* to be the maximum level of all realms containing that node whose boundary is crossed by c. We define the *level of a realm* to be the maximum depth of nesting within that realm; in other words, the height of the tree, whose nodes are realms and whose edges are defined by the "contains" relation, rooted at that realm.

In order for a realm to behave in a manner indistinguishable from an FN or endpoint, it needs to perform the same operations as an FN or endpoint. In general this boils down to the fact that each realm must have all the structural characteristics and all the services of the base-level network described earlier.

The FNs, for example, send out LSA announcements to the TS, implying that a realm must also send LSA announcements to the TS. To accomplish this we assume each realm has a TS that knows the entire topology of the realm including the border channels and can send an LSA on behalf of the realm to the TS in the parent realm. (A TS discovers its parent TS during the initial autoconfiguration process.) The LSA contains information about all the border channels for the realm, much like the LSA for a node contains information about all its directly attached channels.

Like an FN, a realm needs to be able to relay packets from its ingress border channels to its egress border channels. To provide cross-realm relaying, PFRI leverages the path fault handler, routing, and motivation services in a realm. When a packet enters a realm through an ingress border channel, the border FN incurs a path fault because the next channel (i.e., the egress channel) is not directly attached. The packet is then forwarded to the PFH to be filled with a gap-filling segment that will transit the packet to the requested egress channel. As described earlier, PFRI routing services control the mappings in the PFH. We assume each realm has such a service, called the *Realm Routing Service(RRS)*, which knows the entire topology for the realm and can provide the PFH with the gap-filling segments it needs to support cross-realm relaying. Much like MPLS networks [17], PFRI pushes and pops gap-filling segments onto/off-of the packet header as the packet traverses the internal channels within a realm. This enables the packet to follow the user-specified path, but not reveal the internal topology of the realm.

Note that part of handling a path fault involves checking the motivation token to verify that the sender has been authorized to transit packets across the realm. Much like FNs have an associated motivation server, each realm needs an

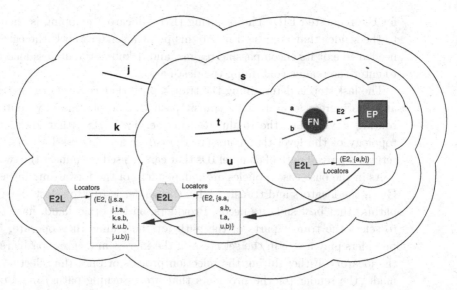

Figure 11.5 A locator accumulates additional channels when registered at higher levels of the hierarchy.

associated motivation server to give out motivation tokens used to transit the realm. Consequently, motivation servers may arrange themselves into a hierarchy (much like the TS servers) to ensure cross-realm motivation tokens are made available to higher level realms.

Realms also need to act like endpoints. Recall that endpoints originate packets and receive packets. As we saw in the base-level example, endpoints must register their locators with their E2L server in order to be "findable." The same is true for realms. A realm, acting like an endpoint must register its locator with its parent realm's E2L server. The parent realm's E2L server must, in turn, register the EID with its parent's E2L server. Consider the hierarchy shown in Figure 11.5. As the EID is registered at successively higher levels of the hierarchy, the associated locator accumulates additional channel IDs. Each time the partial path in the locator is extended based on the parent realm's policies.

It should also be noted that an endpoint that does not want to be "found" need not register its EID with the E2L service. However, it must still discover its locator in order to send packets. To assist in this process, the E2L service can be used without registering an EID. Instead of sending a "register" request to the E2L service, a sender sends a "self locate" request to the E2L service. Like a register request, the self locate request traverses the hierarchy building up the locator. When it reaches the topmost realm, the self locate request returns the locator to the sender indicating an egress path out of the sender's enclosing realms.

A sender can then find the (EID→locator) mapping for the destination EID by invoking the E2L service with a "lookup" request for the destination EID. The lookup request works its way up the E2L hierarchy until it finds a mapping for the requested EID. The resulting (EID→locator) mapping is then returned to the sender that requested it. At this point a sender knows the egress locator needed to exit its encompassing realms, and it knows the ingress locator needed to enter the realms containing the destination.

The last step is determining the transit path that connects the egress locator and the ingress locator. This transit path can be obtained by contacting the routing service (i.e., the Realm Routing Service (RRS)) that knows the entire topology of the level that connects the egress and ingress locators. The RRS returns a partial path of channel IDs that can be used to connect the two locators.

As in the base case, policies are independent of the forwarding infrastructure. Given the locators and transit paths, the sender is free to select the end-to-end path(s) that best suits its needs. Providers, on the other hand, have the ability to select the transit parts of the path for the realms they operate. Moreover, providers play a role in the selection of the ingress and egress paths returned in the locators. Either during the selection process or once the selection has been made, the sender (or the providers that are returning paths) must obtain the motivation tokens needed to relay the packets through each hop along the path.

11.5   Discussion

Historically, particularly in the Internet, policy state has been embedded in routers and switches. In fact, decision-making computations (e.g., routing protocols) are executed directly on the forwarding infrastructure to create the policy state. This embedded policy state and computation is part of what keeps the Internet from being able to adapt to the needs of next-generation networks.

Postmodern Forwarding and Routing Infrastructure on the other hand, separates the basic question "is forwarding this packet in my interest?" into two parts: one in the control plane, and one in the data plane. The policies that determine the answer to the original question involve various considerations – such as the sender's identity, exchange of money or other value, or operational requirements – should not have to be present (implicitly or explicitly) at every forwarding node. Instead, they should be applied out-of-band, separately. The answer, if positive (i.e., the proposed transmission complies with policy), can then be encoded in the form of a capability allowing the source to create a cryptographic seal (i.e., a motivation token) that testifies to the validity of the packet. At forwarding time, the node in the middle of the network need only check the validity of the motivation token. (This idea has been explored in a different incarnation in the Platypus system [15].)

We believe that this structure – which isolates the complexity and dynamics of policy from the infrastructure itself – gives users and providers better and more dynamic control over the behavior of the network. It allows stakeholders (users and providers) to "tussle" and potentially arrive at solutions that are not possible today.

Of course, this approach also has costs. One is obviously increased packet size. As was pointed out earlier, carrying a sequence of (large) CIDs in the packet header represents a significant increase over typical IP packet header sizes. Moreover, the credential placed in the packet for each relay hop must be relatively large (so that guessing is not feasible). This implies that PFRI headers could be large enough to exceed small channel MTUs. A variety of approaches can be applied to this problem, including header compression, and other (typically stateful) methods. In some cases it may be necessary to add a shim layer to do fragmentation and reassembly over the underlying physical channel that implements a virtual channel with large MTUs. In parts of the network where this is not possible (e.g., a network of resource-poor sensor nodes) it may be necessary to abstract the region as a separate realm, which interacts with the rest of the network through a more powerful gateway.

Another cost of removing policy state from the infrastructure is the forwarding-time computation required to verify a motivation token's validity. This typically involves computing a cryptographic function over a portion of the packet's contents using a secret shared between the node and the motivation token's creator. There are several concerns here. The first is how to arrange for a secret to be shared between an arbitrary source and an arbitrary FN – clearly it is not feasible for every node to "pre-share" a secret with every potential (or even actual) sender. The proposed solution is to use a delegation hierarchy. Each FN shares a secret with its responsible motivation server. Through cryptographically secure techniques (such as hashing the secret with an identifier and a timestamp), that secret can be used to derive another, which is provided to the delegatee. The FN can repeat this computation at run time to derive the secret on-the-fly (and then cache it for future use).

Another concern is the expense of cryptographic computation on the forwarding path. Traditional cryptographic hashes are considered to be incompatible with high performance [20], but recently specified primitives such as GMAC-AES [13] are defined specifically for high-performance applications, with up to 40 Gbps throughput advertised by some vendors. However, achieving this performance requires strictly limiting the number of applications of the cryptographic primitive per packet. The simple delegation scheme outlined above requires an amount of cryptographic computation that is linear in the number of delegations. Fortunately, it is possible to leverage precomputation to arrange things so that the linear part of the computation requires only simple bitwise XOR operations, and one cryptographic primitive per packet suffices.

Another potentially challenging aspect of our approach is the need for various components, including the sender, to obtain paths and/or locators before sending/forwarding a packet. In a network where the vast majority of the topology changes slowly, this can be addressed to a large extent through caching. As a realm's E2L service learns (EID→locator) mappings it can cache them to speed future requests. Similarly, FNs are allowed to cache gap-filling segments provided by the PFH to avoid similar path faults in the future. Senders are also expected to cache the paths (and corresponding motivation capabilities) they receive and reuse them on subsequent packets heading for the same destination.

A nice feature of PFRI is that invalid paths are detected quickly. When an FN receives a packet that cannot be forwarded because of a failed channel, the FN immediately returns an ICMP-like error message along the reversed path. A source can immediately select an alternate path (that avoids the failed channel) from among the ones originally received, and begin using it. In the case where a transit realm pushed a gap-filling segment, the packet is returned to the PFH that pushed the gap-filling segment. In other words, route failures/changes in PFRI can be corrected quickly without the need to wait for a set of distributed routing protocols to converge, as is the case in the current Internet.

A final concern is the computational cost of finding paths at the global level. Conventional wisdom holds that "link-state routing doesn't scale." That may be true in the Internet, where all nodes must have consistent views of the topology in order to prevent looping. Moreover, it is far from clear that the current AS-level structure of the Internet would be natural or even suitable for the PFRI architecture. Even if it were, however, several factors make it seem reasonable to contemplate some form of centralized route computation at that scale. The first is the simple observable fact that existing services perform on-demand search computations over much larger datasets, using large-scale computational infrastructures, on timescales that would be acceptable for an initial route-determination step. A second is the ability to postpone some parts of the computation until they are needed. For example, the set of all paths connecting second- and third-tier providers through a fully connected "core" could be computed in advance (and recomputed whenever topology changes), with the "tails" of transit paths computed only on demand. Finally, the fact that paths are strictly local means that the computation can be distributed in ways that maximize the benefits of caching.

Ultimately, the question is whether the benefits of the additional services enabled by our architecture outweigh the aforementioned costs. That question can only be answered by deploying and using the architecture, a process that is now ongoing.

11.6   Experimental evaluation

As a first step to understanding and evaluating the PFRI architecture, we have constructed a simple prototype of the PFRI network layer. The prototype is implemented as an overlay on top of the current Internet using UDP to tunnel PFRI packets between overlay nodes. The overlay nodes are implemented in the *Ruby* scripting language which is useful for quick prototyping, but is not designed for high performance. Although performance is adequate to execute most applications, the goal of the prototype is to evaluate the correctness, flexibility, ease of use, and limitations of our architecture, not its performance. Because the prototype executes as user-level processes, it can be run on most any set of computers, including Emulab [21] and Planetlab [2] which offer the ability to control the network environment or offer real-world traffic loads, respectively.

One of the first things we wanted to evaluate was PFRI's ability to automatically configure and "boot" the network with as little human intervention as possible. We began with a base-level network consisting of a single realm that required no boundary definitions. Booting the network simply involved starting the various components needed by a realm including a TS, PFH, RRS, E2L, MS, and all the FNs, and endpoints. We began by showing that no user configuration is needed to assign addresses to nodes – a feature not present in the current Internet. (Note that the use of DHCP does not completely remove the need to configure addresses, particularly for routers.) Upon booting, each component randomly generated a private/public key for each of its channels, which in turn was used to generate the channel IDs. At this point all channels have a unique CID. The next step was to see if the components could discover one another and exchange the information needed to make the network operate correctly. The TS, PFH, and MS then announce their existence so that FNs and endpoints can discover the path to these crucial network layer services. Once the paths are known, link-state announcements are sent to the TS, which in turn makes the information available to the RRS that is used by senders and the PFH to fill in the FDs carried in packets. In our prototype, FNs send their motivation secret to the MS in an encrypted packet. Having configured themselves with no user intervention, endpoints were now able to select paths and send and receive packets over those paths.

Our next experiment tested whether autoconfiguration works in the inductive case – i.e., with a hierarchy of realms. To test this, we created a multi-level hierarchy of realms with endpoints in the innermost sub-realms. Realm boundaries were specified by assigning channel-levels to each of the channels. Each realm was given a set of network servers (TS, PFH, RRS, E2L, and MS) that were *not configured* to know the hierarchical structure of the network. Like the base-level, components quickly assigned CIDs to their associated channels, and then discovered and began exchanging information with the other components in the local realm. Once the innermost realms had configured themselves, they began to transit the announcement messages of services in the next level up, allowing components at that level to discover one another. The announcements also enabled services in the inner realm to find paths to their parent service at the next level in the hierarchy. Like the base-level, the hierarchical network structure was also able to autoconfigure with no user configuration other than the specification of the boundary channels.

To evaluate our new architecture using real-world applications, we implemented a PFRI tunneling service that intercepts conventional IP traffic, tunnels it across our PFRI network, and then reinjects it on the destination IP network. We used the TUN/TAP virtual network interface [3] to capture IP packets at the sender and also used it to deliver IP packets to the receiver. Our software interfaces with the TUN/TAP to encapsulate/de-encapsulate packets in PFRI packets, mapping IP addresses to PFRI endpoints. Policies (e.g., to select routes) can be defined for IP packets that match certain patterns – say, all packets destined to a certain IP address or originating from a certain TCP port. Using this facility we have been able to execute conventional network applications (e.g., ssh, ftp, and remote desktop apps) running over our PFRI network, with the potential to take different routes based on the type of application being used.

11.7   Other clean-slate approaches

While a clean-slate approach may abandon backward compatibility, it need not consist entirely of new ideas. Indeed, most of the techniques used in PFRI have been proposed before, even in the context of the present Internet architecture. In that sense, PFRI "stands on the shoulders of giants" to leverage a long history of related work – more than can be listed here. The unique aspect of PFRI is the way in which it integrates these techniques. We briefly highlight some of the related architectures and techniques here, but by no means intend this to be an exhaustive list given the massive body of work done on network architectures in the past.

An early proposal for an inter-network protocol based on source routing was Cheriton's proposed Sirpent(tm) system [9]. The design provided for in-band policy-compliance verification, via an analogue of PFRI's motivation tokens. Destinations were specified using hierarchical names stored in the Domain Name System. In addition, routing information was to be distributed via the DNS.

The Nimrod Routing Architecture [7] supported user-directed, service-specific routes tailored to the needs of a particular application, offering control over routes similar to the control that PFRI offers its users. To limit the amount of routing information propagated across the network, Nimrod introduced the concept of hierarchical topology maps that offered multiple levels of abstraction, an approach that has similarities to PFRI's hierarchical realms and link-state topology distribution. Unlike PFRI, Nimrod uses hierarchical addresses/locators to identify nodes rather than channels. A source-routed approach using hierarchical addresses was also used in the ATM Forum's Private Network/Network Interface (PNNI) [18].

The New Internet Routing Architecture (NIRA) [23] also offers users the ability to select routes. As in Nimrod, NIRA nodes must be assigned hierarchical addresses (where the hierarchy is defined by customer–provider relationships between realms, and not by containment as in PFRI). This hierarchy plays an important role in route specification. Basic valley-free routes – that is, routes that consist of two segments, one ascending and one descending in the hierarchy – use common prefixes to find the "turnaround" point in the hierarchy, while valley-free segments in non-canonical (source) routes can reduce the representation size. But PFRI, on the other hand, does not require hierarchical identifiers and assigns identifiers to channels so that hierarchical nodes never need to be configured with an ID, making it possible to autoconfigure the entire network hierarchy. PFRI also supports partial paths that both reduce the path size and allow providers to control their respective part of the path.

The Tesseract [22] architecture takes a four-dimensional view of the network control plane, breaking the components of routing into distinct activities. This separation has similarities to PFRI's goal of separating routing from forwarding, but does not consider separating addressing. Tesseract uses logically centralized decision elements to make policy decisions and push those decisions into the forwarding state of switches. The Internet Engineering Task Force (IETF) Forwarding and Control Element Separation (FORCES) working group [1] also advocated an approach somewhat similar to Tesseract in the sense that the controlling entity was separated from the forwarding entity, pushing decisions made by the controlling entity into the forwarding elements; PFRI, on the other hand, distributes the decision-making process among the sender, receiver, and providers, and carries the decisions along with the packet, allowing policy to change on a per-packet basis.

The Dynamic Host Configuration Protocol (DHCP) [12] commonly used in the Internet today avoids the need to configure addresses on host computers. However, the addresses it hands out are topology-based, and it does not help assign addresses to routers. Network Address Translation (NAT) in the Internet also avoids the need to assign topology-based addresses, but is only useful for nodes that do not wish to be reachable. A variety of other address translation schemes have also been proposed as part of the Internet Research Task Force (IRTF) Routing Research Group (RRG) working group to separate addresses from location. Another approach to avoid topology-based address assignment, Routing on Flat Labels (ROFL) [8], uses flat addresses, with a distributed hash table-style ring. Like PFRI, caching plays an important role in achieving efficiency in ROFL. Unlike PFRI, it uses hop-by-hop routing/forwarding, and suffers from the drawbacks it entails.

Acknowledgements

This work was supported by the National Science Foundation under grants CNS-0626918 and CNS-0435272. The authors would like to thank Bobby Bhattacharjee, Neil Spring, and James Sterbenz, for their helpful insights and suggestions.

References

[1] FORCES Working Group. www.ietf.org/html.charters/forces-charter.html.
[2] Planetlab: An Open Platform for Developing, Deploying, and Accessing Planetary-scale Services. www.planet-lab.org.

[3] Universal TUN/TAP Device Driver. www.kernel.org/pub/linux/kernel/people/marcelo/linux-2.4/Documentation/networking/tuntap.txt.

[4] D. G. Anderson, H. Balakrishnan, M. F. Kaashoek, and R. Morris. Resilient Overlay Networks. In *Symposium on Operating Systems Principles*, pages 131–145, 2001.

[5] O. Ascigil, S. Yuan, J. Griffioen, and K. Calvert. Deconstructing the Network Layer. In *Proceedings of the ICCCN 2008 Conference*, August 2008.

[6] B. Bhattacharjee, K. Calvert, J. Griffioen, N. Spring, and J. P. G. Sterbenz. Postmodern Internetwork Architecture, 2006. Technical Report ITTC-FY2006-TR-45030-01, University of Kansas.

[7] I. Casteneyra, N. Chiappa, and M. Steenstrup. The Nimrod Routing Architecture. RFC 1992, August 1996.

[8] M. Cesar, T. Condie, J. Kannan, *et al.* ROFL: Routing on Flat Labels. In *Proceedings of ACM SIGCOMM 2006*, Pisa, Italy, pages 363–374, August 2006.

[9] D. Cheriton. Sirpent(tm): A High-Performance Internetworking Approach. In *Proceedings of ACM SIGCOMM 1989*, Austin, Texas, pages 158–169, September 1989.

[10] Y. Chu, S. Rao, S. Seshan, and H. Zhang. Enabling Conferencing Applications on the Internet Using an Overlay Multicast Architecture. In *Proceedings of the ACM SIGCOMM 2001 Conference*, August 2001.

[11] D. Clark, J. Wroclawski, K. Sollins, and R. Braden. Tussle in Cyberspace: Designing Tomorrow's Internet. *IEEE/ACM Transactions on Networking*, 13(3):462–475, June 2005.

[12] R. Droms. Dynamic Host Configuration Protocol. RFC 2131, March 1997.

[13] M. Dworkin. Recommendation for Block Cipher Modes of Operation: Galois/Counter Mode (GCM) and GMAC. National Institute of Standards and Technology Special Publication 800-38D, November 2007.

[14] D. Pendarakis, S. Shi, D. Verma, and M. Waldvogel. ALMI: An Application Level Multicast Infrastructure. In *Proceedings of the 3rd USENIX Symposium on Internet Technologies and Systems (USITS)*, pages 49–60, 2001.

[15] B. Raghavan and A. Snoeren. A System for Authenticated Policy-Compliant Routing. In *Proceedings of ACM SIGCOMM 2004*, Portland, Oregon, August 2004.

[16] S. Ratnasamy, P. Francis, M. Handley, R. Karp, and S. Schenker. A Scalable Content-addressable Network. In *SIGCOMM '01: Proceedings of the 2001 Conference on Applications, Technologies, Architectures, and Protocols for Computer Communications*, pages 161–172, New York, 2001. ACM.

[17] E. Rosen, A. Viswanathan, and R. Callon. Multiprotocol Label Switching Architecture. RFC 3031, January 2001.

[18] J. M. Scott and I. G. Jones. The ATM Forum's Private Network/Network Interface. *BT Technology Journal*, 16(2):37–46, April 1998.

[19] I. Stoica, R. Morris, D. Liben-Nowell, *et al.* Chord: A Scalable Peer-to-peer Lookup Protocol for Internet Applications. *IEEE/ACM Transactions on Networking*, 11:17–32, 2003.

[20] J. Touch. Performance Analysis of MD5. In *SIGCOMM '95: Proceedings of the 1995 Conference on Applications, Technologies, Architectures, and Protocols for Computer Communications*, New York, August 1995. ACM.

[21] B. White, J. Lepreau, L. Stoller, *et al.* An Integrated Experimental Environment for Distributed Systems and Networks. In *Proceedings of the Fifth Symposium on Operating Systems Design and Implementation*, pages 255–270, Boston, MA, December 2002. USENIX Association.

[22] Hong Yan, D. A. Maltz, T. S. E. Ng, *et al.* Tesseract: A 4D Network Control Plane. In *Proceedings of USENIX Symposium on Networked Systems Design and Implementation (NSDI '07)*, April 2007.

[23] X. Yang. NIRA: A New Internet Routing Architecture. In *Proceedings of ACM SIGCOMM 2003 Workshop on Future Directions in Network Architecture (FDNA)*, Karlsruhe, Germany, pages 301–312, August 2003.

[24] B. Y. Zhao, L. Huang, J. Stribling, *et al.* Tapestry: A Resilient Global-scale Overlay for Service Deployment. *IEEE Journal on Selected Areas in Communications*, 22:41–53, 2004.

12 Multi-path BGP: motivations and solutions

Francisco Valera[†], Iljitsch van Beijnum[‡], Alberto García-Martínez[†],
Marcelo Bagnulo[†]

[†]Universidad Carlos III de Madrid, Spain [‡]IMDEA Networks, Spain

Although there are many reasons for the adoption of a multi-path routing paradigm in the Internet, nowadays the required multi-path support is far from universal. It is mostly limited to some domains that rely on IGP features to improve load distribution in their internal infrastructure or some multi-homed parties that base their load balance on traffic engineering. This chapter explains the motivations for a multi-path routing Internet scheme, commenting on the existing alternatives, and detailing two new proposals. Part of this work has been done within the framework of the Trilogy[1] research and development project, whose main objectives are also commented on in the chapter.

12.1 Introduction

Multi-path routing techniques enable routers to be aware of the different possible paths towards a particular destination so that they can make use of them according to certain restrictions. Since several next hops for the same destination prefix will be installed in the forwarding table, all of them can be used at the same time. Although multi-path routing has a lot of interesting properties that will be reviewed in Section 12.3, it is important to remark that in the current Internet the required multi-path routing support is far from universal. It is mostly limited to some domains that deploy multi-path routing capabilities relying on Intra-domain Gateway Protocol (IGP) features to improve the load distribution in their internal infrastructure and normally only allowing the usage of multiple paths if they all have the same cost.

---

[1] Trilogy: Architecting the Future (2008–2010). ICT-2007-216372 (http://trilogy-project.org). The research partners of this project are British Telecom, Deutsche Telekom, NEC Europe, Nokia, Roke Manor Research Limited, Athens University of Economics and Business, Universidad Carlos III de Madrid, University College London, Université Catholique de Louvain, and Stanford University.

Next-Generation Internet Architectures and Protocols, ed. Byrav Ramamurthy, George Rouskas, and Krishna M. Sivalingam. Published by Cambridge University Press. © Cambridge University Press 2011.

However, multi-path routing would also present important advantages in the inter-domain routing environment.

In the Internet, for example, the routing system and the congestion control mechanisms which are two of its main building blocks, work in a completely independent manner. That is, the route selection process is performed based on some metrics and policies that are not dynamically related to the actual load of the different available routes. On the other hand, when there is congestion in some parts of the network, the only possible reaction is to reduce the offered load. Current flow control mechanisms cannot react to congestion by rerouting excess traffic through alternative links because typically these alternatives are not known. Clearly, coupling routing, and more specifically multi-path routing, and congestion control has significant potential benefits since it would, for instance, enable routers to spread the traffic through multiple routes based on the utilization of the links.

This kind of coupling and interactions between multi-path and other techniques will be explained in Section 12.2, since they constitute one of the main objectives of the Trilogy project, which is described in this section.

Despite the fact that multi-path alternatives for the inter-domain routing are not available yet in the Internet, some of the existing proposals are described in Section 12.4. Finally, this chapter introduces two additional solutions in Sections 12.4.4.2 and 12.4.4.3. These two solutions are some of the proposals being considered in the Trilogy project to provide non-equal cost multi-path routing at the inter-domain level. The goal of both mechanisms is to enable inter-domain multi-path routing in an incrementally deployable fashion that would result in increased path diversity in the Internet. Unlike the rest of the existing alternatives these new proposals imply minimum changes to the routers and to Border Gateway Protocol (BGP) semantics, are interoperable with current BGP routers, and have as one of their most important objectives an easier adoption of multi-path inter-domain solutions so their advantages can be realized earlier.

12.2   Trilogy project

12.2.1   Objectives

Trilogy is a research and development project funded by the European Commission by means of its Seventh Framework Programme. The main objective of the project is to propose a control architecture for the new Internet that can adapt in a scalable, dynamic, autonomous, and robust manner to local operations and business requirements.

There are two main motivations for this objective. The first one is the traditional limited interaction that has always existed between congestion control, routing mechanisms, and business demands. This separation can be considered as the direct cause of many of the problems which are leading to a proliferation of disperse control mechanisms, fragmentation of the network into private environments, and growing scalability issues. Re-architecting these mechanisms into a more coherent whole is essential if these problems are to be tackled.

The second motivation comes from the observation of the success of the current Internet. More than from its transparency and self-configuration, it comes from the fact that it is architected for change. The Internet seamlessly supports evolution in applications use and adapts to configuration changes; deficiencies have arisen where it is unable to accommodate new types of business relationships. To make the Internet richer and more capable will require more sophistication in its control architecture, but without imposing a single organizational model.

12.2.2   Trilogy technologies

Past attempts to provide joint congestion control and routing have proven that the objective of the Trilogy project is a challenging task. In the late 1980s, a routing protocol that used the delay as the metric for calculating the shortest paths was tried in the ARPANET [16]. While this routing protocol behaved well under mild load conditions, it resulted in severe instabilities when the load was high [16]. Since that experience, it is clear that the fundamental challenge to overcome when trying to couple routing to congestion information is stability. Recent theoretical results [14, 10] have shown that it is indeed possible to achieve stability in such systems. The Trilogy project relies on these recent results in order build a stable joint multi-path routing and congestion control architecture. One key difference between Trilogy's architecture and the previous ARPANET experience is that Trilogy embeds multi-path routing capabilities. Intuitively stability is easier to achieve in a multi-path routing scenario where the load split ratio varies based on the congestion in the different paths than in a single-path routing approach, where all the traffic towards a given destination is shifted to an alternative path when congestion arises in the currently used path. So, multi-path routing capabilities are one of the fundamental components for Trilogy's architecture. In addition, the distribution of traffic among the multiple routes is performed dynamically based on the congestion level of the different paths, as opposed to current multi-path routing schemes. Normal equal cost multi-path practice is to perform round-robin distribution of flows among the multiple routes. It is possible to distribute the flows across the multiple routes in a way that optimizes the traffic distribution for a given traffic matrix [7, 26].

The proposed approach is based on the theoretical results presented in [14]. The basic idea is to define a Multi-Path Transmission Control Protocol (MPTCP) that is aware of the existence of multiple paths. This protocol will then characterize the different paths based on their congestion level. That means that MPTCP will maintain a separate congestion window for each of the available paths and will increase and reduce the congestion window of each path based on the experienced congestion. An MPTCP connection is constituted by multiple subflows associated to the different paths available and each subflow has its own congestion control. By coupling the congestion window of the different

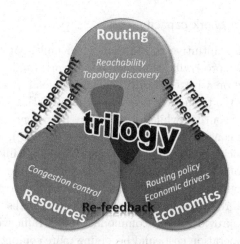

Figure 12.1 Three basic components of the Trilogy project.

subflows, additional benefits may be obtained like the resource pooling benefits, described in [28] (this occurs when the network's resources behave as though they make up a single pooled resource and facilitates increasing reliability, flexibility and efficiency). While the coupling of the congestion windows of the different subflows of MPTCP allows users to move away from congested paths and leave space for flows that have more pressing needs due to the lack of path diversity toward their destination, Trilogy's architecture includes a third component that allows users to provide accountability for the congestion caused in the network, a piece that is missing in the current Internet architecture, but deemed critical for the next-generation Internet. This accountability component, called Re-ECN (Explicit Congestion Notification)[19] would allow users to be accountable for the congestion they generate.

These are the three main components of Trilogy's architecture, see Figure 12.1, and their interaction is detailed in [3].

The rest of the article will detail the multi-path routing component of the architecture, analyzing its most important motivations, different alternatives, and particular proposals.

12.3   Multi-path routing

The adoption of a multi-path routing solution in the Internet will imply changes. Such changes imply costs that need to be assumed by the different business roles and in order to deploy an effective solution it is critical to have the right motivations for the affected parties. In particular, it is critical to have the right incentives, i.e., a scheme where the parties that have to pay for the costs also get some of the resulting benefits. In this section, some motivations to deploy a multi-path routing solution for the Internet are presented from the perspective of each of the stakeholders involved.

12.3.1   Higher network capacity

It is fairly intuitive to see that when multi-path routing is used it is possible to push more traffic through the network (and particularly when it is used in conjunction with congestion-dependent load distribution). This is so basically because the traffic will flow through any path that has available capacity, filling unused resources, while moving away from congested resources. Using a generalized cut constraints approach (see [17] and [15]), it is actually possible to model the capacity constraints for logical paths existing in a network and prove that the set of rates that a multi-path routing capable network that uses logical paths can accommodate is larger than the set of input rates in the same network using uni-path routing directly over the physical paths. This basically means that the network provider can accommodate more traffic with its existing network, reducing its operation costs and becoming more competitive. From the end-users' perspective, they will be able to push more traffic through their existing providers.

12.3.2   Scalable traffic engineering capabilities

The Internet global routing table contains over 300 000 entries and it is updated up to 1 000 000 times a day, according to recent statistics [12], resulting in the scalability challenges identified by the Internet community. There are multiple contributors to the global routing table, but about half of the routing table entries are more specific prefixes, i.e., prefixes that are contained in less specific ones [18]. In addition, they exhibit a much less stable behavior than less specific prefixes, making them major contributors to the BGP churn. Within those more specific prefixes, 40% can be associated with traffic engineering techniques [18] used by the ASs (Autonomous Systems) to change the normal BGP routing. Among the most compelling reasons for doing traffic engineering, we can identify avoiding congested paths. This basically means that ASs inject more specific prefixes to move a subset of traffic from a congested route towards a route with available capacity. In this case, more-specific prefixes act as a unit of traffic sinks that can be moved from one route to another when a path becomes congested. While this is a manual process in BGP, because of its own nature, these more specific prefix announcements tend to be more volatile than less specific prefixes announced to obtain real connectivity. Deploying a multi-path routing architecture would remove the need to use the injection of routes for more specific prefixes in BGP to move traffic away from congested links, especially when used in combination with congestion control techniques.

12.3.3   Improved response to path changes

Logical paths that distribute load among multiple physical paths are more robust than each one of the physical paths, hence, using multiple logical paths would normally result in improved fault tolerance. However, it can be argued that current redundancy schemes manage to use alternative paths when the used path fails without needing to rely on multi-path. Nowadays, there are several mechanisms to provide fault tolerance in the Internet that would allow switching to an alternative path in case the one actually used fails. Notably, BGP reacts to failures, and reroutes packets through alternate routes in case of failures but its convergence times may be measured in minutes and there are a certain amount of failures that are transparent to BGP because of aggregation. There are also other means to provide fault tolerance in the network, such as relying on the IGP, or in local restoration, and although some of them can have good response times, they are not able to deal with all the end-to-end failure modes, since they are not end-to-end mechanisms. On the other hand, end-to-end mechanisms for fault tolerance have been proposed, such as HIP (Host Identity Protocol) [20] or the REAP (REAchability Protocol) [6]. However, in all these cases, only one path is used simultaneously and because they are network layer protocols, it is challenging to identify failures in a transport layer agnostic way, resulting in response times that are measured in seconds [6]. The improved response to path changes that multi-path routing would allow is relevant to the end-users, since they will obtain better resiliency, but it is also a motivation for the network operator, since the path change events would behave in a more congestion friendly manner.

12.3.4   Enhanced security

Logical paths that distribute load among multiple physical paths exhibit superior security characteristics than the physical paths. This is so due to a number of reasons. For instance, man-in-the-middle attacks are much harder to achieve, since the attacker needs to be located along the multiple paths, and a single interception point is unlikely to be enough. The same argument applies to sniffers along the path, resulting in enhanced privacy features. In addition, logical paths are more robust against denial-of-service attacks against any of the links involved in the paths, since attacking any link would simply imply that the traffic will move to alternative physical paths that compose the logical path. The result is that a multi-path routing based architecture results in improved security. This is a benefit for the end-user that would take advantage of the improved security features.

12.3.5   Improved market transparency

Consider the case where a site has multiple paths towards a destination through multiple transit providers. Consider now that it uses the different logical paths that include physical paths through its different transit providers. Since traffic will flow based on congestion pricing, at the end of the day the client may be able to have detailed information about how much traffic has routed through each of its providers. Having more perfect information of the actual quality of the service purchased allows clients to make more informed decisions about their providers, thereby fostering competition and improving the market.

12.4   Multi-path BGP

In the Internet there have already been deployed some alternative means to support the simultaneous usage of multiple paths to reach a certain destination. The best known solutions are the ones being used within the domain of a particular provider (intra-domain routing) since traffic can be conveniently controlled and directed while all the routing devices are under a single management entity, and the multi-path solution is typically common throughout the domain. However, these solutions are not directly applicable to the inter-domain routing framework since there are other important factors beyond the technical ones that must be considered, which are mainly related to policy and economic constraints.

This section provides an overview of the most relevant solutions proposed so far both for the intra-domain and the inter-domain environments, finally focusing on the motivations for other multi-path BGP alternatives and also exposing some of the problems that may arise when designing multi-path inter-domain protocols.

12.4.1   Intra-domain multi-path routing

One of the easiest frameworks to implement multi-path routing would be to use IP source routing, as long as the end systems are provided with enough topological information for them to calculate these multiple paths. However, apart from the security concerns on the use of source routing [4] and the lack of support for IP source routing in current routers, it also has some drawbacks strictly talking from a multi-path practical perspective, such as scalability problems due to the provision of topology maps to the end systems, worse use of resources of the provider since traffic will typically be unbalanced in the network, and the fact that some links may remain unused while others may become congested or form a less flexible routing scheme since IP traffic will normally flow following the same paths. One interesting feature that this scheme would enable is the usage of disjoint paths: since path selection is centrally done by the end systems it can be guaranteed that the selected paths do not partially overlap, improving resiliency that way.

Link-state protocols like OSPF (Open Shortest Path First) [22] explicitly allow equal cost multi-path routing. When multiple paths to the same destination have the same cost, an OSPF router may distribute packets over the different paths. The Dijkstra algorithm makes sure each path is loop-free. A round-robin schedule could be easily used for this, but there are protocols such as TCP that perform better if packets belonging to a certain flow follow the same path and for this more complex techniques are often used (see [11] or [5]). Equal-cost multi-path in general provides a better use of network resources than normal uni-path routing schemes and a better resilience, and this is completely transparent to the end-user. However, sometimes it may not provide enough path diversity so a more aggressive multi-path routing technique may be used. A well-known alternative for the intra-domain routing is the unequal cost multi-path used in EIGRP (Enhanced Interior Gateway Protocol) [1]. Unequal-cost multi-path solutions imply using some other routes in addition to the shortest ones, but these new routes do not guarantee loop freeness in the routing infrastructure. This is solved in most protocols using loop-free conditions like the ones defined in [27]. In essence, this comes down to a router only advertising routes to neighboring routers that have a higher cost than the routes that the router itself uses to reach a destination. See Section 12.4.4.2 for further details.

Open Shortest Path First is also capable of doing multi-topology routing and OSPF type-of-service routing (updated to be more general in [23]) overlays multiple logical topologies on top of a single physical topology. A single link may have different costs in different topologies. As such, the shortest paths will be different for different topologies. However, packets must be consistently forwarded using the same topology to avoid loops. This is different from other types of multi-path routing, where each link that a packet traverses brings the packet closer to its destination, in the sense that the cost for reaching the destination is smaller after each hop. This is also true in multi-topology routing, but only when a packet stays within the same topology, so multi-topology routing requires more complex IP forwarding function than regular hop-by-hop forwarding. If a packet is moved from one topology to another, it could face a higher cost towards its destination after traversing a link. A second topology change then creates a loop. This makes multi-topology routing appropriate for link-state protocols where all routers have the same information, less suitable for distance vector protocols where each router only has a limited view of the network, and unsuitable for policy-based routing protocols such as BGP, where contradictory policies may apply in different parts of the network.

12.4.2   Inter-domain multi-path routing

For the inter-domain environment there are also existing solutions providing limited multi-path routing. For instance, when there are parallel links between two eBGP (external BGP) neighbors, operators may configure a single BGP session between the two routers using addresses that are reachable over each of the links equally. This is normally done by assigning the address used for the BGP session (and thus, the NEXT_HOP address) to a loopback interface, and then having static routes that tell the router that this address is reachable over each of the parallel links. The BGP routes exchanged will now have a next hop address that is not considered directly reachable. Even though the BGP specification does not accommodate this, implementations can typically be configured to allow it. They will then recursively resolve the BGP route's NEXT_HOP address, which will have multiple resolutions in the multi-path case. This will make the IP forwarding engine distribute packets over the different links without involvement from the BGP protocol. For iBGP (internal BGP), the next hop address is not assumed to be directly reachable, so it is always resolved recursively. So in the case of iBGP, the use of multiple paths depends on the interior routing protocol or the configuration of static routes.

Border Gateway Protocol is also capable of explicitly managing equal cost multi-path routing itself. This happens when a BGP router has multiple eBGP sessions, the router is configured to use multiple paths concurrently, and the routes learned over different paths are considered sufficiently equal. The latter condition is implementation specific. In general, if the LOCAL_PREF, AS_PATH, and MED are all equal, routes may be used concurrently. In this case, multiple BGP routes are installed in the routing table and packets are forwarded accordingly. Because all the relevant BGP attributes for the routes over different paths are the same, there is no impact on BGP loop detection or other BGP processing.

Apart from these existing solutions that are currently being applied, there are some other proposals that are worthwhile mentioning.

The source routing alternative is also possible for the inter-domain and similar comments would apply here as the ones made for the intra-domain (see [30] and [13]). In addition, one of the most important considerations now is that lack of flexibility for intermediate providers to apply their policies if packets come to a fixed path from the origin. In intra-domain routing this is not an issue, since it is all related to a single provider, but, for the inter-domain routing, this is critical.

Some other solutions consist of overlays that run on top of the generic Internet routing mechanism. Additional paths are normally obtained by tunneling packets between different nodes that belong to the overlay. The typical problems related to overlays are the additional complexity associated with the tunneling setup mechanisms and the overhead that the tunnels themselves introduce. One of these proposals is MIRO (Multi-path Inter-domain ROuting, [29]) that reduces the overhead during the path selection phase by means of a cooperative path selection involving the different intermediate AS's (additional paths are selected on demand rather than disseminating them all every time). Another alternative is RON (Resilient Overlay Networks, [2]) that builds an overlay on top of the Internet routing layer and continuously probes and monitors the paths between the nodes of the overlay. Whenever a problem is detected, alternate paths are activated using the overlay.

Another recent solution is called path splicing [21] following the multi-topology idea and generating the different paths by running multiple protocol instances to create several trees towards the destination but without sharing many edges in common. While normal multi-topology schemes will just use different topologies for different packets (or flows), the idea here is to allow packets to switch between topologies at any intermediate hop, increasing the number of available paths for a given source–destination pair. The selection of the path is done by the end systems including certain bits in the packets that select the forwarding table that must be used at each hop. This proposal claims for a higher reliability and faster recovery than normal multi-topology alternatives providing less overhead than overlay-based solutions.

12.4.3   Motivations for other solutions

Due to different reasons the previous proposals have still not been promoted into real alternatives. In this chapter two proposals are introduced based on the following motivations and assumptions for an early adoption:

- Change BGP semantics as little as possible.
- Change BGP routers as little as possible.
- Be interoperable with current BGP routers.
- Provide more path diversity than exists today.

In addition, it is worth noting that any solution should comply with the peering/transit Internet model based on economic considerations (see [8]). The rationale for this model is to realize that in most cases a site only carries traffic to or from a neighbor as a result of being paid for this (becoming a provider that serves a customer, or serving a paid peering), or because an agreement exists in which both parties obtain similar benefit (peering). This results in the requirement to enforce two major restrictions:

- Egress route filtering restrictions: customer ASs should advertise their own prefixes and the prefixes of its customers, but they should never advertise prefixes received from other providers. In this way, a site does not offer itself to carry traffic for a destination belonging to a site for which it is not going to obtain direct profit.
- Preferences in route selection: routers should prefer customer links over peering links because sending and receiving traffic over customer links enables them to earn money, and peering over provider links, because peering links at least does not cost them money. According to this, the multi-path route selection process can aggregate routes from many different customer links; or many peering links; or many provider links; but it can never mix links associated to different relationship types. Note that the administrator may even have specific preferences for routes received from neighbors with the same relationship with the site, because of economic reasons, traffic engineering, etc.

As a result of the peering/transit model, paths in the Internet can start going "up" from the originating site to a provider, and up to another provider, many times until it reaches a peering relationship, and then descend to a customer of this site, descend again, many times until it reaches the destination. Since it is impossible to find paths in which descending from a site to a customer and ascending again, or paths in which a peering link is followed by an ascending turn to a provider, it is said that the Internet is "valley-free" [8] as a result of the application of the peering/transit model.

The "valley-free" model suggests that a loop in the advertising process (i.e., a route advertised to a site that already contains in the AS_PATH the AS number of that site) can only occur for a route received by a provider. This is because a customer or peer of a site S, cannot advertise a route that has been previously advertised by S, according to the restrictions stated above. The valley-free condition also assures that a route containing S that is received from a provider P1(S) was advertised by S to another provider. Since S only announces to its providers its own prefixes or customer prefixes, the prefixes received by any provider, whose selection would result in a loop, are its own prefix or customer prefixes. Note that these routes would never be selected because either the destination is already in the site, or because it always prefers customer links to provider links. Consequently, although there is a specific mechanism in BGP for detecting loops in the routes, the application of the peer/transit model by itself would be enough to assure that loops never occur. Of course, loop prevention mechanisms must exist in order to cope with routing instabilities, configuration errors, etc. However, we can extend this reasoning to the multi-path case to state that, if any multi-path BGP strategy complies with the peering/transit model, as requested before, the aggregation of routes with equal condition (just customer routes; if not, just peering routes; and if not just provider-received routes) will not result in route discarding due to loop prevention in the steady state for well-configured networks. However, any multi-path BGP mechanism must provide loop prevention to cope with transient conditions and configuration errors.

In this chapter we present two proposals that share some mechanisms, such as part of the route selection approach, and differ in others, such as the loop prevention mechanism.

12.4.4   mBGP and MpASS

12.4.4.1 Route selection and propagation

Because a router running BGP tends to receive multiple paths to the same destination from different neighboring routers, the modifications to allow for the use of multiple paths can be limited to each individual router and modifications to the BGP protocol are unnecessary. The selection process for multi-path BGP should take as a starting point the rules for uni-path BGP, deactivating the rules that are used for tie-breaking among similar rules to allow the selection of multiple routes instead of just a single one. Note that the more rules that are deactivated, the larger number of routes with the same preference can be selected for multi-path forwarding. However, only routes that are equivalent for the administrator must be selected, arriving at this preference from economic reasons, traffic engineering considerations, or in general any policy that the administrator wants to enforce. So a modified multi-path router first applies normal BGP policy criteria and then selects a subset of the received paths for concurrent use. The attributes and rules through which relevant preferences of the administrator are enforced, in the order in which they are applied, are:

- Discard routes with higher LOCAL_PREF. This rule enforces any specific wish of the administrator, and is the rule used to assure that only routes received from customers are selected; or if no routes from customers exist, only routes received from peers; or if none of the previous exist, routes received from providers.
- Discard routes with higher MED. This rule is used to fulfill the wishes of the customers in order to implement "cold-potato" routing so that customers' costs in terms of transit cost are reduced.
- Discard lowest ORIGIN. This rule is used in some cases as a traffic-engineering tool. If not, the impact of its application is low, since almost all routes should have equal ORIGIN attribute.
- Discard iBGP routes if eBGP routes exist. It is used to deploy "hot-potato" routing, which may be relevant to reduce internal transit costs. In addition, it also eliminates internal loops in route propagation. When applied, routers receiving a route from an external neighbor uses only external neighbors, so internal loops never occur. Routers not receiving a route from an external neighbor selects the router inside the AS that will send the packet out of the AS.
- Discard routes with higher cost to NEXT_HOP. This is also used to enforce "hot-potato" routing. However, some relaxation on this rule can be introduced, provided that prevention of loops in intra-domain forwarding is achieved by means such as some kind of tunneling like MPLS.

The rest of the rules (selecting route received from router with minimum loopback address, etc.) are provided to ensure uniqueness in the result, so they can be removed for multi-path routing.

Therefore, a modified router first applies normal BGP policy criteria and then selects a subset of the received paths for concurrent use. Note that multiple paths mainly come from the possibility of ignoring AS_PATH length (although some conditions on this length could be established for accepting a route), and from accepting routes with different NEXT_HOP distances.

12.4.4.2 LP-BGP: Loop-freeness in multi-path BGP through propagating the longest path

In this particular proposal, after obtaining the different paths that will be installed in the forwarding table for the same destination prefix, the path with longest AS_PATH length to upstream ASs will be disseminated to neighboring routers where allowed by policy. Although disseminating a path that has a larger number of ASs in its AS_PATH seems counterintuitive, it has the property of allowing the router to use all paths with a smaller or equal AS_PATH length without risking loops (see Figure 12.2).

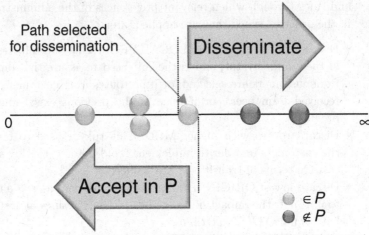

Figure 12.2 Multi-path selection in LP-BGP.

However, this change has the implication that there is no longer a one-to-one relationship between the paths that packets follow through the network and the path that is advertised in BGP. The resulting obfuscation of the network's topology as seen by observers at the edge can either be considered harmful, for those who want to study networks or apply policy based on the presence of certain intermediate domains, or useful, for those intent on hiding the inner workings of their network.

The multi-path BGP modifications allow individual ASs to deploy multi-path BGP and gain its benefits without coordination with other ASs. Hence, as an individual BGP router locally balances traffic over multiple paths, changes to BGP semantics are unnecessary.

Under normal circumstances, the BGP AS_PATH attribute guarantees loop-freeness. Since the changes allow BGP to use multiple paths concurrently, but only a single path is disseminated to neighboring ASs, checking the AS_PATH for the occurrence of the local AS number is no longer sufficient to avoid loops. Instead, the the Vutukury/Garcia-Luna-Aceves LFI (Loop-free Invariant) [27] conditions are used to guarantee loop-freeness.

Intuitively, these conditions are very simple: because a router can only use paths that have a lower cost than the path that it disseminates to its neighbours (or, may only disseminate a path that has a higher cost than the paths that it uses), loops are impossible. A loop occurs when a router uses a path that it disseminated earlier, in which case the path that it uses must both have a higher and a lower cost than the path that it disseminates, situations that can obviously not exist at the same time. When the following two LFI conditions as formulated by Vutukury and Garcia-Luna-Aceves are satisfied, paths are loop-free:

$$FD_j^i \leq D_{ji}^k \quad k \in N^i$$
$$S_j^i = \{k | D_{jk}^i < FD_j^i \wedge k \in N^i\}$$

"where D_{ji}^k is the value of D_j^k reported to i by its neighbor k; and FD_j^i is the feasible distance of router i for destination j and is an estimate of D_j^i, in the sense that FD_j^i equals D_j^i in steady state but is allowed to differ from it temporarily during periods of network transitions" [27]; D_j^k is the distance or cost from router k to destination j, N_i is the set of neighbors for router i, and S_j^i is the successor set that router i uses as next hop routers for destination j.

Our interpretation of the two LFI conditions as they relate to BGP is as follows:

$$cp(p_r) < cp_r(p_r)$$
$$P = \{p | cp(p) \leq cp_r(p_r) \wedge p \in \pi\}$$

where P is the set of paths towards a destination that are under consideration for being used and π is the set of paths towards a destination disseminated to the local router by neighboring routers; p_r is the path selected for dissemination, $cp_r(x)$ the cost to reach a destination through path x that is reported to other routers and the cost $cp(x)$ is taken to mean the AS_PATH length of path x in the case of eBGP and the interior cost for iBGP. The interior cost is the cost to reach a destination as reported by the interior routing protocol that is in use.

Because the local AS is added to the AS_PATH when paths are disseminated to neighboring ASs, the smaller and strictly smaller requirements are swapped between the two conditions.

The BGP-4 specification [24] allows for the aggregation of multiple prefixes into a single one. In that case, the AS numbers in the AS_PATH are replaced with one or more AS_SETs, which contain the AS numbers in the original paths. Should the situation arise where a topology is not valley-free [8] and there is both a router that implements multi-path BGP as described in this chapter as well as, in a different AS, a router that performs aggregation through the use of AS_SETs, then routing loops may be possible. This is so because, depending on the implementation, a router creating an AS_SET could shorten the AS_PATH length and break the limitations imposed by the LFI conditions. To avoid these loops, P may either contain a single path with an AS_PATH that contains an AS_SET, or no paths with AS_PATHs that contain AS_SETs. Note that AS_SETs are rarely used today; a quick look through the Route Views project data reveals that less than 0.02 percent of all paths have one or more AS_SETs in their AS_PATH [25].

All paths that remain in the multi-path set after the previous steps and after applying policy are installed in the routing table and used for forwarding packets.

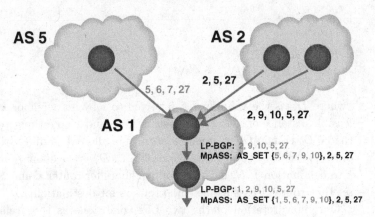

Figure 12.3 BGP propagation in LP-BGP and MpASS.

The determination of traffic split ratios between the available paths is a topic for future work.

At this point, the path with the longest AS_PATH within P is selected for dissemination to BGP neighbors. As a result of the LFI conditions, multi-path-aware ASs will suppress looped paths with a multi-path-aware AS in the looped part of the path, while regular BGP AS_PATH processing suppresses looped paths with no multi-path-aware ASs in the looped part of the path. To avoid loops for non-multi-path-aware iBGP routers, the selected path is also not disseminated over any BGP session through which the router learned a path that is in the multi-path set, and if the router previously disseminated a path over a session towards a neighboring router that supplied a path in the selected multi-path set P, it now sends a withdrawal for the multi-path destination.

12.4.4.3 MpASS: Multi-path BGP with AS_SETs

The main idea behind MpASS is to include in the AS_PATH all the AS numbers resulting from the union of the AS_PATH attributes of the routes aggregated so far. In particular, the AS_PATH is obtained by concatenating an AS_SEQUENCE structure containing the AS_PATH corresponding to the route that the BGP router would select from applying BGP uni-path selection rules, and an AS_SET structure that includes all the AS numbers of the rest of the routes, and the AS number of the site. This particular construction mechanism assures that all AS numbers are included and the length of the AS_PATH structure, as defined for the AS_PATH length comparison rule [24], is equal to the length of the AS_PATH of the best route plus 1 (as it would occur for legacy uni-path BGP routers). In this way, when a legacy route applies the rule of discarding routes with larger AS_PATH length, this multi-path route is not penalized compared to the uni-path route that it would have generated.

Loop prevention is enforced by the check performed by regular uni-path BGP and it is not necessary to define any additional mechanism or particular condition, i.e., discarding routes that contain the AS number of the site of the router receiving the advertisement (see Figure 12.3). An additional characteristic is that the inclusion of all the AS numbers of the sites that may be traversed by a packet sent to the destination allows the application of policies based on the particular AS traversed when selecting a route. Legacy BGP routers receive a route that is indistinguishable to a regular BGP route, and if they select it, packets may benefit from the multiple available paths.

12.5 Conclusions and future work

Multi-path routing presents many advantages when compared with single-path routing: higher network capacity, scalable traffic engineering capabilities, improved response to path changes and better reliability, enhanced security, improved market transparency.

For the intra-domain routing environment there are different solutions that can be applied (and effectively are), and the fact of having the deployment constrained to a single routing domain particularly facilitates this task (only in the interior of a provider's network).

In the inter-domain routing framework, the situation is more complex because most of the different existing proposals imply important changes in the well-established inter-domain communication technology based on BGP, linking different providers, each one with its own interests and requirements.

The European research and development project Trilogy considers multi-path routing as one of its main objectives. In the project, multi-path routing is considered together with congestion control mechanisms and the different Internet economic drivers, so as to try to improve the existing Internet communication mechanisms by means of providing a synergic solution based on the liaison of these three areas.

This chapter is focusing on one of these areas, the multi-path routing, and we have presented two mechanisms for providing multiple routes at the inter-domain level that are being considered in the project. The mechanisms differ in the way routes are selected and how loop prevention is enforced. The first one, LP-BGP, has the potential to reduce the number of BGP updates propagated to neighboring routers, as updates for shorter paths do not influence path selection and are not propagated to neighboring routers. However, in longer paths there is more potential for failures, so the inclusion of long paths in the set of paths that a multi-path router uses may expose it to more updates compared to the situation where only short paths are used. When propagating just the longest path, BGP no longer matches the path followed by all packets. The second proposal (MpASS) allows the selection of routes with any AS_PATH length, since loop prevention relies on transporting the complete list of traversed AS numbers.

One difference between them is that LP-BGP may propagate a route with an AS_PATH larger than the best of the aggregated routes, so that the result of a multi-path aggregation may be a route less attractive to other BGP routers (presenting longer paths to customers may put service providers at a commercial disadvantage). Still, propagating the longest path has robust loop detection properties and operators may limit acceptable path lengths at their discretion, so the second disadvantage is relatively minor (they could require for instance all best routes to be equal length).

On the other hand, MpASS may suffer from excessive update frequency, since each time a new path is aggregated in a router, a new Update must be propagated to all other routers receiving this route, to ensure that loop prevention holds (note that in the uni-path case, BGP only propagates a route if the newly received one improves on the previous one, while in this case many routes may be gradually added to the forwarding route set). This problem can be relieved by setting a rate limit to the aggregation process.

As part of the future work we plan to do a deeper analysis of the stability properties of both protocols, i.e., routing convergence and convergence dynamics. Some intuition around the routing algebra theory developed by Griffin and Sobrinho [9] suggests the LP-BGP is stable and that MpASS is assured to be stable if only routes with equal AS_PATH length are aggregated, although more analysis is required to determine if the use of different lengths may lead to stable solutions.

Finally, an evaluation of the effect of applying these mechanisms in the real Internet is required in order to analyze the path diversity situation: is the current number of available paths too low, or too high? Is it enough to use equal length AS_PATH routes? What is the cost added to the already stressed inter-domain routing system? Work will continue in Trilogy; stay tuned.

References

[1] B. Albrightson, J. J. Garcia-Luna-Aceves, and J. Boyle. EIGRP – A fast routing protocol based on distance vectors. In *Networld/Interop 94, Las Vegas. Proceedings*, pages 136–147, 1994.

[2] D. Andersen, H. Balakrishnan, F. Kaashoek, and R. Morris. Resilient overlay networks. In *ACM SOSP Conference. Proceedings*, 2001.

[3] M. Bagnulo, L. Burness, P. Eardley, *et al.* Joint multi-path routing and accountable congestion control. In *ICT Mobile Summit. Proceedings*, 2009.

[4] S. M. Bellovin. Security problems in the TCP/IP protocol suite. *ACM SIGCOMM Computer Communication Review*, 19(2):32–48, 1989.

[5] T. W. Chim, K. L. Yeung, and K. S. Lui. Traffic distribution over equal-cost-multi-paths. *Computer Networks*, 49(4):465–475, 2005.

[6] A. de la Oliva, M. Bagnulo, A. García-Martínez, and I. Soto. Performance analysis of the reachability protocol for IPv6 multihoming. *Lecture Notes in Computer Science*, 4712:443–454, 2007.

[7] B. Fortz and M. Thorup. Internet traffic engineering by optimizing OSPF weights. In *IEEE INFOCOM 2000. Proceedings*, volume 2, pages 519–528, 2000.

[8] L. Gao and J. Rexford. Stable Internet routing without global coordination. *IEEE/ACM Transactions on Networking*, 9(6):681–692, 2001.

[9] T. G. Griffin and J. L. Sobrinho. Metarouting. *ACM SIGCOMM Computer Communication Review*, 35(4):1–12, 2005.

[10] H. Han, S. Shakkottai, C. V. Hollot, R. Srikant, and D. Towsley. Overlay TCP for multi-path routing and congestion control. *IEEE/ACM Transactions on Networking*, 14(6):1260–1271, 2006.

[11] C. Hopps. Analysis of an Equal-Cost Multi-Path Algorithm. RFC2992, 2000.

[12] G. Huston. Potaroo.net. [Online]. Available: www.potaroo.net/, 2009.

[13] H. T. Kaur, S. Kalyanaraman, A. Weiss, S. Kanwar, and A. Gandhi. BANANAS: An evolutionary framework for explicit and multipath routing in the Internet. *ACM SIGCOMM Computer Communication Review*, 33(4):277–288, 2003.

[14] F. Kelly and T. Voice. Stability of end-to-end algorithms for joint routing and rate control. *ACM SIGCOMM Computer Communication Review*, 35(2):5–12, 2005.

[15] F. P. Kelly. Loss networks. *The Annals of Applied Probability*, 1(3):319–378, 1991.

[16] A. Khanna and J. Zinky. The revised ARPANET routing metric. *ACM SIGCOMM Computer Communication Review*, 19(4):45–56, 1989.

[17] C. N. Laws. Resource pooling in queueing networks with dynamic routing. *Advances in Applied Probability*, 24(3):699–726, 1992.

[18] X. Meng, B. Zhang, G. Huston, and S. Lu. IPv4 address allocation and the BGP routing table evolution. *ACM SIGCOMM Computer Communication Review*, 35(1):71–80, 2005.

[19] T. Moncaster, B. Briscoe, and M. Menth. Baseline encoding and transport of pre-congestion information. IETF draft. draft-ietf-pcn-baseline-encoding-02, 2009.

[20] R. Moskowitz, P. Nikander, P. Jokela, and T. Henderson. Host Identity Protocol. RFC5201, 2008.

[21] M. Motiwala, N. Elmore, M. Feamster, and S. Vempala. Path splicing. In *ACM INFOCOM. Proceedings*, 2008.

[22] J. Moy. OSPF Version 2. RFC2328, 1998.

[23] P. Psenak, S. Mirtorabi, A. Roy, L. Nguyen, and P. Pillay-Esnault. Multi-topology (MT) routing in OSPF. RFC4915, 2007.

[24] Y. Rekhter, T. Li, and S. Hares. A Border Gateway Protocol 4 (BGP-4). RFC4271, 2006.

[25] Routeviews. University of Oregon Route Views Project. [Online] Available: http://routeviews.org/, 2009.

[26] A. Sridharan, R. Guerin, and C. Diot. Achieving near-optimal traffic engineering solutions for current OSPF/IS-IS networks. *IEEE/ACM Transactions on Networking*, 13(2):234–247, 2005.

[27] S. Vutukury and J. J. Garcia-Luna-Aceves. A simple approximation to minimum-delay routing. In *ACM SIGCOMM. Proceedings*, pages 227–238. ACM, 1999.

[28] D. Wischik, M. Handley, and M. Bagnulo. The resource pooling principle. *ACM SIGCOMM Computer Communication Review*, 38(5):47–52, 2008.

[29] W. Xu and J. Rexford. MIRO: Multi-path interdomain routing. In *Proceedings of the 2006 Conference on Applications, Technologies, Architectures, and Protocols for Computer Communications*, volume 36, pages 171–182. ACM, 2006.

[30] D. Zhu, M. Gritter, and D.R. Cheriton. Feedback based routing. *ACM SIGCOMM Computer Communication Review*, 33(1):71–76, 2003.

13 Explicit congestion control: charging, fairness, and admission management

Frank Kelly[†] and Gaurav Raina[‡]

[†] University of Cambridge, UK [‡] Indian Institute of Technology Madras, India

In the design of large-scale communication networks, a major practical concern is the extent to which control can be decentralized. A decentralized approach to flow control has been very successful as the Internet has evolved from a small-scale research network to today's interconnection of hundreds of millions of hosts; but it is beginning to show signs of strain. In developing new end-to-end protocols, the challenge is to understand just which aspects of decentralized flow control are important. One may start by asking how should capacity be shared among users? Or, how should flows through a network be organized, so that the network responds sensibly to failures and overloads? Additionally, how can routing, flow control, and connection acceptance algorithms be designed to work well in uncertain and random environments?

One of the more fruitful theoretical approaches has been based on a framework that allows a congestion control algorithm to be interpreted as a distributed mechanism solving a global optimization problem; for some overviews see [1, 2, 3]. Primal algorithms, such as the Transmission Control Protocol (TCP), broadly correspond with congestion control mechanisms where noisy feedback from the network is averaged at endpoints, using increase and decrease rules of the form first developed by Jacobson [4]. Dual algorithms broadly correspond with more explicit congestion control protocols where averaging at resources precedes the feedback of relatively precise information on congestion to endpoints. Examples of explicit congestion control protocols include the eXplicit Control Protocol (XCP) [5] and the Rate Control Protocol (RCP) [6, 7, 8].

There is currently considerable interest in explicit congestion control. A major motivation is that it may allow the design of a fair, stable, low-loss, low-delay, and high-utilization network. In particular, explicit congestion control should allow short flows to complete quickly, and also provides a natural framework for charging. In this chapter we review some of the theoretical background on explicit congestion control, and provide some new results focused especially on admission management.

Next-Generation Internet Architectures and Protocols, ed. Byrav Ramamurthy, George Rouskas, and Krishna M. Sivalingam. Published by Cambridge University Press. © Cambridge University Press 2011.

In Section 13.1 we describe the notion of proportional fairness, within a mathematical framework for rate control which allows us to reconcile potentially conflicting notions of fairness and efficiency, and exhibits the intimate relationship between fairness and charging. The Rate Control Protocol uses explicit feedback from routers to allow fast convergence to an equilibrium and in Section 13.2 we outline a proportionally fair variant of the Rate Control Protocol designed for use in a network where queues are small. In Section 13.3 we focus on admission management of flows where we first describe a step-change algorithm that allows new flows to enter the network with a fair, and high, starting rate. We then study the robustness of this algorithm to sudden, and large, changes in load. In particular, we explore the key tradeoff in the design of an admission management algorithm: namely the tradeoff between the desired utilization of network resources and the scale of a sudden burst of newly arriving traffic that the network can handle without buffer overload. Finally, in Section 13.4, we provide some concluding remarks.

13.1   Fairness

A key question in the design of communication networks is just how should available bandwidth be shared between competing users of a network? In this section we describe a mathematical framework which allows us to address this question.

Consider a network with a set J of *resources*. Let a *route* r be a non-empty subset of J, and write $j \in r$ to indicate that route r passes through resource j. Let R be the set of possible routes. Set $A_{jr} = 1$ if $j \in r$, so that resource j lies on route r, and set $A_{jr} = 0$ otherwise. This defines a $0 - 1$ incidence matrix $A = (A_{jr}, j \in J, r \in R)$.

Suppose that route r is associated with a *user*, representing a higher-level entity served by the flow on route r. Suppose if a rate $x_r > 0$ is allocated to the flow on route r then this has *utility* $U_r(x_r)$ to the user. Assume that the utility $U_r(x_r)$ is an increasing, strictly concave function of x_r over the range $x_r > 0$ (following Shenker [9], we call traffic that leads to such a utility function *elastic* traffic). To simplify the statement of results, we shall assume further that $U_r(x_r)$ is continuously differentiable, with $U_r'(x_r) \to \infty$ as $x_r \downarrow 0$ and $U_r'(x_r) \to 0$ as $x_r \uparrow \infty$.

Assume further that utilities are additive, so that the aggregate utility of rates $x = (x_r, r \in R)$ is $\sum_{r \in R} U_r(x_r)$. Let $U = (U_r(\cdot), r \in R)$ and $C = (C_j, j \in J)$. Under this model the system optimal rates solve the following problem.

$SYSTEM(U, A, C)$:

$$\text{maximize} \quad \sum_{r \in R} U_r(x_r)$$
$$\text{subject to } Ax \leq C$$
$$\text{over} \qquad x \geq 0.$$

While this optimization problem is mathematically fairly tractable (with a strictly concave objective function and a convex feasible region), it involves utilities U that are unlikely to be known by the network. We are thus led to consider two simpler problems.

Suppose that user r may choose an amount to pay per unit time, w_r, and receives in return a flow x_r proportional to w_r, say $x_r = w_r/\lambda_r$, where λ_r could be regarded as a charge per unit flow for user r. Then the utility maximization problem for user r is as follows.

$USER_r(U_r; \lambda_r)$:

$$\text{maximize } U_r \left(\frac{w_r}{\lambda_r} \right) - w_r$$
$$\text{over} \qquad w_r \geq 0.$$

Suppose next that the network knows the vector $w = (w_r, r \in R)$, and attempts to maximize the function $\sum_r w_r \log x_r$. The network's optimization problem is then as follows.

$NETWORK(A, C; w)$:

$$\text{maximize } \sum_{r \in R} w_r \log x_r$$
$$\text{subject to } Ax \leq C$$
$$\text{over} \qquad x \geq 0.$$

It is known [10, 11] that there always exist vectors $\lambda = (\lambda_r, r \in R)$, $w = (w_r, r \in R)$ and $x = (x_r, r \in R)$, satisfying $w_r = \lambda_r x_r$ for $r \in R$, such that w_r solves $USER_r(U_r; \lambda_r)$ for $r \in R$ and x solves $NETWORK(A, C; w)$; further, the vector x is then the unique solution to $SYSTEM(U, A, C)$.

A vector of rates $x = (x_r, r \in R)$ is *proportionally fair* if it is feasible, that is $x \geq 0$ and $Ax \leq C$, and if for any other feasible vector x^*, the aggregate of proportional changes is zero or negative:

$$\sum_{r \in R} \frac{x_r^* - x_r}{x_r} \leq 0. \tag{13.1}$$

If $w_r = 1, r \in R$, then a vector of rates x solves $NETWORK(A, C; w)$ if and only if it is proportionally fair. Such a vector is also the natural extension of Nash's bargaining solution, originally derived in the special context of two users [12], to an arbitrary number of users, and, as such, satisfies certain natural axioms of fairness [13, 14].

A vector x is such that the *rates per unit charge* are proportionally fair if x is feasible, and if for any other feasible vector x^*,

$$\sum_{r \in R} w_r \frac{x_r^* - x_r}{x_r} \leq 0. \tag{13.2}$$

The relationship between the conditions (13.1) and (13.2) is well illustrated when $w_r, r \in R$, are all integral. For each $r \in R$, replace the single user r by w_r identical sub-users, construct the proportionally fair allocation over the resulting $\sum_r w_r$ users, and provide to user r the aggregate rate allocated to its w_r sub-users; then the resulting rates *per unit charge* are proportionally fair. It is straightforward to check that a vector of rates x solves $NETWORK(A, C; w)$ if and only if the rates per unit charge are proportionally fair.

13.1.1   Why proportional fairness?

RCP approximates the processor-sharing queueing discipline when there is a single bottleneck link, and hence allows short flows to complete quickly [15, 7]. For the processor-sharing discipline at a single bottleneck link, the mean time to transfer a file is proportional to the size of the file, and is insensitive to the distribution of file sizes [16, 15]. Proportional fairness is the natural network generalization of processor-sharing, with a growing literature showing that it has exact or approximate insensitivity properties [17, 18] and important efficiency and robustness properties [19, 20].

In their study of multi-hop wireless networks, Le Boudec and Radunovic [20] highlight that proportional fairness achieves a good tradeoff between efficiency and fairness, and recommend that metrics for the rate performance of mobile ad hoc networking protocols be based on proportional fairness. We also highlight the two-part paper series [21] that studies the use of proportional fairness as the basis for resource allocation and scheduling in multi-channel multi-rate wireless networks. Among numerous aspects of their study, the authors conclude that the proportional fairness solution simultaneously achieves higher system throughput, better fairness, and lower outage probability with respect to the default solution given by today's 802.11 commercial products.

Briscoe [22] has eloquently made the case for *cost fairness*, that is, rates per unit charge that are proportionally fair. As Briscoe discusses, it does not necessarily follow that users should pay according to the simple model described above; for example, if users prefer ISPs to offer flat rate subscriptions. But to avoid perverse incentives, accountability should be based on cost fairness. For example, ISPs might want to limit the congestion costs their users can cause, not just charge them for whatever unlimited costs they cause.

In the next section we show that the Rate Control Protocol may be adapted to achieve cost fairness, and further that it is possible to show convergence, to equilibrium, on the rapid timescale of round-trip times.

13.2   Proportionally fair rate control protocol

In this section we recapitulate the proportionally fair variant of RCP introduced in [23]. The framework we use is based on fluid models of packet flows where the dynamics of the fluid models allows the machinery of control theory to be used to study stability on the fast timescale of round-trip times.

Buffer sizing is an important issue in the design of end-to-end protocols. In rate controlled networks, if links are run close to capacity, then buffers need to be large, so that new flows can be given a high starting rate. However, if links are run with some spare capacity, then this may be sufficient to cope with new flows, and allow buffers to be small. Towards the goal of a low delay and low loss network, it is imperative to strive to keep queues small. In such a regime, the queue size fluctuations are very fast – so fast that it is impossible to control the queue size. Instead, as described in [24, 25], protocols act to control the *distribution* of queue size. Thus, on the timescale relevant for convergence of the protocol it is then the *mean* queue size that is important. This simplification of the treatment of queue size allows us to obtain a model that remains tractable even for a general network topology. Next we describe our network model of RCP with small queues, designed to allow buffers to be small.

Recall that we consider a network with a set J of resources and a set R of routes. A route r is identified with a non-empty subset of J, and we write $j \in r$ to indicate that route r passes through resource j. For each j, r such that $j \in r$, let T_{rj} be the propagation delay from the source of flow on route r to the resource j, and let T_{jr} be the return delay from resource j to the source. Then

$$T_{rj} + T_{jr} = T_r \quad j \in r, r \in R, \tag{13.3}$$

where T_r is the round-trip propagation delay on route r: the identity (13.3) is a direct consequence of the end-to-end nature of the signaling mechanism, whereby congestion on a route is conveyed via a field in the packets to the destination, which then informs the source. We assume queueing delays form a negligible component of the end-to-end delay – this is consistent with our assumption of the network operating with small queues.

Our small queues fair RCP variant is modeled by the system of differential equations

$$\frac{d}{dt} R_j(t) = \frac{a R_j(t)}{C_j \overline{T}_j(t)} \left(C_j - y_j(t) - b_j C_j p_j(y_j(t)) \right), \tag{13.4}$$

where

$$y_j(t) = \sum_{r:j \in r} x_r(t - T_{rj}) \tag{13.5}$$

is the aggregate load at link j, $p_j(y_j)$ is the mean queue size at link j when the load there is y_j, and

$$\overline{T}_j(t) = \frac{\sum_{r:j \in r} x_r(t) T_r}{\sum_{r:j \in r} x_r(t)} \tag{13.6}$$

is the average round-trip time of *packets* passing through resource j. We suppose the flow rate $x_r(t)$ leaving the source of route r at time t is given by

$$x_r(t) = w_r \left(\sum_{j \in r} R_j(t - T_{jr})^{-1} \right)^{-1}. \tag{13.7}$$

We interpret these equations as follows. Resource j updates $R_j(t)$, the nominal rate of a flow which passes through resource j alone, according to Equation (13.4). In this equation the term $C_j - y_j(t)$ represents a measure of the rate mismatch, at time t, at resource j, while the term $b_j C_j p_j(y_j(t))$ is proportional to the mean queue size at resource j. Equation (13.7) gives the flow rate on route r as the product of the weight w_r and reciprocal of the sum of the reciprocals of the nominal rates at each of the resources on route r. Note that Equations (13.5) and (13.7) make proper allowance for the propagation delays, and the average round-trip time (13.6) of packets passing through resource j scales the rate of adaptation (13.4) at resource j.

The computation (13.7) can be performed as follows. If a packet is served by link j at time t, $R_j(t)^{-1}$ is added to the field in the packet containing the indication of congestion. When an acknowledgement is returned to its source, the acknowledgement feedbacks the sum, and the source sets its flow rate equal to the returning feedback to the power of -1.

A simple approximation for the mean queue size is as follows. Suppose that the workload arriving at resource j over a time period τ is Gaussian, with mean $y_j \tau$ and variance $y_j \tau \sigma_j^2$. Then the workload present at the queue is a reflected Brownian motion [26], with mean under its stationary distribution of

$$p_j(y_j) = \frac{y_j \sigma_j^2}{2(C_j - y_j)}. \tag{13.8}$$

The parameter σ_j^2 represents the variability of resource j's traffic at a packet level. Its units depend on how the queue size is measured: for example, packets if packets are of constant size, or kilobits otherwise.

At the equilibrium point $y = (y_j, j \in J)$ for the dynamical system (13.4)–(13.8) we have

$$C_j - y_j = b_j C_j p_j(y_j). \tag{13.9}$$

From Equations (13.8)–(13.9) it follows that at the equilibrium point

$$p_j'(y_j) = \frac{1}{b_j y_j}. \tag{13.10}$$

Observe that in the above model formulation there are two forms of feedback: rate mismatch and queue size.

13.2.1   Sufficient conditions for local stability

For the RCP dynamical system, depending on the form of feedback that is incorporated in the protocol definition one may exhibit *two* simple sufficient conditions for local stability; for the requisite derivations and associated analysis see [23].

Local stability with feedback based on rate mismatch and queue size. A sufficient condition for the dynamical system (13.4)–(13.8) to be locally stable about its equilibrium point is that

$$a < \frac{\pi}{4}. \tag{13.11}$$

Observe that this simple decentralized sufficient condition places *no* restriction on the parameters $b_j, j \in J$, provided our modeling assumptions are satisfied.

The parameter a controls the speed of convergence at each resource, while the parameter b_j controls the utilization of resource j at the equilibrium point. From (13.8)–(13.9) we can deduce that the utilization of resource j is

$$\rho_j \equiv \frac{y_j}{C_j} = 1 - \sigma_j \left(\frac{b_j}{2} \cdot \frac{y_j}{C_j} \right)^{1/2},$$

and hence that

$$\rho_j = \left(\left(1 + \frac{\sigma_j^2 b_j}{8} \right)^{1/2} - \left(\frac{\sigma_j^2 b_j}{8} \right)^{1/2} \right)^2$$

$$= 1 - \sigma_j \left(\frac{b_j}{2} \right)^{1/2} + O(\sigma_j^2 b_j). \tag{13.12}$$

For example, if $\sigma_j = 1$, corresponding to Poisson arrivals of packets of constant size, then a value of $b_j = 0.022$ produces a utilization of 90 percent.

Local stability with feedback based only on rate mismatch. One may also derive an alternative sufficient condition for local stability. If the parameters b_j are all set to zero, and the algorithm uses as C_j not the actual capacity of resource j, but instead a target, or virtual, capacity of say 90 percent of the actual capacity, then this too will achieve an equilibrium utilization of 90 percent. In this case the equivalent sufficient condition for local stability is

$$a < \frac{\pi}{2}. \tag{13.13}$$

Although the presence of a queueing term is associated with a smaller choice for the parameter a — note the factor two difference between conditions (13.11) and (13.13) — nevertheless the local responsiveness is comparable, since the queueing term contributes roughly the same feedback as the term measuring rate mismatch.

13.2.2   Illustrative simulation

Next we illustrate our small queue variant of the RCP algorithm with a simple packet-level simulation in the case where there is feedback based only on rate mismatch.

The network simulated has a single resource, of capacity one packet per unit time and 100 sources that each produce Poisson traffic. Let us motivate a simple calculation. Assume that the round-trip time is 10 000 units of time. Then assuming a packet size of 1000 bytes, this would translate into a service rate of 100 Mbytes/s, and a round-trip time of 100 ms, or a service rate of 1 Gbyte/s and a round-trip time of 10 ms. The figures bearing observations or traces from packet-level simulations were produced using a discrete event simulator of packet flows in RCP networks where the links are modeled as FIFO queues. The round-trip times that are simulated are in the range of 1000 to 100 000 units of time. In our simulations, as the queue term is absent from the feedback, i.e., $b = 0$, we set $a = 1$ and replace C with γC for $\gamma \in [0.7, \ldots, 0.90]$ in the protocol definition. The simulations were started close to equilibrium.

Figure 13.1 shows the comparison between theory and the simulation results, when the round-trip times are in the range of 1000 to 100 000 units of time. Observe the variability of the utilization, measured over one round-trip time, for shorter round-trip times. This is to be expected, since there would remain variability in the empirical distribution of queue size. This source of variability decreases as the bandwidth-delay product increases, and in such a regime there is excellent agreement between theory and simulations.

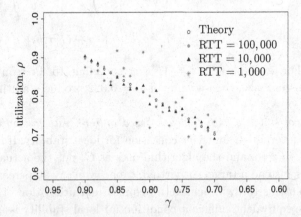

Figure 13.1 Utilization, ρ, measured over one round-trip time, for different values of the parameter γ with 100 RCP sources that each produce Poisson traffic.

13.2.3   Two forms of feedback?

Rate controlled communication networks may contain two forms of feedback: a term based on rate mismatch and another term based on the queue size.

It has been a matter of some debate whether there is any benefit in including feedback based on rate mismatch *and* on queue size. The systems with and without feedback based on queue size give rise to different nonlinear equations; but, notwithstanding an innocuous-looking factor of two difference, they both yield decentralized sufficient conditions to ensure local stability.

Thus far, as methods based on linear systems theory have not offered a preferred design recommendation – note the simple factor two difference between conditions (13.11) and (13.13) – for further progress it is quite natural to employ nonlinear techniques. For a starting point for such an investigation see [27], where the authors investigate some nonlinear properties of RCP with a conclusion that favors the system whose feedback is based only on rate mismatch.

13.2.4 Tatonnement processes

Mechanisms by which supply and demand reach equilibrium have been a central concern of economists, and there exists a substantial body of theory on the stability of what are termed tatonnement processes. From this viewpoint, the rate control algorithm described in this section is just a particular embodiment of a *Walrasian auctioneer* searching for market clearing prices. The Walrasian auctioneer of tatonnement theory is usually considered a rather implausible construct; however, we showed that the structure of a communication network presents a rather natural context within which to investigate the consequences for a tatonnement process.

In this section, we showed how the proportionally fair criteria could be implemented in a large-scale network. In particular, it was highlighted that a simple rate control algorithm can provide stable convergence to proportional fairness per unit charge, and be stable even in the presence of random queueing effects and propagation time delays.

A key issue, however, is how new flows should be admitted to such a network, a theme that we pursue in the next section. The issue of buffer sizing in rate controlled networks is a topical one and the reader is referred to [28], and references therein, for some recent work in this regard. However, our focus in this chapter will be on developing the admission management process of [23].

13.3   Admission management

In explicit congestion controlled networks when a new flow arrives, it expects to learn, after one round-trip time, of its starting rate. So an important aspect in the design of such networks is the management of new flows; in particular, a key question is the scale of the step-change in rate that is necessary at a resource to accommodate a new flow. We show that, for the variant of RCP considered here, this can be estimated from the aggregate flow through the resource, without knowledge of individual flow rates.

We first describe, in Section 13.3.1, how a resource should estimate the impact upon it of a new flow starting. This suggests a natural step-change algorithm for a resource's estimate of its nominal rate. In the remainder of this section we explore the effectiveness of the admission management procedure based on the step-change algorithm to large, and sudden, variations in the load on the network.

13.3.1   Step-change algorithm

In equilibrium, the aggregate flow through resource j is y_j, the unique value such that the right-hand side of (13.4) is zero. When a new flow, r, begins transmitting, if $j \in r$, this will disrupt the equilibrium by increasing y_j to $y_j + x_r$. Thus, in order to maintain equilibrium, whenever a flow, r, begins, R_j needs to be decreased, for all j with $j \in r$.

According to (13.5)

$$y_j = \sum_{r:j\in r} w_r \left(\sum_{k\in r} R_k^{-1} \right)^{-1}$$

and so the sensitivity of y_j to changes in the rate R_j is readily deduced to be

$$\frac{\partial y_j}{\partial R_j} = \frac{y_j \bar{x}_j}{R_j^2}, \tag{13.14}$$

where

$$\bar{x}_j = \frac{\sum_{r:j\in r} x_r \left(\sum_{k\in r} R_k^{-1} \right)^{-1}}{\sum_{r:j\in r} x_r}.$$

This $\bar{x}_j$ is the average, over all packets passing through resource j, of the unweighted fair share on the route of a packet.

Suppose now that when a new flow r, of weight w_r, arrives, it sends a request packet through each resource j on its route, and suppose each resource j, on observation of this packet, immediately makes a step-change in R_j to a new value

$$R_j^{new} = R_j \cdot \frac{y_j}{y_j + w_r R_j}. \tag{13.15}$$

The purpose of the reduction is to make room at the resource for the new flow. Although a step-change in R_j will take time to work through the network, the scale of the change anticipated in traffic from existing flows can be estimated from (13.14) and (13.15) as

$$(R_j - R_j^{new}) \cdot \frac{\partial y_j}{\partial R_j} = w_r \bar{x}_j \cdot \frac{y_j}{y_j + w_r R_j}.$$

Thus the reduction aimed for from existing flows is of the right scale to allow one extra flow at the average of the w_r-weighted fair share through resource j. Note that this is achieved without knowledge at the resource of the individual flow rates through it, $(x_r, r : j \in r)$: only knowledge of their equilibrium aggregate y_j is used in expression (13.15), and y_j may be determined from the parameters C_j and b_j as in (13.9).

We now describe an important situation of interest.

Large and sudden changes in the number of flows. It is quite natural to ask about the robustness of any protocol to sudden, and large, changes in the number of flows. A network should be able to cope sensibly to local surges in traffic. Such surges in traffic could simply be induced by a sudden increase in the number of users wishing to use a certain route. Or, such a surge may be induced by the failure of a link, where a certain fraction, or all, of the load is transferred to a link which is still in operation.

13.3.2 Robustness of the step-change algorithm

In this subsection we briefly analyze the robustness of the admission control process based on the above step-change algorithm against large, and sudden, increases in the number of flows.

Consider the case where the network consists of a single link j with equilibrium flow rate y_j. If there are n identical flows, then at equilibrium $R_j = y_j/n$. When a new flow begins, the step-change (13.15) is performed and R_j becomes $R_j^{new} = y_j/(n+1)$. Hence equilibrium is maintained. Now suppose that m new flows begin at the same time. Once the m flows have begun, R_j should approach $y_j/(n+m)$. However, each new flow's request for bandwidth will be received one at a time. Thus the new flows will be given rates $y_j/(n+1), y_j/(n+2), \ldots, y_j/(n+m)$. So when the new flows start transmitting, after one round-trip time, the new aggregate rate through j, y_j^{new} will approximately be

$$y_j^{new} \approx n\frac{y_j}{n+m} + \int_n^{n+m} \frac{y_j}{u}du.$$

If we let $\epsilon = m/n$, we have

$$y_j^{new} \approx y_j \left(\frac{1}{1+\epsilon} + \log(1+\epsilon) \right). \tag{13.16}$$

For the admission control process to be able to cope when the load is increased by a proportion ϵ, we simply require y_j^{new} to be less than the capacity of link j. Direct calculation shows that if the equilibrium value of y_j is equal to 90% of capacity, then (13.16) allows an increase in the number of flows of up to 66%. Furthermore, if at equilibrium y_j is equal to 80% of capacity, then the increase in the number of flows can be as high as 122% without y_j^{new} exceeding the capacity of the link.

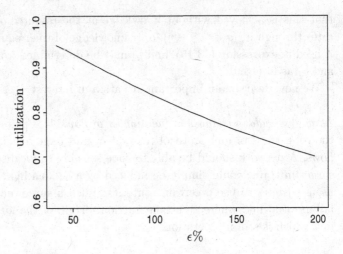

Figure 13.2 Utilization one can expect to achieve and still have the system be robust against an $\epsilon\%$ sudden increase in load; numerical values computed from (13.16).

13.3.3   Guidelines for network management

Figure 13.2 highlights the tradeoff between the desired utilization of network resources and the scale of a sudden burst of newly arriving traffic that the resource can absorb. The above analysis and discussion revolves around a single link, but it does provide a simple rule of thumb guideline for choosing parameters such as b_j or C_j. If one takes ϵ to be the largest plausible increase in load that the network should be able to withstand, then from (13.16), one can calculate the value of y_j which gives y_j^{new} equal to capacity. This value of y_j can then be used to choose b_j or C_j, using the equilibrium relationship $C_j - y_j = b_j C_j p_j(y_j)$. There are two distinct regimes that are possible after a sudden increase in the number of flows:

1. If, after the increase, the load y_j remains less than the capacity C_j, then we are in a regime where the queue remains stable. Its stationary distribution (13.8) will have an increased mean and variance, but will not depend on the bandwidth-delay product.
2. If, after the increase, the load y_j exceeds C_j, then we are in a regime where the queue is unstable, and in order to prevent packet drops it is necessary for the buffer to store an amount proportional to the excess bandwidth times the delay.

The approach we advise is to select buffer sizes and utilizations to cope with the first regime, and to allow packets to be dropped rather than stored in the second regime. The second regime should occur rarely if the target utilization is chosen to deal with plausible levels of sudden overload.

13.3.4 Illustrating the utilization–robustness tradeoff

We first recapitulate the processes involved in admitting a new flow into an RCP network. A new flow first transmits a request packet through the network. The links, on detecting the arrival of the request packet, perform the step-change algorithm to make room at the respective resources for the new flow. After one round-trip time the source of the flow receives back acknowledgement of the request packet, and starts transmitting at the rate (13.7) that is conveyed back. This procedure allows a new flow to reach near equilibrium within *one* round-trip time. We now illustrate, via some simulations, the admission management procedure for dealing with newly arriving flows.

We wish to exhibit the tradeoff between a target utilization, and the impact at a resource of a sudden and large increase in load. Consider a simple network, depicted in Figure 13.3, consisting of five links where we do not include feedback based on queue size in the RCP definition and the end-systems produce Poisson traffic. In our simulations, as the queue term is absent from the feedback, i.e., $b = 0$, we replace C_j with $\gamma_j C_j$ for $\gamma_j < 1$, in the protocol definition, in order to aim for a target utilization. The value of a was set at $0.367 \approx 1/e$, to ensure that the system is well within the sufficient condition for local stability. In our experiments, links A, B, C and D each start with 20 flows operating in equilibrium. Each flow uses link X and one of links A, B, C or D. So, for example, a request packet originating from flows entering link C, would first go through link C and then link X before returning back to the source.

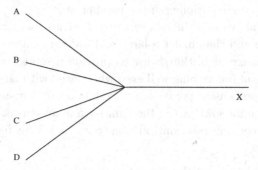

Figure 13.3 Toy network used, in packet-level simulations, to illustrate the process of admitting new flows into a RCP network. The links labeled A, B, C, D and X have a capacity of $1, 10, 1, 10$ and 20 packets per unit time, respectively. The physical transmission delays on links A, B and X are 100 time units and on links C and D are 1000 time units.

The experiment we conduct is as follows. The target utilization at all the links is set at 90%, and the scenarios we consider are a 50%, a 100%, and then a 200% instantaneous increase in the number of flows. The choice of these numbers is guided by the robustness analysis above, which is illustrated in Figure 13.2. Since our primary interest is to explore the impact at the resource of a sudden, and instantaneous, increase in load we shall exhibit the impact at one of the ingress links, i.e., link C.

When dealing with new flows there are two quantities that we wish to observe at the resource: the impact on the rate, and the impact on the queue size. In Figures 13.4(a)–(c), the necessary step-change in rate required to accommodate the new flows is clearly visible. The impact on the queue sizes is, however, more subtle. In Figure 13.4(a), which corresponds to a 50% increase in the number of flows, observe the minor spike in the queue at approximately 4000 time units. The spike in queue size gets more visible when we have a 100% increase in the number of flows; see Figure 13.4(b). The spike lasts for approximately 2200 time units, which is twice the sum of the physical propagation delays along links C and X; the round-trip time of flows originating at link C. With a 200% increase in the number of flows, this spike is extremely pronounced and in fact pushes the peak of the queue close to 300 packets; see Figure 13.4(c). However, the queue does return to its equilibrium state, approximately one round-trip time later.

Figure 13.4(a) illustrates the first regime described in Section 13.3.3: after the increase in load the queue remains stable, albeit with an increased mean and variance. Figures 13.4(b) and (c) illustrate the second regime, where after the increase the load y_j exceeds the capacity C_j. In Figure 13.4(b) the excess load is relatively small, and there is only a gentle drift upwards in the queue size, with random fluctuations still prominent. In Figure 13.4(c) the excess load, $C_j - y_j$, causes an approximately linear increase in the queue size over a period of length one round-trip time. Recall that these two cases correspond respectively to a doubling and a tripling of the number of flows.

The above experiments serve to illustrate the tradeoff between a target utilization and the impact a large and sudden load would have at a resource. The step-change algorithm helps to provide a more resilient network; one that is capable of functioning well even when faced with large surges in localized traffic. A comprehensive performance evaluation of the step-change algorithm, which forms an integral part of the admission management process, to demonstrate its effectiveness in rate controlled networks is left for further study.

13.3.5   Buffer sizing and the step-change algorithm

The Transmission Control Protocol is today the de facto congestion control standard that is used in most applications. Its success, in part, has been due to the fact that it has mainly operated in wired networks where losses are mainly due to the overflow of a router's buffer. The protocol has been designed to react to, and cope with, losses; the multiplicative decrease component in the congestion avoidance phase of TCP provides a risk averse response when it detects the loss of a packet. Losses, however, constitute damage to packets. This concern is expected to get compounded in environments where bit error rates may not be negligible; a characteristic usually exhibited in wireless networks.

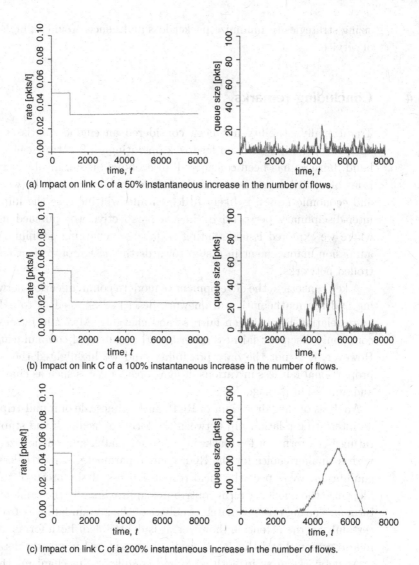

(a) Impact on link C of a 50% instantaneous increase in the number of flows.

(b) Impact on link C of a 100% instantaneous increase in the number of flows.

(c) Impact on link C of a 200% instantaneous increase in the number of flows.

Figure 13.4 Illustration of a scenario with a 50%, 100% and then a 200% increase in the flows which instantaneously request to be admitted into the network depicted in Figure 13.3. The target utilization for all the links in the simulated network is 90%.

In rate controlled networks, loss gets decoupled from flow control and it is possible to maintain small queues in equilibrium and also in challenging situations. A consequence of this is that buffers, in routers, can be dimensioned to be much smaller than the currently used bandwidth-delay product rule of thumb [25] without incurring losses. The role played by the step-change algorithm in ensuring that the queue size remains bounded is exhibited, and so it forms a rather natural component of system design; for example, in developing buffer sizing strategies to minimize packet loss and hence provide a high-grade quality of service.

13.4　Concluding remarks

Traditionally, stability has been considered an engineering issue, requiring an analysis of randomness and feedback operating on fast timescales. On the other hand, fairness has been considered an economic issue, involving static comparisons of utility. In networks of the future this distinction, between engineering and economic issues, is likely to lessen and will increase the importance of an inter-disciplinary perspective. Such a perspective was pursued in this chapter where we explored issues relating to fairness, charging, stability, feedback and admission management in a step towards the design of explicit congestion controlled networks.

A key concern in the development of modern communication networks is charging and the mathematical framework described enabled us to exhibit the intimate relationship between fairness and charging. Max-min fairness is the most commonly discussed fairness criteria in the context of communication networks. However, it is not the only possibility, and we highlighted the role played by proportional fairness in various design considerations such as charging, stability, and admission management.

Analysis of the fair variant of RCP on the timescale of round-trip times reveals an interesting relationship between the forms of feedback and stability. Incorporating both forms of feedback, i.e., rate mismatch and queue size, is associated with a smaller choice for the RCP control parameter. Nevertheless, close to the equilibrium we expect the local responsiveness of the protocol to be comparable, since the queueing term contributes approximately the same feedback as the term measuring rate mismatch. Analysis of the system far away from equilibrium certainly merits attention; however, it is debatable if both forms of feedback are indeed essential and this issue needs to be explored in greater detail.

As networks grow in both scale and complexity, mechanisms that may allow the self regulation of large-scale communication networks are especially appealing. In a step towards this goal, the automated management of new flows plays an important role in rate controlled networks and the admission management procedure outlined does appear attractive. The step-change algorithm that is invoked at a resource to accommodate a new flow is *simple*, in the sense that the requisite computation is done without knowledge of individual flow rates. It is also *scalable*, in that it is suitable for deployment in networks of any size. Additionally, using both analysis and packet-level simulations, we developed insight into a fundamental design aspect of an admission management process: there is a tradeoff between the desired utilization and the ability of a resource to absorb, and hence be robust towards, sudden and large variations in load.

In the design of any new end-to-end protocol there is considerable interest in how simple, local and microscopic rules, often involving random actions, can produce coherent and purposeful behavior at the macroscopic level. Towards the quest for desirable macroscopic outcomes, the architectural framework described in this chapter may allow the design of a fair, stable, low-loss, low-delay, high-utilization, and robust communication network.

Acknowledgement

Gaurav Raina acknowledges funding by the Defense Research Development Organisation (DRDO)-IIT Madras Memorandum of Collaboration.

References

[1] M. Chiang, S. H. Low, A. R. Calderbank, and J. C. Doyle. Layering as optimization decomposition: a mathematical theory of network architectures. *Proceedings of the IEEE*, **95** (2007) 255–312.

[2] F. Kelly. Fairness and stability of end-to-end congestion control. *European Journal of Control*, **9** (2003) 159–176.

[3] R. Srikant. *The Mathematics of Internet Congestion Control* (Boston: Birkhauser, 2004).

[4] V. Jacobson. Congestion avoidance and control. *Proceedings of ACM SIG-COMM* (1988).

[5] D. Katabi, M. Handley, and C. Rohrs. Internet congestion control for future high bandwidth-delay product environments. *Proceedings of ACM SIGCOMM* (2002).

[6] H. Balakrishnan, N. Dukkipati, N. McKeown, and C. Tomlin. Stability analysis of explicit congestion control protocols. *IEEE Communications Letters*, **11** (2007) 823–825.

[7] N. Dukkipati, N. McKeown, and A. G. Fraser. RCP-AC: congestion control to make flows complete quickly in any environment. *High-Speed Networking Workshop: The Terabits Challenge*, Spain (2006).

[8] T. Voice and G. Raina. Stability analysis of a max-min fair Rate Control Protocol (RCP) in a small buffer regime. *IEEE Transactions on Automatic Control*, **54** (2009) 1908–1913.

[9] S. Shenker. Fundamental design issues for the future Internet. *IEEE Journal on Selected Areas of Communication*, **13** (1995) 1176–1188.

[10] R. Johari and J. N. Tsitsiklis. Efficiency of scalar-parameterized mechanisms. *Operations Research*, articles in advance (2009) 1–17.

[11] F. Kelly. Charging and rate control for elastic traffic. *European Transactions on Telecommunications*, **8** (1997) 33–37.

[12] J. F. Nash. The bargaining problem. *Econometrica*, **28** (1950) 155–162.

[13] R. Mazumdar, L. G. Mason, and C. Douligeris. Fairness and network optimal flow control: optimality of product forms. *IEEE Transactions on Communications*, **39** (1991) 775–782.

[14] A. Stefanescu and M. W. Stefanescu. The arbitrated solution for multi-objective convex programming. *Revue Roumaine de Mathématiques Pures et Appliquées*, **20** (1984) 593–598.

[15] N. Dukkipati, M. Kobayashi, R. Zhang-Shen, and N. McKeown. Processor sharing flows in the Internet. *Thirteenth International Workshop on Quality of Service*, Germany (2005).

[16] S. Ben Fredj, T. Bonald, A. Proutière, G. Régnié, and J. W. Roberts. Statistical bandwidth sharing: a study of congestion at flow level. *Proceedings of ACM SIGCOMM* (2001).

[17] L. Massoulié. Structural properties of proportional fairness: stability and insensitivity. *The Annals of Applied Probability*, **17** (2007) 809–839.

[18] J. Roberts and L. Massoulié. Bandwidth sharing and admission control for elastic traffic. *ITC Specialists Seminar*, Yokohama (1998).

[19] T. Bonald, L. Massoulié, A. Proutière, and J. Virtamo. A queueing analysis of max-min fairness, proportional fairness and balanced fairness. *Queueing Systems*, **53** (2006) 65–84.

[20] J.-Y. Le Boudec and B. Radunovic. Rate performance objectives of multihop wireless networks. *IEEE Transactions on Mobile Computing*, **3** (2004) 334–349.

[21] S. C. Liew and Y. J. Zhang. Proportional fairness in multi-channel multi-rate wireless networks – Parts I and II. *IEEE Transactions on Wireless Communications*, **7** (2008) 3446–3467.

[22] B. Briscoe. Flow rate fairness: dismantling a religion. *Computer Communication Review*, **37** (2007) 63–74.

[23] F. Kelly, G. Raina, and T. Voice. Stability and fairness of explicit congestion control with small buffers. *Computer Communication Review*, **38** (2008) 51–62.

[24] G. Raina, D. Towsley, and D. Wischik. Part II: control theory for buffer sizing. *Computer Communication Review*, **35** (2005) 79–82.

[25] D. Wischik and N. McKeown. Part I: buffer sizes for core routers. *Computer Communication Review*, **35** (2005) 75–78.

[26] J. M. Harrison. *Brownian Motion and Stochastic Flow Systems* (New York: Wiley, 1985).

[27] T. Voice and G. Raina. Rate Control Protocol (RCP): global stability and local Hopf bifurcation analysis, preprint (2008).

[28] A. Lakshmikantha, R. Srikant, N. Dukkipati, N. McKeown, and C. Beck. Buffer sizing results for RCP congestion control under connection arrivals and departures. *Computer Communication Review*, **39** (2009) 5–15.

14 KanseiGenie: software infrastructure for resource management and programmability of wireless sensor network fabrics

Mukundan Sridharan, Wenjie Zeng, William Leal, Xi Ju, Rajiv Ramnath, Hongwei Zhang, and Anish Arora
The Ohio State University, USA

This chapter describes an architecture for slicing, virtualizing, and federating wireless sensor network (WSN) resources. The architecture, which we call KanseiGenie, allows users – be they sensing/networking researchers or application developers – to specify and acquire node and network resources as well as sensor data resources within one or more facilities for launching their programs. It also includes server-side measurement and management support for user programs, as well as client-side support for experiment composition and control. We illustrate KanseiGenie architectural concepts in terms of a current realization of KanseiGenie that serves WSN testbeds and application-centric fabrics at The Ohio State University and at Wayne State University.

14.1 Introduction

Deployed wireless sensor networks (WSN) have typically been both small-scale and focused on a particular application such as environmental monitoring or intrusion detection. However, recent advances in platform and protocol design now permit city-scale WSNs that can be deployed in such a way that new, unanticipated sensing applications can be accommodated by the network. This lets developers focus more on leveraging existing network resources and less on individual nodes.

Network abstractions for WSN development include APIs for scheduling tasks and monitoring system health as well as for in-the-field programming of applications, network components, and sensing components. As a result, WSN deployments have in several cases morphed from application-specific custom solutions to "WSN fabrics" that may be customized and reused in the field. In some cases,

Next-Generation Internet Architectures and Protocols, ed. Byrav Ramamurthy, George Rouskas, and Krishna M. Sivalingam. Published by Cambridge University Press. © Cambridge University Press 2011.

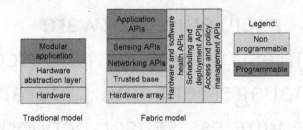

Figure 14.1 Traditional network model and fabric model.

these fabrics support and manage the concurrent operation of multiple applications. Figure 14.1 compares the traditional WSN model with the emerging fabric model of WSNs.

Why programmable WSN fabrics?   A primary use case of programmable WSN fabrics is that of testbeds. High-fidelity validation of performance-sensitive applications often mandates rigorous end-to-end regression testing at scale. Application developers need to evaluate candidate sensing logics and network protocols in the context of diverse realistic field conditions. This can be done in the field, but in many cases locating the testbed in the field is inconvenient, so field conditions can be emulated by collecting data from relevant field environments and injecting those datasets into the testbed. Application developers also need to configure and tune system performance to meet application requirements. System developers, in contrast, need to gain insight and concept validation by examining phenomena such as link asymmetry; interference, jamming behavior, node failure, and the like.

Sensing platforms that focus on people and on communities are another important use case for WSN fabrics. It is now economically feasible to deploy or technically feasible to leverage a number of connected devices across campus, community, and city spaces. In several cases, the cost of the facility is shared by multiple organizations and individuals; consider, for example, a community of handheld users who are willing to share information sensed via their privately owned devices. Assuming a programmable WSN fabric with support for these devices, hitherto unanticipated applications, missions, and campaigns can be launched upon demand, exploiting the existence of multiple sensor modalities and the benefits of fusion. Programs can also be refined based on data obtained from the field. Not least, a range of users must be supported – from non-expert clients/application developers that will access and use fabric resources and data, to expert domain owners with whom the responsibility of managing and maintaining the network will lie.

GENI.   The Global Environment for Network Innovation project [1] concretely illustrates an architecture in which a WSN fabric is a key component. GENI is a next-generation experimental network research infrastructure currently in its prototyping phase. It includes support for control and programming

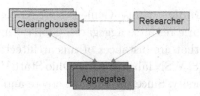

Figure 14.2 GENI federated fabric overview.

of resources spanning facilities with next-generation fiber optics and switches, novel high-speed routers, city-wide experimental urban radio networks, high-end computational clusters, and sensor grids. It intends to support large numbers of individuals and large and simultaneous experiments with extensive instrumentation designed to make it easy to collect, analyze, and share real measurements and to test load conditions that match those of current or projected Internet usage. To this end, it is characterized by the following features.

- Programmability – researchers may download software into GENI-compatible nodes to control how they behave.
- Slice-based experimentation – each GENI experiment will be on a designated slice that consists of an interconnected set of reserved resources on platforms in diverse locations. Researchers, represented by slices in the GENI context, will remotely discover, reserve, configure, program, debug, operate, manage, and teardown distributed systems established across parts of the GENI suite.
- Virtualization – when feasible, researchers can run experiments concurrently on the same set of resources as if each experiment were running alone.
- Federation – different resources are owned and operated by different organizations.

Figure 14.2 depicts the GENI architecture from a usage perspective. In a nutshell, GENI consists of three entities: researchers, clearinghouses, and the resource aggregates. A researcher queries the clearinghouse for the set of available resources at one or more aggregates and requests reservations for those that she requires.

Users and aggregates in GENI establish trust relationships with GENI clearinghouses. Aggregates and users authenticate themselves via the clearinghouse. The clearinghouse keeps track of the authenticated users, resource aggregates, slices, and reservations. Each resource provider may be associated with its own clearinghouse but there are also central GENI clearinghouses for federated discovery and management of all resources owned by participating organizations. GENI also relies on a standard for all entities to describe the underlying resource. This resource description language serves as the glue for the three entities because all interactions involve some description of resource, be it a physical resource such as routers and clusters or a logical resource such as CPU time or wireless frequency.

KanseiGenie.   KanseiGenie is a GENI-compatible architecture that focuses on WSN fabric resource aggegates. Global Environment for Network Innovation aggregates that are instances of this architecture include the Kansei [3] and the PeopleNet [8] WSN fabrics at The Ohio State University and NetEye [6] at Wayne State University. Since the effort to create and deploy a WSN fabric can be high, KanseiGenie installer packages are being developed to enable rapid porting of GENI-compatible programmability to other resource aggregates. The modular design of the KanseiGenie software suite lets testbed and fabric developers customize the package for any combination of sensor arrays already supported by KanseiGenie or new sensor arrays.

Goals of this chapter.   In this chapter, we overview requirements of next-generation sensing infrastructures and motivate the various elements of KanseiGenie architecture for WSN fabrics. We then describe how the architecture can be extended and how it supports integration more broadly with other programmable networks.

14.2   Features of sensing fabrics

A fabric is an independent, decoupled, programmable network, capable of sensing, storing, and communicating a set of physical phenomena [14]. A programmable WSN fabric provides different services depending on its policy and hardware capability. Services are provided in the form of application programming interface (API). These services may be classified as horizontal and vertical services. Horizontal services are generic services serving as building blocks for more complex services. Vertical services are domain specific and are optimized for specific application goals. Standardizing vertical services is desirable, so that applications can be readily composed and ported across fabrics geared to support a particular application domain. Note that, in general, a fabric need not make guarantees about its quality of service, delivering its results based only on a "best effort."

14.2.1   Generic services

We identify four types of generic services that future WSN fabrics should provide:

- **Resource management services:** help researchers discover, reserve, and configure the resource for experimentations.
- **Experiment management services:** provide basic data communication and control between experiments.
- **Operation and management services:** enable administrators to manage the resources.
- **Instrumentation and measurement services:** enable the fabric to make measurements of physical phenomena, store the measurements and make them securely available.

14.2.1.1 Resource management

All fabrics provide a set of shareable resources. To utilize the resource, a researcher needs to discover, reserve, and configure the resource. Resources are allocated to a slice consisting of some sliver(s) of the fabric(s) that will serve as the infrastructure in which she runs experiments. For consistency, we follow the definitions of researcher, slice, and sliver from GENI [1]. We distinguish between experiment creators and the experiment participants by calling the creators "researchers" and the participants "end-users" building their experiments on top of long running experiments. A slice is an empty container into which experiments can be instantiated and to which researchers and resources may be bound. All resources, be they physical or logical, are abstracted as components. A component can be (for example) a sensor node, a Linux machine, or a network switch. When possible, a component should be able to share its resource among multiple slices. Thus, a subset of the fabric occupied by a slice is called a sliver, which is isolated from other slivers. Only researchers bound to a slice are eligible to run experiments on it and a slice can only utilize the resource (slivers) bound to it. Resource management features include at least the following and possibly others.

- Resource publishing
- Resource discovery
- Resource reservation
- Resource configuration

For short-term experiments, resource discovery, reservation, and configuration may be performed at the beginning of the experiment in a one-shot fashion and assuming the underlying environment is relatively static in the short time window. However, for long-term experiments, resource management has to be an ongoing process to adapt to network changes. For example, nodes involved in the experiment could crash or more suitable nodes might join the network.

Resource publishing. A fabric shares its resources by publishing information about some subset to a clearinghouse. A clearinghouse may be held by some well-known site or the fabric itself. To promote utilization, a fabric can publish the same set of resources to multiple clearinghouses. This could result in multiple reservation requests to the same resource; in the end, the fabric decides which reservation(s) to honor, if any. Note that uncoordinated clearinghouses could make inconsistent resource allocations resulting in deadlock or livelock situations. Global clearinghouses hierarchy and interaction architecture, as well as clearinghouse and resource provider policy should explicitly address this problem.

Resource discovery. Sensor fabrics provide heterogeneous types of resources with diverse sensors, storage space, and computing and communication capabilities. Resource discovery is two-fold. First, resource providers must be able to accurately describe the resource, the associated attributes, and the relationships between the resources. Second, researchers need to search for the resource they need with descriptions at different levels of details. Central to this discovery service is a resource description language for both the provider and the researcher. The resource request provided by the researcher can be concrete, such as which physical node and router should be selected, or abstract, such as a request for a 100 by 100 grid of fully connected wireless sensor nodes. In this case the discovery service has to map the request onto a set of physical resources by finding out the most suitable set of resources to fulfill the request.

Resource reservation. Once the desired resource is discovered, it needs to be requested directly from the resource provider or a third-party broker to which the resource has been delegated. Both the set of available resources and the permitted operations on it will vary according to the researcher's privileges. Note that if a researcher reserves the resource from a third-party clearinghouse instead of directly from the provider, the researcher only has a promise of the resource rather than a guaranteed allocation of the resource. The reserved resource is allocated to the researcher's slice only after the researcher claims it from the provider. The success of the resource claim depends on the resource availability at the instant when the claim is issued as well as the provider's local policy.

Resource configuration. The reserved resource slice needs to be configured in the beginning to meet the application's specification. Configurations range from the software and runtime environment setups to setting the transmission power or the network topology. The resource configuration service needs to expose for each device the set of configurable parameters and the values these parameters can take on. Eliminating redundancy and performing other optimizations could be important. For example, if different experiments running on the same device require the same software library, this would result in wasted storage if duplicates of the library are installed.

14.2.1.2 Experiment interaction

Experiments in wireless sensor networks take on many forms. Some run without human intervention whereas some adapt to human inputs and new sensing needs. Some run for months while some are short. We identify a set of features for experiment interaction as a basis for standardizing a common set of ways that researchers interact with their deployed experiments.

Debugging and logging. In some cases, a network is so resource poor (in terms of memory, bandwidth and reliability) that it is difficult or impossible to exfiltrate debugging data. But when it is possible to do so, standard services that permit the experimenter to pause part of the experiment, inspect state, and single-step the code should be provided. Logging is a related capability, which, when the resources permit, provides output of the experiment and can give after-the-fact debugging information. Typically a WSN application involves tens to hundreds of nodes, so the logging service should provide a central point of access to all the logging data.

Exfiltration and visualization. Exfiltration, either in push or pull mode, allows researchers to collect the application output, possibly aggregated or transformed as instructed by users. Advanced exfiltration patterns such as

publish-subscribe should be supported. Standard visualizations of data exfil-
trated by an experiment, such as a node map with connectivity information,
should be provided, along with customization "hooks," providing the option to
the researcher to build application-specific visualizations.

Data injection. Data injection may be of two forms. In compile-time injection
the content (i.e., when, where and what to inject) of the injection is already
determined at compile-time. During run-time injection a researcher generates
the data and injects it into the desired devices at run-time through client-side
software that provides data channels to the devices.

Key and frequency change. Long-running experiments evolve over time.
To ensure security and reduce interference, a researcher may change the shared
or public key or the communication frequency of the experiment every now and
then.

Pause and resume. As intuitive as it sounds, the difficulty to provide pause
and resume services varies dramatically depending on the semantics of an experiment. The service boils down to the problem of taking a snapshot of the global
system state and then reestablishing it later. Ensuring consistency of the system
state after a pause just before a resume is a matter of research interest.

Move. The move service is an extension on top of the "pause and resume"
service. It enables the researcher to continue the experiment even when the
resource allocated to the experiment's slice becomes unavailable or when a better
resource is found. A researcher can pause her experiment, remove it from the
current resource, acquire and configure other available resources for the slice,
and finally migrate and resume the experiment.

Experiment composition. Sometimes the fabric owner might not want to
directly run user executables on the fabric for security or other reasons. Experiments on such fabrics can be viewed as black boxes that take certain end-user
input, such as sensor parameters, to compute certain outputs. On such fabric the
designer might want to provide experiment composition libraries, which can be
statically or dynamically recomposed to satisfy the user's needs. Also to promote
the reuse of existing WSN applications, fabrics should provide a way to redirect
the output from existing fabric applications to the input of another experiment,
possibly the user application on another fabric. The redirected output can be
viewed as a data-stream resource. In other words, the user rather than utilizing
the physical resource of the fabric, uses configurable virtual (data) resources for
experimentation.

Batching. This service enables a researcher to schedule a series of experiments, possibly with a number of controlled parameters. Instead of scheduling
and configuring these experiments individually, the researcher simply gives the
ranges that each parameter will iterate through and the step size between con-
secutive iterations. Such a batching service is especially useful when a researcher
is trying to find out the set of optimal parameters for her experiment or to study
different parameters' impacts on experiment performance.

14.2.1.3 Management and operations services

Operational global view. In many cases, managing the fabrics requires a global operational view of the fabric or multiple federated fabrics. This service provides a portal through which researchers can find operational information such as node health and resource usage statistics. Since sensor nodes are much less robust than PC class computing devices, a user must know which set of sensor nodes worked correctly during the experiment in order to interpret the results. Thus generic node health information such as whether a node and its communicating interfaces function correctly during an experiment should be made available by this service. Resource usage auditing is necessary to check whether the user abides by policies and does not abuse privileges. The auditing includes such things as CPU, network, and storage consumption. However, for mote class devices, detailed usage information on a per-node basis may be unavailable due to resource limitations.

Emergency stop. When a misbehaving experiment is identified, the administrator must be able to either isolate or stop the experiment so that its negative impact on the network is minimized.

Resource usage policy. It must be possible to limit the privileged operations that an experiment may invoke over the underlying fabric. Example privileges include the right to access certain sensors and the right to read and write fabric states. Both compile-time and run-time support should be provided for enforcing resource usage policies. For the run-time case, an administrator should be able to grant or revoke a slice's privileges for the set of reserved resources.

Reliable configuration. The fidelity of the experiment results depends on how well a researcher's experiment configuration specification was executed by the resource provider. If there was any difference or error while configuring the experiment, the researcher could possibly filter out the abnormalities from the results, if a service is provided by the fabric that specifies the set of resources that are correctly configured. Providing this service in a wireless network can be quite a challenge where the process is subject to noise and interference and thus has a higher failure rate. For example, wireless sensor nodes need to be programmed before the experiment can be run on them. In many cases, the sensing program is transferred to the sensor nodes in wireless channels. The fabric should provide some service to make sure that every sensor node is programmed correctly.

14.2.1.4 Instrumentation and measurement services

These services will support multiple, concurrent, diverse experiments from physical to application layer that require measurements. Inexperienced researchers can build their sensing applications by writing simple scripts on top of these services instead of learning a complex programming language, such as nesC, to write their sensing programs. Ideally, each measurement should carry the time and space information for future analysis. Also, given the nature of the sensor network, each experiment can potentially generate a huge amount of measurements. As a result, services for measurement storage, aggregation, and query of measurement data are required. Another critical requirement is that the instrumentation and measurements should be isolated from the execution of the experiments so that results of the experiments are not affected.

Traffic monitoring. Traffic monitoring in the context of WSN includes both application traffic and interfering traffic. One unique property of wireless sensing fabrics is the complex interference phenomenon in their communications. Noise and interference can have a substantial impact on the experiment result. Therefore, collecting noise and interference data is necessary for the researchers to correctly interpret their experiment results.

Sensing. Wireless sensor nodes are equipped with multiple sensors, each of which are capable of measuring different physical phenomena with different accuracies. The sensing service should allow for controlling existing sensors as well as adding new ones. Usually, wireless sensor nodes have very limited local storage. Thus, storage of sensing data should be explicitly addressed by the sensing service.

14.2.2   Domain-specific services

In the previous section we described four classes of generic (horizontal) services that comprise the basic building blocks for most wireless sensing fabrics. In this section, we shift our attention to domain-specific or vertical services, which are tailored to the specific requirements of a given application. We give two examples of vertical API, one a search service designed in the security context and the other a rapid program development API for the KanseiGenie testbed.

14.2.2.1 Search API for security networks

In security networks (such as an intrusion detection network), a designer might want to provide the end-user with a flexible search interface [14] that can search for objects of interest and can re-task the network as necessary. These objects can be various types of physical targets (such as humans, animals, cars, security agents) or sensor and network data objects (such as failed nodes or nodes with low battery life). These searches can be temporary (one-shot) or persistent (returns data at periodic intervals). While such an interface would be useful, it would be specific only to a security-oriented network and hence we call it a vertical API. The key to providing such a search API is to design the sensor fabric with a service, that will interpret the user queries and retask the network with appropriate programs and parameters.

14.2.2.2 DESAL for KanseiGenie

DESAL [13] is a state-based programming language developed specifically for sensor networks that is convenient for rapid prototyping of applications for sensor networks. While DESAL is primarily a programming language, a platform-specific DESAL compiler can be viewed as a domain-specific service. Current DESAL compilers produce programs in nesC [5], which can be complied and executed on testbeds like Kansei. Thus DESAL is a vertical API for KanseiGenie, which allows users to write their applications in high-level DESAL code rather than in more laborious TinyOS/nesC.

14.3   KanseiGenie architecture

14.3.1   The fabric model

As we have noted, the architecture for KanseiGenie is based on a the idea of a fabric, a general networking model that takes the view that the network designer is not responsible for writing applications but for providing a core set of services that can be used by the application developer to create applications. The fabric model is agnostic about whether the application runs inside or outside of the fabric. In fact, depending on the capabilities of the fabric, it can be a combination of both. The generic services exposed by the fabric are analogous to those of a modern operating system on a PC. The user interacts through well-defined APIs and commands to develop sophisticated applications, while leaving the low-level management functions to the OS.

The fabric model clearly separates the responsibilities of the network designer from those of the application developer (who is less likely to be a sensor network expert). A fabric manager, also called the Site Authority (SA), takes care of implementing the services and exposing the APIs to the user, who might access the fabric APIs through any of the available communication networks (even possibly through another fabric leased to the user). Figure 14.3 shows the interactions in the fabric model involving the user, site authority, and clearinghouse. Users could access the fabric through a specially designed user portal that implements and automates a number of user functions like resource discovery, reservation, configuration, and visualization. A clearinghouse (CH) arbitrates the user access to the fabric and users might be able discover the user APIs and fabric resources from the CH. The CH could potentially reside anywhere (even on the same machine as that of the SA) so long as it can securely communicate with the site authority. A single CH could manage access to multiple sensor fabrics and a user could seamlessly compose an application using one or multiple sensor fabrics. This aspect of the fabric model is especially suitable for the emerging layered-sensing applications [4].

Depending on the policy or on platform ability, a fabric need not in general share state or cooperate with other fabrics. A fabric designer must attend to two key aspects:

Isolation. The designer must make sure that, programs, queries, and applications from one user do not interfere with the applications of other users using the same fabric. Depending on the nature of the fabric this task might be hard, as in

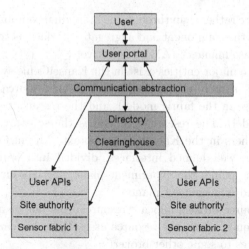

Figure 14.3 Fabric model.

the case of sensor networks. Nevertheless, coexistence of multiple user applications is especially important in fabrics, which are full-fledged deployments where production applications are likely to be running at all times. While it is important not to disrupt the data from the production application, it is also important to let users test newer versions of the production application, short high-value applications, and other applications

Sharing. In some fabrics, the designer might want to allow user applications to interact, while providing isolation where necessary. A classic example is an urban security network, where a federal agency might want to access data generated by a state agency. But sometimes the trust relationships between two users is not straightforward, such as when users who do not trust each other might want to interact for achieving a mutually beneficial goal – like that of a peer-to-peer file sharing service in the Internet. The challenge of fabric designers is to allow dynamic trust relationships to enable such interaction, without compromising the security of user applications or the fabric itself.

14.3.2   KanseiGenie architecture

KanseiGenie is designed for a multi-substrate sensor testbed. Each of these substrates (also called arrays) can contain a number of sensor motes of the same type. This physical architecture is abstracted by a software architecture comprised of components and aggregates. Each sensor device is represented as a component that defines a uniform set of interfaces for managing that sensor device. An aggregate contains a set of components of the same type and provides control over the set. It should provide at least the set of interfaces provided by the contained components and possibly other internal APIs needed for inter-component interactions. Given the collaborative nature of WSN applications, we believe that most users will interact with a sensor fabric through the aggregate interface rather than through the individual component interfaces. We henceforth denote the component and aggregate interface as component manager (CM) and aggregate manager (AM), respectively.

Three major entities exist in the KanseiGenie system (analogous to the fabric model), namely the site authority (SA), resource broker (RB) (aka the clearinghouse in the fabric model), and the reseacher portal (RP) (user in the fabric model). The researcher portal is the software component representing the researchers in the KanseiGenie context. All entities interact with one another through well-defined interfaces, divided into four logical planes, namely the resource plane, the experiment plane, the measurement plane, and the operation and management plane.

Resource brokers allocate resource from one or more site authorities and process requests from the resource users. A broker can also delegate a subset of its resources to some other broker(s).

Given that each sensor array is an aggregate, the KanseiGenie SA is conceptually an aggregate of aggregates managers (AAM) that provides access to all the arrays. The AAM provides an AM interface for each sensor array through parameterization. Externally, the AAM (i) administrates the usage of the resource provided by the site according to local resource management policies, (ii) provides the interface through which the SA advertises its shared resource to one or more authenticated brokers, and (iii) provides a programming interface through which a researcher (using RP) can schedule, configure, deploy, monitor, and analyze their experiments. Internally, the AAM provides mechanisms for inter-aggregate communications and coordination.

The RP is the software that interacts with the SA to run experiments on behalf of the researcher. It contains a suite of tools that simplifies the life cycle of an experiment from resource reservation to experiment cleanup. The RP could provide a GUI, command line or a raw programming interface depending on the targeted user. A GUI user interface will be the most user-friendly and the easiest to use, though the kinds of interactions might be limited in their functionality because of its graphical nature. Command line and programming interface would be for more experienced users that need more control and feasibility than what a GUI can offer. For example, a researcher can write scripts and programs to run a batch of experiments using the provided command-line – something which a GUI may not provide.

All these operations are based on trust relationships, and all entities must implement appropriate authentication and authorization mechanisms to support the establishment of trust. For example, the SA may either directly authorize and authenticate resource users, or delegate such authorization and authentication to other trusted third-party brokers. To invoke the AAM functions, a user will need to present forms of authentication and authorization during the interaction with the AAM. Since all AM interfaces in the AAM are the same except for the parameter used to instantiate the AM interface for a specific array, it thus suffices to describe just one AM interface to understand all the functionality of the AAM. The APIs provided by an AM are organized into the four planes mentioned in Section 14.2.

14.3.3   GENI extension to KanseiGenie

Figure 14.4 illustrates the KanseiGenie architecture in the GENI context. The KanseiGenie SA implements the interface for the aggregates. The SA interacts with the clearinghouse for resource publishing while the researcher interacts with the clearinghouse to reserve resources. The researcher also interacts with SA to redeem their reservations and run experiments. The SA's interface is implemented as Web-service for extensibility and compatibility reasons.

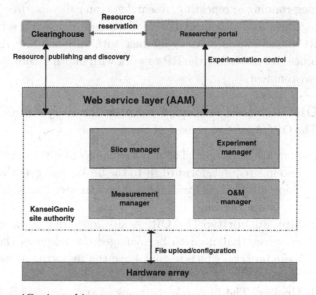

Figure 14.4 KanseiGenie architecture.

KanseiGenie provides a Web-based researcher portal implementation for easy access to its resource. A downloadable IDE-software (much like Eclipse) version of the RP is under development and will be available for future users. The researcher portal provides the most common interaction patterns with the KanseiGenie substrate so that most researchers do not need to implement their client-side software to schedule experiments in KanseiGenie.

KanseiGenie implements the clearinghouse utilizing the ORCA resource management framework [7]. Utilizing a third-party framework is motivated by the need of federated experimentation in the GENI. Although the interface for each entity is well defined, using a common and stable resource management framework helps resolve many incompatibility issues and helps to decouple the concurrent evolutions of the SA and the broker.

14.3.4   Implementation of KanseiGenie

The KanseiGenie software suite consist of the three entities: the researcher portal software, the Orca-based resource management system, and the KanseiGenie site authority.

14.3.4.1 Researcher portal (RP)

The researcher portal is implemented using the PHP programming language. The PHP Web front-end interacts with the AAM through a Web service layer. Such decoupling enables concurrent developments and evolvement of the RP and SA. The Web-based RP abstracts the most common interactions with the testbed, that of one-time, short experiments. To conduct experiments that are long-running or repeating, researchers can gain more fine-grained control by writing their own programs based on KanseiGenie Web service APIs. The current RP portal allows users to interact with only one Site Authority. We are in the process of extending the RP to work with multiple Site Authorities in a federated environment.

14.3.4.2 ORCA-based resource management system

The ORCA system consist of three actors.

1. **Service manager.** The service manager interacts with the user and gets the resource request, forwards it to the broker and gets the lease for the resources. Once a lease is received, the service manager forwards it to the Site Authority to redeem the lease.
2. **Site authority.** The ORCA site authority keeps an inventory of all the resources that need to be managed. It delegates these resources to one or more brokers, which in turn lease the resources to users through the Service Manager.
3. **Broker.** The broker keeps track of the resources delegated by various SAs. It receives the resource request from the SM and if the resource is free, it leases the resource to the SM. A number of different allocation policies can be implemented using a policy plug-in.

To integrate ORCA for resource management, we modified the ORCA service manager to include an XML-RPC server that receives the resource requests from the RP. Similarly the ORCA SA was suitably modified to make Web service calls to the KanseiGenie AAM for experiment setup and teardown. Figure 14.5 shows this integration architecture.

Shared resource and trust relationships among the ORCA broker, SA, and RP are configured beforehand within ORCA.

14.3.4.3 KanseiGenie site authority

The KanseiGenie site authority in turn has three components, the Web service layer (WSL), the KanseiGenie aggregate of aggregate manager (KG AAM), and

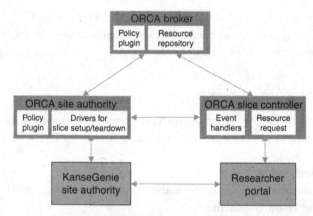

Figure 14.5 KanseiGenie integration with ORCA.

the individual component managers (CM) for each device type supported by KanseiGenie.

Web service layer. The WSL acts as a single point external interface for the KanseiGenie Site Authority. The WSL provides a programmatic, standards-based interface to the AAM. We utilize the Enterprise Java Bean framework to wrap the four functional GENI planes. Each of the manager interfaces are implemented as a SessionBean. We utilize the JBoss application server as the container for the EJBs partly because JBoss provides mechanisms for users to conveniently expose the interface of SessionBeans as Web services. Other reasons that motivated us to choose JBoss include the good support from the community, wide adoption, open-source licence, and stability.

KanseiGenie aggregate of aggregate manager. The KG AAM implements the APIs for Kansei substrates. It keeps track of the overall resources available at a site, monitoring their health and their allocation to users. It also keeps track of individual experiments, their status, deployment, and clean up. The status of the resources and experiments are stored in a MySQL database. A scheduler daemon (written in the Perl) accomplishes experiment deployment, clean up, and retrieval of log files (after an experiment is complete) using the sub-strate manager of the individual substrates. Most of the generic APIs discussed in Section 14.2 are supported by the KanseiGenie AAM.

Component manager. Each of the devices in Kansei has its own manager (but for some primitive devices – such as motes – the manager might be implemented on other more capable devices). The component manager implements the same APIs as that of the KG AAM and is responsible for executing the APIs on the individual devices. The logical collection of all the managers of devices belonging to the same substrate form as the aggregate manager of that substrate. Currently KG supports Stargates [9], TelosBs [11], and XSMs [12]. The AMs for Imote2 [2] and SunSpots [10] are under development. The TelosBs and XSMs being mote-class devices without a presistent operating system, their Managers are implemented on the Stargates (in Perl). The CMs are implemented using Perl. A number of tools for user programming the motes and interacting with them are written in the Python and C programming languages.

14.3.5   KanseiGenie federation

The KanseiGenie software architecture is designed for sensor network federation. We discuss below the use cases and key issues in KanseiGenie federation.

14.3.5.1   Use cases

Federated WSN fabrics enable regression testing, multi-array experimentation, and resource sharing.

Regression testing.   WSNs introduce complex dynamics and uncertainties in aspects such as wireless communication, sensing, and system reliability. Yet it is desirable to have predictable system behavior especially when sensors are used to support mission-critical tasks such as safety control in alternative energy power grids. That is, it is desirable for system services and applications to behave according to certain specifications in a wide range of systems and environmental settings.

Existing measurement studies in WSNs are mostly based on a single fabric such as Kansei and NetEye, which only represents a single system and environmental setting. To understand the predictability and sensitivity of WSN system services and applications, however, we need to evaluate their behavior in different systems and environmental settings. Seamlessly integrating different WSN fabrics together to provide a federated measurement infrastructure will enable us to experimentally evaluate system and application behavior in a wide range of systems and environmental settings, thus providing an essential tool for evaluating the predictability and sensitivity of WSN solutions.

Multi-array experimentation.   Next-generation WSN applications are expected to involve different WSN substrates. These include the traditional, resource-constrained WSN platforms as well as the emerging, resource-rich WSN platforms such as the Imote2 and the Stargate. Federation also enables experimentation with multiple WSN fabrics at the same time, thus enabling evaluation of WSN system services and applications that involve multiple WSN substrates.

Resource sharing.   Federating different WSN fabrics together also enables sharing resources between different organizations. Resource sharing can enable system-wide optimization of resource usage, which will improve both resource utilization and user experience. Resource sharing also helps enable experiment predictability, predicting the behavior of a protocol or application in a target network based on its behavior in testbed. This is because federated WSN fabrics improve the probability of finding a network similar to the target network of a field deployment. Resource sharing is also expected to expedite the evolution of WSN applications by enabling more users to access the heterogeneous WSN fabrics at different organizations.

14.3.5.2 Key issues

To enable secure, effective WSN federation, we need to address issues in clearinghouse architecture, access control, resource discovery and allocation, as well as network stitching.

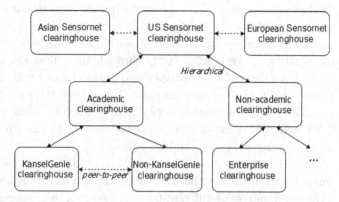

Figure 14.6 Clearinghouse architecture in the KanseiGenie federation.

Clearinghouse architecture. For effective resource management among federated but autonomous WSN fabrics, we expect the clearinghouses to be organized in a hierarchical manner. As shown in Figure 14.6, for instance, several WSNs in the USA may form one cluster, and this cluster interacts with a cluster of European WSNs. The exact form of the clearinghouse hierarchy depends on factors such as trust relations, resource management policies, and information consistency requirements. We are currently working with our domestic and international partners, including Los Alamos National Lab and Indian Institute of Science, to develop the conceptual framework for the federation hierarchy as well as the implementation mechanisms for supporting flexible adaptation of federation hierarchy as policy and technology evolve. This hierarchical federation architecture will be reflected by a hierarchy of clearinghouses in charge of the resource management within some part of the hierarchy.

Within the hierarchy of federated WSN fabrics, we also expect peer-to-peer interaction between different fabrics. For instance, the WSNs from US academia may form a sub-cluster of the US cluster, and these fabrics may share resources in a peer-to-peer manner without involving the top-level clearinghouse of the US cluster.

Access control. One basic security requirement of GENI is to only allow legitimate users to access their authorized resources. Therefore, one basic element of federated WSN fabrics is the infrastructure for authentication, authorization, and access control. For authentication, we can use the cryptographic authentication methods via public key cryptography, and, in this context, adopt the hybrid trust model where the monopolistic and anarchic models are tightly integrated. In this hybrid trust model, each fabric maintains its own certificate authority (CA) which manages the public keys for users directly associated with the fabric, and the distributed, PGP-style, anarchic trust model is used among different fabrics to facilitate flexible policies on federation and trust management. (Note that a fabric may also maintain an internal hierarchy of CAs depending on its local structure and policy.) Using this public key infrastructure (PKI), a fabric CA can vouch for the public key of every entity (e.g., a user or a software service such as a component manager) within its fabric, and every entity can verify the identity of every other entity through the public key certified by the local fabric-CA. The chain of trust formed across fabrics will enable authentication across fabrics. For instance, an entity B can trust the public key Key-A certified by a CA CA-A for another entity A in a different fabric Fabric-A, if the local CA CA-B of B trusts certificates issued by CA-A. The PKI will also be used in secure resource discovery and allocation.

For each legitimate user, the slice manager (SM) at its associated clearinghouse generates the necessary slice credential through which the user can request tickets to access resources of different fabrics. Based on their access control policies, the component managers (CMs) of the involved fabrics interact with the user to issue the ticket and later to allocate the related resources according to the authorized rights carried in the ticket.

Network stitching. After an experiment has acquired the resources from the federated KanseiGenie fabrics, one important task is to stitch together slices from different WSN fabrics. To this end, the KanseiGenie research portal will coordinate with one or multiple clearinghouses to get tickets for accessing the allocated resources. Then the research portal will coordinate with the involved site authorities to set up the communication channels between slices in different fabrics, after which the experiment will have access to the connected slices within the KanseiGenie federation.

14.4   KanseiGenie customization and usage

14.4.1   How to customize KanseiGenie

KanseiGenie SA software uses a distributed architecture with hierarchical design. The KanseiGenie AAM implements the APIs of the four functional planes. The AAM redelegates the API calls to a specific AM depending on what substrate is being used for an experiment. The APIs of the AM are well defined and can be implemented on any platform/technology.

KanseiGenie currently supports the Stargate, XSM, and TelosB sensor platforms. The AMs for Imote2 and SunSpots are under development. The architecture of KanseiGenie supports both a flat or a hierarchical arrangement of substrates and using platform neutral language like Perl makes its customization quite easy. To use KanseiGenie to manage sensor substrates already supported, an administrator only needs to populate the MySQL database tables regarding the number and the physical topology of substrates. Also a testbed administrator can configure KanseiGenie for any combination of the already supported sensor platforms.

Customization of the KanseiGenie to a new sensor substrate involves three steps. The first step is implementing the AM APIs for that substrate (either on those devices directly or another substrate, which can in turn control the new one being added). The second is updating the AAM resource and policy database about the new substrates topology and resources. The third is modifying and/or adding new GUI interfaces to the RP to support the configuration parameters for the new platform.

14.4.2   Vertical APIs and their role in customization

The KanseiGenie architecture supports additional APIs apart from the standardized (generic) APIs of the four functional planes. Basing the architecture on a Service Oriented Architecture (SOA) and dividing them into vertical (domain specific) and horizontal (generic) APIs provides a basis for customizing for different substrates. The vertical APIs is a convenient way of providing for specific application domains, while at the same time standardizing them.

Two of the customizations of the KanseiGenie architecture are the Peoplenet and the Search API for intrusion detection system, which we describe below.

PeopleNet. PeopleNet [8] is a mobility testbed at The Ohio State University composed of about 35 cell phones and the Dreese building fabric. The Dreese building fabric provides services like the elevator localization, building temperature control, and light monitoring, along with a fabric for generic experimentation. The PeopleNet cell phone fabric supports APIs for instant messaging and buddy search services.

Search API. Search API [14] consist of an interface for an intrusion detection network. The interface lets the user query the network for target objects such as single shot queries or persistent queries. More over, the queries can be about real physical targets or logical data objects in the network and they can be confined to a geographical area in the network or the user can query the entire network.

The above two examples provide insight into the customization of the fabric architecture and how it can support multiple dissimilar fabrics. The flexibility of the architecture is enabled by the separation of the horizontal APIs from the vertical APIs and because we leave the implementation of APIs to the specific Aggregate Managers, the same API can be implemented differently by different substrates.

14.4.3   KanseiGenie usage step-by-step run-through

The KanseiGenie Researcher Portal is designed to be an intuitive and easy way for a user to access the testbed resources. Here we give a short step-by-step run-through of a typical usage scenario.

1. **Get access.** The first thing a user needs to do is to get access to the testbed resources. A user can do this by contacting the KanseiGenie administrator (say by e-mail) or by getting a login from the GENI clearinghouse.
2. **Create a slice.** A user will first create one or more slices (if a user wants to run more than one experiment concurrently she will need more than one slice). A slice represents the user inside the fabric. It is a logical container for the resources of the user.
3. **Choose the substrate.** The portal displays all the different substrates available in the KanseiGenie federation. The user needs to decide which substrate she is going to use for this experiment.
4. **Upload the executable.** The user will next prepare the executable and/or scripts that needs to be tested for the particular substrate.
5. **Create the resource list.** A user might want to test her program on a particular topology. The portal provides a service that lets the user create any topology they want from the available nodes from a particular substrate.
6. **Get lease.** The user will next have to get a lease for the resources she wants to use. The portal interacts with the ORCA resource management system and gets the lease for the resources.
7. **Configure experiment.** Once the user has the lease, she needs to configure the experiment she wants to run on these resources. She will choose parameters such as how long the experiment should run, which executable should be run, what logging and exfiltration service to use, what injection is required, etc.
8. **Run experiment.** Finally, once the configuration is done, the user can start the experiment from the experiment dashboard.
9. **Interact with experiment.** The portal also provides services through which a user can interact with the experiment, while it is running. A user can inject prerecorded sensor data into the slice, view logs in real time, visualize the network in real time, view health data of resources. To enable most of these services, the user should specify/select the services in the configuration step.
10. **Download results.** Once an experiment is complete, the results and logs from the experiment are available for download from the portal.

14.5   Evolving research issues in next-generation networks

In this section we will look at some of the emerging research issues in the fabric model.

14.5.1   Resource specifications for sensor fabrics

The sensor fabric resource specifications act as the language that all the entities in the architecture understand. It is important for the designers to come up with an ontology that is detailed enough for the users of the domain to take advantage of the fabric services and features, and broad enough to enable interaction and joint experimentation with other programmable fabrics (such as other wireless networks and core networks).

Much of the complexity of sensor networks need to be embedded in resource specifications (RSpecs). Resource specifications will also be an extensible part of the federated KanseiGenie interface. As new resources and capabilities are added, these specifications will inevitably need to be extended. We expect the extensions to leverage a hierarchical name space. This will allow new communities that federate with KanseiGenie to extend the resource specification within their own partition of the name space. Components that offer specialized resources will similarly extend the resource specification in their own name space. Additionally, we need to consider the granularity of resource specification, which decides the level of detail in the resource description. In federated resource management, there is no unique solution and the implementation strategy is subject to both technical and administrative constraints. For instance, whether and how much information about resource properties should be maintained by clearinghouses will depend on the trust relations among the entities involved and may be encoded in resource specifications at different levels of granularity.

To enable reliability and predictability in experimentation, resource specification also needs to characterize precisely the reliability and predictability properties of WSN testbeds, including the external interference from 802.11 networks, the stability of link properties, and failure characteristics of nodes in a testbed, so that an experiment will also use reliability- and predictability-oriented specifications to define its requirements on the allocated resources.

For the same experiment there may be different ways of specifying the actual resources needed. For an experiment requiring two TelosB motes and a link of 90 percent reliability connecting these two motes, for instance, we may define the resource specification to request two motes six meters away with the necessary power level for ensuring a 90 percent link reliability between these two motes, or we may define the resource specification to request any two motes connected by a link of 90 percent reliability. Both methods will give the user the desired resources, but the second method will allow for more flexibility in resource allocation and thus can improve overall system performance.

14.5.2   Resource discovery

For federated resource management, different clearinghouses need to share resource information with one another according to their local resource sharing policies. The two basic models of resource discovery are the push and pull models. In the push model, a clearinghouse periodically announces to its peering or upper-level clearinghouses the available resources at its associated fabrics that can be shared. In the pull model, a clearinghouse requests from its peers or upper-level clearinghouses their latest resource availability. We expect the pull model to be mainly used in an on-demand manner when a clearinghouse cannot find enough resources to satisfy a user request. Note that this interaction between clearinghouses also needs to be authenticated using, for example, the PKI discussed earlier.

14.5.3   Resource allocation

We expect that federated WSN infrastructures will support a large number of users. Hence effective experiment scheduling will be critical in ensuring high system utilization and in improving user experience. Unlike scheduling computational tasks (e.g., in Grid computing), scheduling wireless experiments introduces unique challenges due to the nature of wireless networking. For instance, the need for considering physical spatial distributions of resources such as sensor nodes affects how we should schedule experiments. To give an example, let's consider two fabrics S1 and S2 where both fabrics have 100 TelosB motes, but the motes are deployed as a 10×10 grid in S1 whereas the motes are deployed as a 5×20 grid in S2. Now suppose that we have two jobs J1 and J2, where J1 arrives earlier than J2, and J1 and J2 request a 5×10 and a 5×12 grid, respectively. If we only care for the number but not the spatial distribution of the requested motes, whether J1 is scheduled to run on S1 and S2 does not affect the schedulability of J2 while J1 is running. But spatial distribution of nodes does matter in wireless networks, and allocating J1 to S2 will prevent J2 from running concurrently, whereas allocating J1 to S1 will allow the concurrent execution of J2 by allocating it to S2, improving system utilization and reducing waiting time.

Wireless experiments in federated GENI may well use resources from multiple fabrics in a concurrent and/or evolutional manner. Scheduling concurrently used resources from multiple fabrics is similar to scheduling resources within a single fabric even though we may need to consider the interconnections between fabrics. For experiments that use multiple fabrics in an evolutional manner, we can schedule their resource usage based on techniques such as "task clustering," where sequentially requested resources are clustered together and each cluster of requests is assigned to the same fabric to reduce coordination overhead and to maximize resource utilization. To reduce deadlock and contention, we need to develop mechanisms so that an experiment can choose to inform the clearinghouse scheduler of its temporal resource requirement so that subsequent experiments do not use resources that may block previously scheduled experiments.

14.5.4   Data as resource

One consequence of the fabric model is that the network is hidden behind a collection of interfaces and as long as the interfaces are standardized and known, a user can programmatically access it. In other words, it does not matter if the interface is implemented by a sensor network or by a single PC. Thus, dataHubs – which are databases that can annotate and store results of experiments and replay the data for similar future queries – can now be viewed as a sensor resource. Alternately, a sensor network can be viewed as a source for a data stream and the user as a data transformation program. Under this unified view a dataHub which can interpret queries and transform the stored data accordingly can fake a sensor fabric. Hence, in the fabric model, data (properly annotated and qualified) and sensing resources are interchangeable and provides for interesting hybrid experimentation scenarios.

The architecture provides much research opportunities and challenges, as a number of questions need to answered before the architecture can be used beyond the most simplistic scenarios. Challenges include the following:

- How to automatically annotate and tag data coming from sensor networks to create a credible dataHub?
- It is common for the same experiment to produce multiple similar datasets in wireless networks. How does a user decide which dataset to use as representative of an experiment?
- Does the RSpec ontology need to be extended to represent data?
- What range of queries can be answered with the current data? Should data be preprocessed to decided acceptable querries?

14.5.5   Network virtualization

The fundamental aim of the fabric architecture is to virtualize and globalize the resources in a sensor network, so that in principle a user anywhere in the world can request, reserve, and use the resources. However, the more resource is virtualized, the less control the user has over it. Thus there is a tradeoff between the level of access and virtualization. The challenge for modern network designers is to provide as much control (as low in the stack as possible) to the users, while retaining the ability to safely recover the resource and also making sure the resource might be shareable.

In a fabric multiple researchers will run their experiments concurrently on different subsets of an array of sensors of the same type. Usually, sensors are densely deployed over space. Such density provides the means for different experimenters to share the same geographical space and sensor array to conduct concurrent experiments that are subject to very similar, if not statistically identical, physical phenomena. In such environments, interference is inherent between users due to the broadcast nature of the wireless communications; its effect is more prominent when the communicating devices are close to one another. The virtualization of wireless networks imposes a further challenge for the sensor fabric providers to ensure isolation between concurrently running experiments. Such interference isolation is usually achieved by careful frequency or time slot allocations. However, these solutions are quite primitive in nature and do not provide optimum network utilization. Even more important, these solutions are not suitable for sensing infrastructures where multiple applications from different users need to be run concurrently in a production mode.

Our recent research in this area using statistical multiplexing as the basis of virtualization is promising to provide better solutions, enabling much better network utilization and external noise isolation.

14.6   Conclusion

The KanseiGeni architecture for wireless sensor network fabrics supports a wide range of experimentation by enabling slicing, virtualization, and federation among diverse networks. Not restricted to just sensor network fabrics, the architecture can be readily introduced into a more general programmable network such as GENI. The architectural model centers on network fabrics rather than on nodes, resulting in a level of abstraction that supports a wide range of services and enables applications that were not anticipated by the fabric designer. KanseiGeni can be customized for domain-specific applications via vertical APIs, allowing researchers to add a rich functionality that goes well beyond the basic set of services mentioned here.

There are many open areas, including resource specification, discovery and allocation, as well as the question of data as a resource.

References

[1] Global environment for network innovation. www.geni.net.
[2] Intelmote2: High-performance wireless sensor network node. http://docs.tinyos.net/index.php/Imote2.
[3] Kansei wireless sensor testbed. kansei.cse.ohio-state.edu.
[4] Layered sensing. www.wpafb.af.mil/shared/media/document/AFD-080820-005.pdf.
[5] Nested c: A language for embedded sensors. www.tinyos.net.
[6] NetEye wireless sensor testbed. http://neteye.cs.wayne.edu.
[7] Open resource control architecture. https://geni-orca.renci.org/trac/wiki/.

[8] Peoplenet mobility testbed. `http://peoplenet.cse.ohio-state.edu`.

[9] Stargate gateway devices.   `http://blog.xbow.com/xblog/stargate_xscale_platform/`.

[10] Sunspots: A java based sensor mote. `www.sunspotworld.com/`.

[11] Telosb sensor motes. `http://blog.xbow.com/xblog/telosb/`.

[12] Xsm: Xscale sensor motes. `www.xbow.com/Products/Product_pdf_files/Wireless_pdf/MSP410CA_Da%tasheet.pdf`.

[13] Arora A., Gouda M., Hallstrom J. O., Herman T., Leal W. M., Sridhar N. (2007). A state-based language for sensor-actuator networks. *SIGBED Rev.* **4**, 3, 25–30.

[14] Kulathamani V., Sridharan M., Arora A., Ramnath R. (2008). Weave: An architecture for tailoring urban sensing applications across multiple sensor fabrics. MODUS, International Workshop on Mobile Devices and Urban Sensing.

Part IV

Theory and models

15 Theories for buffering and scheduling in Internet switches

Damon Wischik

University College London, UK

15.1   Introduction

In this chapter we argue that future high-speed switches should have buffers that are much smaller than those used today. We present recent work in queueing theory that will be needed for the design of such switches.

There are two main benefits of small buffers. First, small buffers means very little queueing delay or jitter, which means better quality of service for interactive traffic. Second, small buffers make it possible to design new and faster types of switches. One example is a switch-on-a-chip, in which a single piece of silicon handles both switching and buffering, such as that proposed in [7]; this alleviates the communication bottleneck between the two functions. Another example is an all-optical packet switch, in which optical delay lines are used to emulate a buffer [3]. These two examples are not practicable with large buffers.

Buffers cannot be made arbitrarily small. The reason we have buffers in the first place is to be able to absorb fluctuations in traffic without dropping packets. There are two types of fluctuations to consider: fluctuations due to end-to-end congestion control mechanisms, most notably TCP; and fluctuations due to the inherent randomness of chance alignments of packets.

In Section 15.2 we describe queueing theory which takes account of the interaction between a queue and TCP's end-to-end congestion control. The Transmission Control Protocol tries to take up all available capacity on a path, and in particular it tries to fill the bottleneck buffer. Yet it is self-evidently absurd to build a large buffer just so that TCP can maintain a large queue. The analysis in Section 15.2 shows that a small buffer can still allow TCP to get high utilization, by providing appropriate feedback about congestion. As part of this analysis we explain how TCP synchronization arises, and also the consequences of bursty TCP traffic.

In Section 15.3 we describe queueing theory for analyzing the impact of chance fluctuations in traffic. This matter has been the domain of queueing theory for

Next-Generation Internet Architectures and Protocols, ed. Byrav Ramamurthy, George Rouskas, and Krishna M. Sivalingam. Published by Cambridge University Press. © Cambridge University Press 2011.

decades, but what is new is the development of theory that takes account of the architecture of the switch and its scheduling algorithm. For example, in an input-queued switch, there may be several inputs that all have packets to send to the same output; giving service to one will deny service to the others. Section 15.3 gives an analysis of the interplay between queueing and scheduling.

In Section 15.4 we synthesize the lessons that may be drawn from these two branches of theory, and propose an architecture for the future Internet's packet-level design. We suggest a form of active queue management that is suited to high-speed switches. We describe the sorts of performance analysis that are appropriate for new switch architectures. We discuss the design of replacements for TCP, and how they can be "kinder" to the network.

Before continuing, we should draw attention to the fact that these theories are all based on statistical regularities that emerge in large systems. They are offered as models that might be applied to core Internet routers, but not to small-scale switches. More experimental work is needed to determine the extent of applicability.

15.2   Buffer sizing and end-to-end congestion control

One of the functions of the buffer in an Internet router is to keep a reserve of packets, so that the link rarely goes idle. This relies on there being enough traffic to keep the link busy.

Router vendors today typically use a rule of thumb for buffer sizing: they ensure that routers provide at least one round-trip time's worth of buffering, often taken to be around 250 ms. There is a simple heuristic argument which justifies this, when the traffic consists of a single TCP flow. The first serious challenge to the rule of thumb came in 2004 from Appenzeller, Keslassy, and McKeown [1], who pointed out that core routers serve many thousands of flows, and that large aggregates behave quite differently to single flows. Their insight prompted the theoretical work described in this section, which is taken from [9]. For a recent perspective on different approaches see [15].

In Section 15.2.1 we give four heuristic arguments about buffer sizing. The first is a justification for the rule of thumb in use today; the others are a simple introduction to the in-depth model in this section. In section 15.2.2 we derive a traffic model for the aggregate of many TCP flows sharing a single bottleneck link. In Section 15.2.3 we explore the behavior of this model: we present a rule for buffer sizing, and we explain the cause of synchronization between TCP flows. In Section 15.2.4 we discuss the impact of burstiness at sub-*RTT* timescales. In the conclusion, Section 15.4, we will draw design lessons about how packet-level burstiness should be handled in the future Internet, including the design of replacements for TCP.

15.2.1   Four heuristic arguments about buffer sizing

Heuristic 1

The Transmission Control Protocol controls the number of packets in flight in the network by limiting the congestion window, i.e., the number of packets that the source has sent but for which it has not yet received an acknowledgement. It increases its congestion window linearly until it detects a packet drop, whereupon it cuts the window by half.

Consider a single TCP flow using a single bottleneck link. Here is a simple heuristic calculation of how big the buffer needs to be in order to prevent the link going idle. Let B be the buffer size, let C be the service rate, and let PT be the round-trip propagation delay, i.e., the round-trip time excluding any queueing delay. When the buffer is full, immediately before a drop, the congestion window will be $w = B + CPT$, where B is the number of packets queued and CPT is the number of packets in flight. Immediately after a drop the congestion window is cut to $w' = (B + CPT)/2$. We want $w' \geq CPT$ so that there are still enough packets in flight to keep the link busy. Therefore we need $B \geq CPT$. This simple heuristic can be rigorized to take proper account of how TCP controls its window size; see for example [1]. An experimental study from 1994 [14] confirmed that this heuristic applies for up to 8 TCP flows on a 40 Mbit/s link.

At 40 Mbit/s the rule of thumb recommends a buffer of 10 Mbit; today 10 Gbit/s links are common, and the rule of thumb recommends 2.5 Gbit of buffering.

Heuristic 2

Here is a crude queueing model that suggests that tiny buffers are sufficient. Consider a queue with arrival rate λ packets/s, and service rate μ packets/s, and suppose that packets are all the same size, that the buffer is B packets, that arrivals are a Poisson process, and that service times are exponential, i.e. that this is an $M/M/1/B$ queue. Classic queueing theory says that the packet drop probability is

$$\frac{(1-\rho)\rho^B}{1-\rho^{B+1}}, \quad \text{where} \quad \rho = \lambda/\mu.$$

A buffer of 80 packets $\approx$ 1 Mbit should be sufficient to achieve packet drop probability of less than 0.5 percent even at 98 percent utilization. This means 24 ms of buffering for a 40 Mbit/s link, and 0.1 ms of buffering on a 10 Gbit/s link.

The striking feature of the equation is that packet drop probability depends only on utilization ρ, rather than on absolute link speed μ, so a 10 Gbit/s link has exactly the same buffer requirement (measured in Mbit) as a 10 Mbit/s link.

This heuristic neglects the fact that TCP has closed-loop congestion control. The next two heuristics show how it may be remedied.

Heuristic 3

Consider N TCP flows with common round-trip time RTT. Suppose these flows share a single bottleneck link with packet drop probability p and service rate C. The TCP throughput equation (derived in Section 15.2.2 below) says that the average throughput is

$$x = \frac{\sqrt{2}}{RTT\sqrt{p}}.$$

We know however that TCP seeks to fully utilize any available capacity; if the buffer is large enough to allow near full utilization then

$$Nx \approx C.$$

Now, round-trip time consists of $RTT = PT + QT$ where PT is propagation delay and QT is queueing delay. Rearranging,

$$p = \frac{2}{(C/N)^2(PT + QT)^2}.$$

The larger the buffer size, the larger QT, and the smaller the packet drop probability. In particular, heuristic 1 recommends $B = CPT$, hence $QT = B/C = PT$, whereas heuristic 2 recommends $QT \approx 0$, hence the packet drop probability will be four times higher under heuristic 2.

Note however that both recommendations keep the link nearly fully utilized. In the model we have described here, large buffers act as delay pipes, so TCP flows experience larger delay and TCP is coerced into being less aggressive, which keeps packet drop probability low. We believe that in the future Internet, routers should not introduce artificial delay merely to slow down end-systems. End-systems can cope with the packet loss by retransmission or forward error correction, but they can never undo the damage of latency.

Heuristic 4

Another heuristic that has received a great deal of attention [6] comes from assuming that TCP flows are limited by a maximum window size imposed by the operating system at the end-node. Specifically, suppose there are N flows, with common round-trip time RTT, and suppose each flow's window w is limited to $w \leq w_{max}$. Suppose further that the link in question has service rate $C > Nw_{max}/RTT$. If there were no packet drops in the network, each flow would send at a steady rate of $x_{max} = w_{max}/RTT$. But every packet drop causes the window size to drop by $w_{max}/2$, whereafter the window size increases by 1 packet per RTT until it reaches $w = w_{max}$ again; the total number of packets that could have been sent but weren't is $w_{max}^2/2$. To see the impact this has on utilization at a link with packet drop probability p, consider the total number of packets sent over some time period Δ: there could have been $Nx_{max}\Delta$ packets sent, but in fact there were around $Nx_{max}\Delta p$ packet drops, hence there were only $Nx_{max}\Delta(1 - pw_{max}^2/2)$ packets sent. In order that this link should not impair utilization too much, bearing in mind that it is not after all the bottleneck, we might require $p \leq 2\varepsilon/w_{\max}^2$ for some suitably small ε. Based on heuristic 2, if ρ is the actual traffic intensity and $\rho < \rho_{\max} = N x_{\max}/C < 1$ then

$$p = \frac{(1 - \rho)\rho^B}{1 - \rho^{B+1}} \leq \rho^B \leq \rho_{\max}^B,$$

which suggests using a buffer of size

$$B \geq \frac{\log(2\varepsilon/w_{\max}^2)}{\log \rho_{\max}}.$$

We will not pursue this analysis further, since the underlying assumption (that flows are window-limited at the end-system) is not a reliable basis for designing the next-generation Internet.

15.2.2   Fluid traffic model and queue model

In order to make queueing problems tractable, one must typically consider a limiting sequence of queueing systems and apply some sort of probabilistic limit theorem. Here we shall consider a sequence of queueing systems, indexed by N, where the Nth system is a bottleneck link fed by N TCP flows, with common round-trip time RTT, and where the link speed is NC. We are aiming here to derive a model for TCP congestion control: if we took the link speed to be any smaller, e.g., $\sqrt{N}C$, then the flows would mostly be in timeout, and if we took it to be any larger then we would need some high-speed TCP modification to make use of all the capacity.

TCP model
Suppose all the N TCP flows are subject to a common packet drop probability. Let the average congestion window size of all N flows at time t be $w(t)$, measured in packets. Let $x(t) = w(t)/RTT$; this is the average transmission rate at time t, averaged across all the flows. Let the packet drop probability experienced by packets sent at time t be $p(t)$. Then a reasonable approximation for how the total window size $Nw(t)$ evolves over a short interval of time is

$$Nw(t + \delta) \approx Nw(t) + \delta N x(t - RTT)(1 - p(t - RTT))\frac{1}{w(t)} - \delta N x(t)p(t)\frac{w(t)}{2}.$$

$$(15.1)$$

The term $Nx(t - RTT)(1 - p(t - RTT))$ is the total rate of sending packets at time $t - RTT$, times the probability that the packet was not dropped, i.e., it is the rate at which acknowledgement packets (ACKs) are received at time t. Multiplying by δ gives the total number of ACKs that are received in $[t, t + \delta]$. TCP specifies that a flow should increase its congestion window w by $1/w$ on receipt of an ACK. Each flow has its own congestion window size, but we shall approximate each of them by the average window size $w(t)$. Thus the middle term on the right-hand side of (15.1) is the total amount by which the flows increase their windows in $[t, t + \delta]$. The final term in the equation is obtained by similar reasoning: it is the total amount by which congestion windows are decreased due to packet drops in that interval, based on TCP's rule that the congestion window should be decreased by $w/2$ on detection of a dropped packet. The equation may be further approximated by

$$\frac{d\,w(t)}{dt} \approx \frac{1}{RTT} - \frac{w(t)}{2}\Big[x(t - RTT)p(t - RTT)\Big], \qquad (15.2)$$

an equation which has been widely used in the literature on TCP modeling.

There are many approximations behind this equation: all flows are in congestion avoidance; all packets experience the same packet loss probability; $p(t)$ is small so that $1 - p(t) \approx 1$; the average window increase may be approximated by $1/w(t)$; flows may be treated as though they detect a drop immediately when an ACK is missing; flows respond to every drop, not just once per window; queueing delay is negligible, i.e., the round-trip time is constant. Nonetheless, the equation seems to be faithful enough to predict the outcome of simulations.

Observe that if the system is stable then $w(t)$ and $p(t)$ are by definition constant, and (15.2) reduces to

$$0 = \frac{1}{RTT} - \frac{w}{2}xp \quad \Longrightarrow \quad x = \frac{\sqrt{2}}{RTT\sqrt{p}},$$

which is the classic TCP throughput equation.

Queue model for small buffers
Suppose that the N TCP flows share a common bottleneck link with link speed NC and that the buffer size is B, i.e., buffer size does not depend on N, as heuristic 2 from Section 15.2.1 proposed. Let the total arrival rate at time t be $Nx(t)$.[1] By how much might the queue size change over a short interval $[t, t + \delta]$? We will build up the answer in layers, first by considering an open-loop queue fed by Poisson arrivals of rate $Nx(t)$, then by justifying the Poisson assumption. In Section 15.2.3 we close the loop.

First, suppose the queue to be fed by Poisson arrivals of rate $Nx(t)$. We know from (15.2) that over a short interval of time $[t, t + \delta]$ the traffic intensity hardly changes. We also know from classic Markov chain analysis that an $M/D/1/B$ queue with arrival rate Nx and service rate NC has exactly the same distribution as if it had arrival rate x and service rate C; it just runs N times faster. Therefore the queue will see rapid fluctuations, busy cycles of length $O(1/N)$, and it will rapidly attain its equilibrium distribution much quicker than the timescale over which $x(t)$ changes.

---

[1] In the previous section, we let $Nx(t)$ be the total transmission rate at time t, so strictly speaking the arrival rate to the queue at time t is $x(t - T)$, where T is the time it takes packets to reach the queue. We could subtract T from all the time terms in this section, but it would make the notation messy. Instead, it is simpler to picture the queue as if it is immediately next to the source, and the return path has delay RTT.

Why did we assume that the aggregate traffic was a Poisson process? It is a standard result that the aggregate of N independent point processes converges over a timescale of length $O(1/N)$ to a Poisson process, as long as there is some minimum time-separation between the points [4]. We need to justify the assumptions of (i) independence and (ii) minimum time-separation. See [6, Appendix II of extended version] for explicit calculations. Alternatively, observe that (i) the timescale of queue fluctuations is $O(1/N)$, whereas the round-trip time is $RTT = O(1)$, so the queue has long forgotten its state by the time packets have completed one round trip, and any minor jitter or difference in round-trip times should be enough to make the flows seem independent over short timescales (though see the discussion in Section 15.4 point 3). Furthermore (ii) if the packets from each TCP flow experience some minimum serialization delay, or if they pass through some relatively slow access-point queue, then the time-separation is satisfied (though see Section 15.2.4).

In summary, we expect the equilibrium distribution of queue size to be a function of the instantaneous arrival rate $x(t)$, and as $x(t)$ changes slowly so too does the equilibrium distribution. In particular, the packet drop probabilty $p(t)$ is a function of the instantaneous arrival rate $x(t)$, and this function may be calculated using classic Markov process techniques; call it

$$p(t) = D_{C,B}(x(t)). \tag{15.3}$$

For large B a simpler expression is available:

$$p(t) \approx \max(1 - C/x(t), 0) = \begin{cases} (x(t) - C)/x(t) & \text{if } x(t) > C \\ 0 & \text{if } x(t) \le C. \end{cases}$$

The $x(t) < C$ case assumes B is large enough that the queue rarely overflows. The $x(t) > C$ case is derived from applying Little's Law to the free space at the tail of the buffer, assuming the queue rarely empties.

We have made several approximations in this derivation, the most questionable of which is the assumption that packets are sufficiently spaced in time for the Poisson limit to apply. In Section 15.2.4 we discuss how to cope with bursty traffic.

15.2.3   Queueing delay, utilization, and synchronization

We have derived a dynamical model for the system of N TCP flows sharing a bottleneck link with link speed NC and buffer size B. If $x(t)$ is the average transmission rate at time t and $p(t)$ is the packet drop probability for packets sent at time t, then

$$\frac{dx(t)}{dt} = \frac{1}{RTT^2} - \frac{x(t)}{2}\left[x(t - RTT)p(t - RTT)\right]$$
$$p(t) = D_{C,B}(x(t)). \tag{15.4}$$

Remember that there is no term for queue size because the queue size fluctuates over timescale $O(1/N)$, where N is the number of flows, and so over a short interval $[t, t + \delta]$ all we see of the queue size is a blur. The equilibrium *distribution* of queue size is the quantity that varies smoothly, as a function of $x(t)$, and it is the queue size distribution not the queue size that gives us $p(t)$.

The first step in analyzing the system is to find the fixed point, i.e., the values x and p at which the dynamical system is stable. These are

$$x = \frac{\sqrt{2}}{RTT\sqrt{p}} \quad \text{and} \quad p = D_{C,B}(x). \tag{15.5}$$

If the buffer size is large then $p = \max(1 - C/x, 0)$ and the fixed point is at $x > C$, i.e., the system is perpetually running with the queue full or near-full, so the utilization is 100 percent. If the buffer size is small, and we approximate $D_{C,B}(x)$ by $(x/C)^B$, the formula for tail queue size distribution in an $M/M/1/\infty$ queue, we get utilization $x/C = (\sqrt{2}/wnd)^{1/(1+B/2)}$, where $wnd = CRTT$ is the congestion window that each flow would have if the link were completely utilized. If $B = 250$ packets, then utilization is at least 97 percent up to $wnd = 64$ packets. It is clear that small buffers entail some loss of utilization, but it is perhaps surprising that this loss is so little.

The next step is to analyze the stability of the dynamical system. One could simulate the TCP flows at the packet level, or solve the differential equations numerically, or linearize (15.4) about the fixed point and calculate algebraically whether it is locally stable, or perform a power-series expansion about the fixed point and if the system is unstable estimate the amplitude of oscillations.

To calculate whether (15.4) is locally stable, take a first-order approximation of (15.4) about the fixed point x, guess the solution $x(t) = x + e^{\omega t}$, and solve for ω. It is locally stable if ω has negative real part. It turns out that a simple sufficient condition for local stability is

$$RTT < \frac{\pi}{2} \frac{2}{x\sqrt{(p + xp')^2 - p^2}}, \tag{15.6}$$

where $p = D(x)$ and $p' = dD_{C,B}(x)/dx$. Raina and Wischik [9] calculate the amplitude of oscillations when the system goes unstable, and using an $M/D/1/B$ queueing model to define $D_{C,B}(\cdot)$ they recommend that buffers should be around 20–60 packets: any smaller and the utilization is too low, any larger and the system has large oscillations.

Figure 15.1 shows packet-level simulations of a buffer of 70 packets (left) and a buffer of 15 packets (right). We have chosen the other parameters so that the algebraic theory predicts oscillations for a buffer of 70 packets, and it predicts stability for a buffer of 15 packets. There are 1000 flows sharing a 480-Mb/s link (i.e., the available bandwidth per flow is $C = 40$ packet/s). Round-trip times are chosen uniformly at random from $[120, 280]$ ms. Also, each flow has an ingress link of capacity $3C$, and the reverse path is loaded with 1000 TCP flows with similar parameters. The top panel shows the mean throughput of the flows $x(t)$,

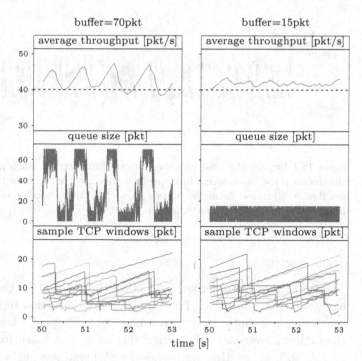

Figure 15.1 Traces from a packet-level simulation of a single bottleneck link with 1000 flows, round-trip times uniform in [120, 280] ms, capacity 480 Mb/s, and buffer of either 70 or 15 packets.

estimated by dividing the average window size by the average round-trip time. The dotted line shows the available bandwidth per flow C. The middle panel shows the queue size. For a buffer of 70 packets, when $x(t)$ oscillates around C, the queue size fluctuates markedly: when $x(t) > C$ the queue size bounces around full, and the packet drop probability is $p(t) \approx 1 - C/x(t)$; when $x(t) < C$ the queue size bounces around empty and $p(t) \approx 0$. It doesn't take much change in $x(t)$ to change the queue size dramatically. For a buffer of 15 packets, $x(t)$ has small oscillations, and they do not have such a big impact; instead, the queue has small and very fast fluctuations. The bottom panel shows a sample of TCP window sizes.

The sample TCP window plots show that in the unstable case, there are periods of time when the queue becomes full, many flows receive drops at the same time, and they become synchronized. In the stable case this does not happen. The lesson to be drawn is simply that "unstable dynamical system" has the interpretation "synchronized TCP flows." In this particular example the synchronization does not harm utilization, though the bursty drops may be harmful for, e.g., voice traffic that is sharing the link.

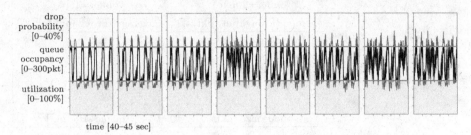

drop
probability
[0–40%]

queue
occupancy
[0–300pkt]

utilization
[0–100%]

time [40–45 sec]

Figure 15.2 Impact of access link speed on synchronization. Each plot shows a simulation trace, with access link speeds increasing from 5 times the core bandwidth per flow to 40 times. As access speeds grow, TCP flows become burstier, which reduces the impact of synchronization.

15.2.4   Traffic burstiness

The traffic model we derived in Section 15.2.2 relied on the assumption that aggregate traffic flows look Poisson over short timescales. In this section we use simulation and simple models to explore what happens when traffic is burstier than Poisson over short timescales; this analysis is taken from [16]. In section 15.4 we suggest practical ways to deal with burstiness in the future Internet.

Consider a TCP flow that has very fast links along its entire route. It will typically send a packet, get an ACK back, send two packets back-to-back, get two ACKs back back-to-back, send three packets back-to-back, and so on, i.e., every round-trip time it will send a single burst of packets. The problem is likely to be exacerbated if there are many flows in slow-start, which is inherently bursty. The empirical outcome is that the traffic does not look Poisson on sub-*RTT* timescales; the modeling problem is that it does not satisfy the "minimum spacing" requirement from Section 15.2.2, and so the Poisson limit does not apply.

The plots in Figure 15.2 are from simulations with a bottleneck link with total capacity $NC = 460\,\text{Mb/s}$ and $N = 1000$ long-lived TCP flows with *RTT* drawn uniformly from $[150, 200]$ ms, and a buffer of 300 packets. We smoothed the traffic by making each flow pass through its own rate-limiting access link, and changed the speed of this from $5C$ (left-hand panel) to $40C$ (right-hand panel). When the access link is relatively slow there is clear synchronization. As the access link gets faster the queue size still fluctuates over its entire range, but these fluctuations are now random and there is no synchronization.

Here is a crude theoretical model to explain this behavior. Consider two extreme cases. In the "smooth traffic" case, we assume that all packets are well spaced and the Poisson limit holds, so that all the previous theory applies, and $D_{C,B}(x)$ is the packet drop probability for a queue fed by Poisson traffic of rate x. In the "bursty traffic" case, suppose that the source emits back-to-back clumps of m packets, and they arrive in a clump to the bottleneck link. The total traffic is still the aggregate of many independent point processes, but now every point refers to a clump of back-to-back packets, and so all the previous theory applies except that now $D_{C,B}(x)$ is the packet drop probability for a queue fed by a Poisson process of rate x/m, where each Poisson arrival signifies the arrival of a burst of m packets. To get a qualitative idea of the impact of burstiness, we shall approximate $D(\cdot)$ by the queue length distribution for an $M/M/1/\infty$ queue, and treat packet clumps as indivisible; this gives $D_{C,B}(x) \approx (x/C)^B$ in the smooth case and $D_{C,B}(x) \approx (x/C)^{B/m}$ in the bursty case. In the bursty case, $D'_{C,B}(x)$ is m times smaller, and (15.6) indicates that the system should therefore be more stable.

15.3 Queueing theory for switches with scheduling

Switching is an integral function in a packet-switched data network. An Internet router has several input ports and several output ports, and its function is to receive packets at input ports, work out which output port to send them to, and then switch them to the correct output port. The physical architecture of the switch may place restrictions on which packets may be switched simultaneously, e.g., in an input-queued switch the restriction is that in any clock tick no input port may contribute more than one packet, and no output port may be sent more than one packet. The switch must therefore incorporate a scheduling algorithm, to decide which packets should be switched at what time.

A switch presents two questions: given a physical architecture, what scheduling algorithm should be used; and what is the resulting performance, as measured in terms of mean queuing delay etc.?

This chapter explains some tools that have recently been developed for performance analysis of switches. It is worth stressing two points. First, the tools apply to a general class of switched systems, not just input-queued switches. Second, the tools can as yet only deal with a small class of scheduling algorithms, namely algorithms derived from the maximum-weight (MW) algorithm. The tools we describe here are just a few years old; they are taken from [10, 13, 12]. There is likely to be significant development especially in the design and performance analysis of easily implementable distributed scheduling algorithms.

In Section 15.3.1 we describe the general model of a switch. In Section 15.3.2 we specify a fundamental optimization problem that is used to define the capacity region of a switch. In Section 15.3.3 we summarize the theory of performance analysis for switches that are overloaded, underloaded, and critically loaded.

15.3.1 Model for a switched network

The abstract model we will consider is as follows. Consider a collection of N queues. Let time be discrete, indexed by $t \in \{0, 1, \dots\}$. Let $Q_n(t)$ be the amount of work in queue $n \in \{1, \dots, N\}$ at time t; write $\mathbf{Q}(t)$ for the vector of queue sizes. In each timeslot, the scheduling algorithm chooses a schedule $\boldsymbol{\pi}(t) = (\pi_1(t), \ldots, \pi_n(t))$, and queue n is offered service $\pi_n(t)$. The schedule $\boldsymbol{\pi}(t)$ is required to be chosen from a set $\mathcal{S} \subset \mathbb{R}_+^N$ where $\mathbb{R}_+$ is the set of non-negative real numbers. After the schedule has been chosen and work has been served, new work may arrive; let each of the N queues have a dedicated exogenous arrival process, and let the average arrival rate at queue n be λ_n units of work per timeslot. For simplicity, assume the arrivals are Bernoulli or Poisson. We shall be interested in the max-weight scheduling algorithm, which chooses a schedule $\boldsymbol{\pi}(t) \in \mathcal{S}$ such that

$$\boldsymbol{\pi}(t) \cdot \mathbf{Q}(t) = \max_{\rho \in \mathcal{S}} \boldsymbol{\rho} \cdot \mathbf{Q}(t), \qquad (15.7)$$

where $\boldsymbol{\pi} \cdot \mathbf{Q} = \sum_n \pi_n Q_n$. Ties are broken arbitrarily. We might also apply some sort of weighting, e.g., replace $Q_n(t)$ in this equation by $w_n Q_n(t)^\alpha$ for some fixed weights $\mathbf{w} > 0$ and $\alpha > 0$. All the analyses in this chapter require $\mathcal{S}$ to be finite.

For example, a 3×3 input-queued switch has $N = 9$ queues in total, three of them located at every input port. The set $\mathcal{S}$ of possible schedules is

$$\mathcal{S} = \left\{ \begin{bmatrix} 1 & 0 & 0 \\ 0 & 1 & 0 \\ 0 & 0 & 1 \end{bmatrix}, \begin{bmatrix} 1 & 0 & 0 \\ 0 & 0 & 1 \\ 0 & 1 & 0 \end{bmatrix}, \begin{bmatrix} 0 & 1 & 0 \\ 1 & 0 & 0 \\ 0 & 0 & 1 \end{bmatrix}, \begin{bmatrix} 0 & 1 & 0 \\ 0 & 0 & 1 \\ 1 & 0 & 0 \end{bmatrix}, \begin{bmatrix} 0 & 0 & 1 \\ 1 & 0 & 0 \\ 0 & 1 & 0 \end{bmatrix}, \begin{bmatrix} 0 & 0 & 1 \\ 0 & 1 & 0 \\ 1 & 0 & 0 \end{bmatrix} \right\}.$$

Here we have written out the schedules as 3×3 matrices rather than as 1×9 vectors. These schedules are all possible schedules that satisfy the constraint "one packet from any given input, one packet to any given output."

The switched network model can be modified to apply to wireless base stations [11]. For this purpose, assume that the set of possible schedules depends on some "state of nature" that changes randomly. Assume that the scheduling algorithm knows the state of nature in any given timeslot, and it chooses which node to transmit to based on the backlog of work for that node and on the throughput it will get in the current state of nature.

The model can also be modified to apply to a flow-level model of bandwidth allocation by TCP [8]. Take $Q_n(t)$ to be the number of active TCP flows on route n through the network at time $t \in \mathbb{R}_+$, and let $\pi_n(t)$ be the throughput that flows on route n receives, according to TCP's congestion control algorithm rather than (15.7).

15.3.2   The capacity region, and virtual queues

The study of switched networks is all based on optimization problems. The most fundamental is called PRIMAL($\boldsymbol{\lambda}$), and it is

$$\text{minimize} \quad \sum_{\pi \in \mathcal{S}} \alpha_\pi \quad \text{over} \quad \alpha_\pi \in \mathbb{R}_+ \text{ for all } \boldsymbol{\pi} \in \mathcal{S}$$

$$\text{such that} \quad \boldsymbol{\lambda} \leq \sum_{\pi \in \mathcal{S}} \alpha_\pi \boldsymbol{\pi} \quad \text{componentwise.}$$

This problem asks whether it is possible for an offline scheduler, which knows the arrival rates λ, to find a combination of schedules that can serve λ; α_π is the fraction of time that should be spent on schedule π. If the solution of PRIMAL(λ) is ≤ 1 we say that λ is admissible. Define the capacity region Λ to be the set of admissible λ. The dual problem DUAL(λ) is

$$\text{maximize} \quad \boldsymbol{\xi} \cdot \boldsymbol{\lambda} \quad \text{over} \quad \boldsymbol{\xi} \in \mathbb{R}^N_+$$

$$\text{such that} \quad \max_{\pi \in \mathcal{S}} \boldsymbol{\xi} \cdot \boldsymbol{\pi} \leq 1.$$

Interpret $\boldsymbol{\xi}$ as queue weights, and $\boldsymbol{\xi} \cdot \mathbf{q}(t)$ as a virtual queue. The arrival rate of work to the virtual queue is $\boldsymbol{\xi} \cdot \boldsymbol{\lambda}$. Since $\boldsymbol{\xi} \cdot \boldsymbol{\pi} \leq 1$, no service action can drain more than one unit of work from the virtual queue. Clearly if the solution of DUAL(λ) is > 1 then the switch is overloaded.

Both optimization problems are soluble, though the solutions may not be unique. The set of dual feasible variables is a convex polyhedron, so we might as well restrict attention to the extreme points, of which there are finitely many. We call these the *principal virtual queues*.

For example, in a 3×3 input-queued switch there are six principal virtual queues, namely

$$\left\{ \begin{bmatrix} 1 & 1 & 1 \\ 0 & 0 & 0 \\ 0 & 0 & 0 \end{bmatrix}, \begin{bmatrix} 1 & 0 & 0 \\ 1 & 0 & 0 \\ 1 & 0 & 0 \end{bmatrix}, \begin{bmatrix} 0 & 0 & 0 \\ 1 & 1 & 1 \\ 0 & 0 & 0 \end{bmatrix}, \begin{bmatrix} 0 & 1 & 0 \\ 0 & 1 & 0 \\ 0 & 1 & 0 \end{bmatrix}, \begin{bmatrix} 0 & 0 & 0 \\ 0 & 0 & 0 \\ 1 & 1 & 1 \end{bmatrix}, \begin{bmatrix} 0 & 0 & 1 \\ 0 & 0 & 1 \\ 0 & 0 & 1 \end{bmatrix} \right\}.$$

These represent "all work at input port 1," "all work for output port 1," etc.

15.3.3 Performance analysis

Queueing theory for switches is divided into three parts: for overloaded switches, i.e., PRIMAL(λ) > 1; for underloaded switches; i.e., PRIMAL(λ) < 1, and for critically loaded switches, i.e., PRIMAL(λ) $= 1$. The key to understanding them all is the notion of the fluid model.

Fluid model

The fluid model is a collection of differential equations, which describe the operation of the switch. These equations relate the amount of work $\mathbf{a}(t)$ that has arrived in $[0, t]$, the amount of time $s_\pi(t)$ spent on action π in the interval, the idleness $\mathbf{z}(t)$ incurred at each queue over that interval, and the queue sizes $\mathbf{q}(t)$ at time $t \in \mathbb{R}_+$. The equations are

$$\mathbf{a}(t) = \boldsymbol{\lambda} t$$

$$\mathbf{q}(t) = \mathbf{q}(0) + \mathbf{a}(t) - \sum_{\pi \in \mathcal{S}} s_\pi(t) \boldsymbol{\pi} + \mathbf{z}(t)$$

$$\sum_{\pi \in \mathcal{S}} s_\pi(t) = t$$

$$\dot{z}_n(t) = 0 \text{ if } q_n(t) > 0$$
$$\dot{s}_{\boldsymbol{\pi}}(t) = 0 \text{ if } \boldsymbol{\pi} \cdot \mathbf{q}(t) < \max_{\boldsymbol{\rho} \in \mathcal{S}} \boldsymbol{\rho} \cdot \mathbf{q}(t).$$

The first four equations are straightforward expressions of queue dynamics, and the final equation is a way to express the scheduling rule (15.7).

These differential equations arise as a limiting description of a large-scale system. Let $\mathbf{A}(t)$ be the actual amount of work that has arrived to each queue over time $[0, t]$, let $\mathbf{Q}(t)$ be the actual queue size at timeslot t, let $\mathbf{Z}(t)$ be the cumulative idleness in $[0, t]$, and $S_{\boldsymbol{\pi}}(t)$ be the cumulative time spent on action $\boldsymbol{\pi}$. Then define rescaled versions: $\mathbf{a}^r(t) = \mathbf{A}(rt)/r$, $\mathbf{q}^r(t) = \mathbf{Q}(rt)/r$, $\mathbf{z}^r(t) = \mathbf{Z}(rt)/r$ and $s_{\boldsymbol{\pi}}^r(t) = S_{\boldsymbol{\pi}}(rt)/r$, and extend these functions to all $t \in \mathbb{R}$ by linear interpolation. In words, we are zooming out in time and space by a factor of r. It may be shown that these rescaled versions "nearly" solve the fluid model equations. Specifically, let $x^r(\cdot)$ consist of the rescaled processes, let FMS consist of all solutions to the fluid model equations, and let FMS_ε be the ε-fattening of the paths over an interval $[0, T]$, i.e.,

$$\text{FMS}_{\varepsilon, T} = \left\{ x : \sup_{0 \le t \le T} |x(t) - y(t)| < \varepsilon \text{ for some } y \in \text{FMS} \right\},$$

where $|\cdot|$ is the maximum over all components. Then, for a wide range of arrival processes, including Bernoulli arrivals and Poisson arrivals,

$$\mathbb{P}\big(x^r(\cdot) \in \text{FMS}_{\varepsilon, T}\big) \to 1 \quad \text{for any } \varepsilon > 0 \text{ and } T > 0. \tag{15.8}$$

Stability analysis
The fluid model is said to be *stable* if there is some draining time $H > 0$, such that every fluid model solution with bounded initial queue size $|\mathbf{q}(0)| \le 1$ ends up with $\mathbf{q}(t) = \mathbf{0}$ for all $t \ge H$. Define

$$L(\mathbf{q}) = \left(\sum_{1 \le n \le N} q_n^2 \right)^{1/2}.$$

It is easy to use the fluid model equations to show that

$$\frac{d\,L(\mathbf{q}(t))}{dt} = \frac{\boldsymbol{\lambda} \cdot \mathbf{q}(t) - \max_{\boldsymbol{\rho} \in \mathcal{S}} \boldsymbol{\rho} \cdot \mathbf{q}(t)}{L(\mathbf{q}(t))}.$$

Suppose the switch is underloaded, i.e., $\text{PRIMAL}(\boldsymbol{\lambda}) < 1$, so we can write $\boldsymbol{\lambda} \le \sum_{\boldsymbol{\pi}} \alpha_{\boldsymbol{\pi}} \boldsymbol{\pi}$ for some $\sum_{\boldsymbol{\pi}} \alpha_{\boldsymbol{\pi}} < 1$. After some more algebra,

$$\frac{d\,L(\mathbf{q}(t))}{dt} < \left(\sum_{\boldsymbol{\pi}} \alpha_{\boldsymbol{\pi}} - 1 \right) \frac{S^{\min} |\mathbf{q}(t)|}{N^{1/2} |\mathbf{q}(t)|} = -\eta < 0,$$

as long as $\mathbf{q}(t) \ne \mathbf{0}$. Here $S^{\min}$ is the smallest nonzero amount of service it is possible to give to a queue, and $|\mathbf{q}|$ is $\max_n q_n$. Since $L(\mathbf{q}(t))$ is decreasing at least at rate η, it follows that the system drains completely within time $L(\mathbf{q}(0))/\eta$. This proves stability.

It may be shown that if the fluid model is stable, and if arrivals are independent across timeslots, then the queue size process is a positive-recurrent Markov chain. This is the conventional notion of stability of a queueing system.

Overloaded switches

Suppose that the switch is overloaded, i.e., that DUAL(λ) > 1. We know that there is some virtual queue for which the arrival rate exceeds the maximum possible service rate, hence this virtual queue must grow indefinitely. In fact it grows in a very precise manner. By showing that $dL(\mathbf{q}(t)/t)/dt \leq 0$ one can prove that for any initial queue size $\mathbf{q}(0)$, the scaled queue size approaches a limit i.e., $\mathbf{q}(t)/t \to \mathbf{q}^\dagger$. If the queues start empty then the solution to the fluid model is $\mathbf{q}(t)/t = \mathbf{q}^\dagger$. We can identify $\mathbf{q}^\dagger$: it is the unique solution to the optimization problem ALGP$^\dagger(\boldsymbol{\lambda})$:

$$\text{minimize} \quad L(\mathbf{r}) \quad \text{over} \quad \mathbf{r} \in \mathbb{R}_+^N$$
$$\text{such that} \quad \mathbf{r} \cdot \boldsymbol{\xi} \geq \boldsymbol{\lambda} \cdot \boldsymbol{\xi} - 1 \quad \text{for all } \boldsymbol{\xi} \in \mathcal{S}^\dagger(\boldsymbol{\lambda})$$

where $\mathcal{S}^\dagger(\boldsymbol{\lambda})$ is the set of overloaded principal virtual resources, i.e., extreme solutions to DUAL($\boldsymbol{\lambda}$) for which $\boldsymbol{\xi} \cdot \boldsymbol{\lambda} > 1$.

The MW algorithm shows some surprising behavior in overload: it turns out that an increase in load can actually cause the departure rate to decrease. Here is an example, in a 2×2 input-queued switch. The two possible service actions are

$$\pi^1 = \begin{bmatrix} 1 & 0 \\ 0 & 1 \end{bmatrix} \quad \text{and} \quad \pi^2 = \begin{bmatrix} 0 & 1 \\ 1 & 0 \end{bmatrix}.$$

Consider two possible arrival rate matrices

$$\lambda^{\text{critical}} = \begin{bmatrix} 0.3 & 0.7 \\ 0.7 & 0.3 \end{bmatrix} \quad \text{and} \quad \lambda^{\text{overload}} = \begin{bmatrix} 0.3 & 1.0 \\ 0.7 & 0.3 \end{bmatrix}.$$

In the critically loaded case, the switch can only be stabilized by $\sigma^{\text{crit}} = 0.3\pi^1 + 0.7\pi^2$, and it is straightforward using the fluid model equations to check that MW will eventually achieve this service rate, for any initial queue size. However, in the overloaded case, starting from $\mathbf{q}(0) = \mathbf{0}$, MW will achieve

$$\mathbf{q}(t) = \begin{bmatrix} 0.1t & 0.2t \\ 0 & 0.1t \end{bmatrix}$$

by serving at rate $0.2\pi^1 + 0.8\pi^2$, which means that queue $q_{2,1}$ is idling, and the total departure rate is 1.9. A different scheduling algorithm might have chosen to serve at rate σ^{crit}, which would result in a higher total departure rate, namely 2.

Underloaded switches

Consider an underloaded switch. Since the switch is stable, we know that the queue size does not grow unboundedly, and in fact we can use the fluid limit result (15.8) to deduce

$$\mathbb{P}\big(L(\mathbf{Q}) \geq r\big) \to 0 \quad \text{as} \quad r \to \infty.$$

In this section we will use large deviations theory to estimate the speed of convergence; in other words, we will find the tail of the distribution of $L(\mathbf{Q})$ under the MW algorithm. The method also allows us to find lower bounds for the tail of other quantities such as the total amount of work $\mathbf{Q} \cdot \mathbf{1}$ or the maximum queue size $|\mathbf{Q}|$, though not upper bounds. See also Section 15.4 point 4.

Here is an heuristic argument for obtaining the tail of the distribution. Suppose the queue starts empty at time 0, and that T is large enough that $\mathbf{Q}(T)$ has the stationary distribution of queue size. Then for large r,

$$\frac{1}{r} \log \mathbb{P}\big(L(\mathbf{Q}(rT)) \approx r\big) = \frac{1}{r} \log \mathbb{P}\big(L(\mathbf{q}^r(T)) \approx 1\big)$$

$$= \frac{1}{r} \log \mathbb{P}\Big(\mathbf{a}^r(\cdot) \in \{\mathbf{a} : L(\mathbf{q}(T; \mathbf{a})) \approx 1\}\Big)$$

$$\approx \sup_{\mathbf{a} : \mathbf{q}(T; \mathbf{a}) = 1} \frac{1}{r} \log \mathbb{P}\big(\mathbf{a}^r \approx \mathbf{a}\big)$$

$$\approx - \inf_{\mathbf{a} : \mathbf{q}(T; \mathbf{a}) = 1} \int_0^T l\big(\dot{\mathbf{a}}(t)\big)\, dt. \qquad (15.9)$$

The first step is a simple rescaling. In the second step, we emphasize that the underlying randomness is in the arrival process, and we want to find the probability that the arrivals are such as to cause large queues. We have written $\mathbf{q}(T; \mathbf{a})$ for the queue size at time T, to emphasize that it depends on the arrival process. The third step is called the principle of the largest term, and it says that the probability of a set is roughly the probability of its largest element. The final step is an estimate of the probability of any given sample path; this sort of estimate holds for a wide range of arrival processes. The function l is called the local rate function; if $\mathbf{X}$ is the distribution of the amount of work arriving to each queue in a timeslot, so that the arrival rate is $\boldsymbol{\lambda} = \mathbb{E}\mathbf{X}$, and if arrivals are independent across timeslots, then

$$l(\mathbf{x}) = \sup_{\boldsymbol{\theta} \in \mathbb{R}^N} \boldsymbol{\theta} \cdot \mathbf{x} - \log \mathbb{E}e^{\boldsymbol{\theta}\mathbf{X}}.$$

Note that $l(\boldsymbol{\lambda}) = 0$, and that l is non-negative. Large deviations theory is concerned with making all these heuristic steps rigorous.[2]

It remains to calculate (15.9). This problem is known as "finding the cheapest path to overflow." It is easy to bound the answer, by simply guessing a solution $\mathbf{a}(\cdot)$ such that $\mathbf{q}(T; \mathbf{a}) = 1$. In this case, a good guess is that the most likely path is made up of two linear pieces: run at $\dot{\mathbf{a}}(t) = \boldsymbol{\lambda}$ for time $T - U$, keeping $\mathbf{q}(t) = 0$, and then pick some arrival rates $\mathbf{x} \geq 0$ and run at $\dot{\mathbf{a}}(t) = \mathbf{x}$ for time

---

[2] A caution: the results for underloaded switches that we present here are the natural extension of results proved for a wireless base station [13, 12]. They have not yet been formally proved for the general switch model presented here.

U, choosing U so that $L(\mathbf{q}(T)) = 1$. The first phase has cost 0, and the second phase has cost $Ul(\mathbf{x})$. Note that in the second phase the switch is running in overload, and from our previous results for switches in overload we know that the queue will grow linearly, hence $L(\mathbf{q})$ will grow linearly; indeed it must take time $U = 1/\text{ALGP}^\dagger(\mathbf{x})$ to reach $L(\mathbf{q}) = 1$. We might as well choose the overload rates $\mathbf{x}$ so as to minimize the cost of the path. This tells us

$$\inf_{\mathbf{a}\,:\,\mathbf{q}(T;\mathbf{a})=1} \int_0^T l\big(\dot{\mathbf{a}}(t)\big)\, dt \leq \inf_{\mathbf{x}\geq 0} \frac{l(\mathbf{x})}{\text{ALGP}^\dagger(\mathbf{x})}.$$

In fact it is possible to prove that this is an equality. The proof uses the dual problem to $\text{ALGP}^\dagger$.

Critically loaded switches

Suppose the switch is critically loaded, i.e., suppose $\text{DUAL}(\boldsymbol{\lambda}) = 1$. Recall that feasible solutions to the $\text{DUAL}(\boldsymbol{\lambda})$ problem can be interpreted as virtual queues; since the switch is critically loaded there must be some virtual queues for which the arrival rate is exactly equal to the maximum possible service rate.

For example, consider an input-queued switch, and suppose that the virtual queue "all work for output 2" is critically loaded. This means that the arrival rate of work destined for output 2, summed across all the input ports, is equal to 1 packet per timeslot. Now, the contents of the virtual queue, i.e., the aggregate of all work destined for output 2, is split across the input ports. By devoting more service effort to input port 1 the virtual queue might be shifted onto the other input ports; or by giving priority to the longest of these queues the virtual queue might be spread evenly across the input ports.

We will see that a scheduling algorithm can in effect choose where to store the contents of the virtual queues. To make this precise we shall introduce another optimization problem called $\text{ALGD}(\boldsymbol{\lambda}, \mathbf{q})$:

$$\begin{aligned} &\text{minimize} &&L(\mathbf{r}) \quad \text{over} \quad \mathbf{r} \in \mathbb{R}_+^N \\ &\text{such that} &&\boldsymbol{\xi}\cdot\mathbf{r} \geq \boldsymbol{\xi}\cdot\mathbf{q} \quad \text{for all } \boldsymbol{\xi} \in \mathcal{S}^*(\boldsymbol{\lambda}) \end{aligned}$$

where $\mathcal{S}^*(\boldsymbol{\lambda})$ consists of all the principal virtual queues that are critically loaded. This problem says: shuffle work around between the actual queues so as to minimize cost (as measured by the Lyapunov function L), subject to the constraint that all the work in critically loaded virtual queues $\boldsymbol{\xi} \in \mathcal{S}^*(\boldsymbol{\lambda})$ has to be stored somewhere. The problem has a unique solution, call it $\Delta^{\boldsymbol{\lambda}}(\mathbf{q})$. (The solution is unique because the feasible set is convex and non-empty, and the objective function is an increasing function of $\sum r_n^2$ which is convex.)

It can be shown that the MW scheduling algorithm effectively solves this optimization problem, in that it seeks out a queue-state $\mathbf{q}$ that solves $\text{ALGD}(\boldsymbol{\lambda}, \mathbf{q})$. In other words, the queue-state lies in or near the set

$$\mathcal{I} = \{\mathbf{q} : \mathbf{q} = \Delta^{\boldsymbol{\lambda}}(\mathbf{q})\}.$$

We can use this for explicit calculations. For example, in an input-queued switch in which say only output port 2 is critically loaded, it is easy to check that $\Delta^\lambda(\mathbf{q})$ spreads all the work for port 2 evenly across the input ports, and that all the other queues are empty. More generally, we believe that by understanding the geometry of $\mathcal{I}$ we can learn something about the performance of the algorithm. For example, $\mathcal{I}$ depends on the scheduling algorithm, and, in all the cases we have looked at so far, the bigger $\mathcal{I}$ the lower the average queueing delay.

Here are two ways to formalize the claim that MW effectively solves ALGD($\lambda, \mathbf{q}$). (i) It can be shown that $\mathbf{q}$ is a fixed point of the fluid model if and only if $\mathbf{q} = \Delta^\lambda(\mathbf{q})$. Furthermore, for any fluid model, $|\mathbf{q}(t) - \Delta^\lambda(\mathbf{q}(t))| \to 0$ as $t \to 0$. (ii) There is also a probabilistic interpretation of this convergence, for which we need to define another rescaling: let $\hat{\mathbf{q}}^r(t) = \mathbf{Q}(r^2 t)/r$. Then

$$\mathbb{P}\left(\frac{\|\hat{\mathbf{q}}^r(t) - \Delta^\lambda(\hat{\mathbf{q}}^r(t))\|}{\max(1, \|\hat{\mathbf{q}}^r(t)\|)} > \varepsilon\right) \to 0 \quad \text{as } r \to 0$$

where $\|x(\cdot)\| = \sup_{0 \le t \le T} |x(t)|$, and the limit holds for any $\varepsilon > 0$ and $T > 0$. (It seems likely that the denominator on the left-hand side is not necessary, but the result has not yet been proved without it.)

15.4   A proposed packet-level architecture

In this section we propose an architecture that pulls together the strands of theory discussed in Sections 15.2 and 15.3. Briefly, buffers in core routers should be small, in order to keep queueing delay and jitter small. This has the side benefit of permitting new switch architectures, such as switch-on-a-chip. In order for this to work well:

1. Switches should use explicit congestion notification (ECN) marks to signal congestion, in order to keep utilization to a reasonable level, e.g., no more than 95 percent. For flows which are not ECN-capable, packets should be dropped rather than marked.
2. Congestion should be signaled for each virtual queue. A virtual queue consists of a weighted collection of actual queues; the virtual queues for a given switch architecture are derived from the capacity region of that architecture.
3. A packet might be given an ECN mark when it causes a virtual queue to exceed some small threshold, e.g., 30 packets. The response function, i.e., the probability of an ECN mark given a utilization level, should be tuned to achieve stability (i.e., desynchronization between flows) and decent utilization.
4. The actual buffer should be somewhat larger, and it should be dimensioned so that actual packet loss is very rare. This may be done using large deviations theory for underloaded switches.
5. Any future high-speed replacement for TCP should be designed so that it responds smoothly to congestion, and it produces well-spaced packets.

1. ECN marks

Heuristic 3 in Section 15.2.1 explains that TCP reduces its transmission rate in response to both queueing delay and packet drops. If buffers are small then queueing delay will be small, and therefore packet drop probability must be higher in order to regulate TCP flows. Explicit congestion notification is an alternative to packet drops: when a switch is congested it can set the ECN bit in packet headers, and this tells TCP to back off as if the packet had been dropped but without actually dropping it.

2. Virtual queues

Virtual queues, as defined in Section 15.3.2, express the capacity constraints of a switch. For example, a 3×3 input-queued switch has nine queues in total, namely "work at input 1 for output 1," etc., but only six virtual queues, namely "all work at input 1," "all work for output 1," etc. The loading on the virtual queues specifies whether the switch as a whole is underloaded, critically loaded or overloaded. The performance analysis in Section 15.3.3 shows that the scheduling algorithm can shift work between the actual queues but it cannot drain a virtual queue if that virtual queue is critically loaded or overloaded. Therefore the virtual queues are the resources that need to be protected from congestion, therefore ECN marks should be generated based on virtual queues.

3. Response function

Section 15.2.3 shows how to calculate utilization and stability, using Equations (15.5) and (15.6). These depend on the function $D(x)$, the packet drop probability (or ECN marking probability) when the traffic arrival rate is x. However, the analysis in Section 15.2.4 shows that for a simple queue that drops packets when it is full, $D(x)$ depends on the burstiness of traffic, which is influenced by access speeds and indeed any other form of packet spacing. The consequence is that there is no single buffer size that can balance utilization and stability uniformly across a range of packet spacings. Furthermore, if there is significant traffic shaping by upstream queues then flows will not be independent, so the Poisson approximation on page 308 will not hold, and this will further affect $D(x)$.

Instead, we propose that the switch should enforce a response function $D(x)$ that is broadly consistent across a range of packet spacings and traffic statistics. A crude solution would simply be to measure x and then to mark packets with probability $D(x)$, where D is a predefined function. A more elegant solution might be to simulate a virtual queue with exponential service times, and mark packets based on queue size in this virtual queue. An alternative approach [2] is to build traffic shapers at the edge of the network to make traffic Poisson, and to build switches that explicitly delay packets just long enough to ensure that the Poisson nature of the traffic is preserved.

4. Buffer dimensioning and management

Suppose the virtual queue response function is designed to ensure that the offered arrival rates to a switch lie within some bound, e.g., 95 percent of the full capacity region. Section 15.3.3 suggests the approximation

$$\mathbb{P}(L(\mathbf{Q}) > b) \approx e^{-bI}, \tag{15.10}$$

where L is the Lyapunov function for the switch scheduling algorithm and $\mathbf{Q}$ is the vector of queue sizes. The formula for I is based on an optimization problem revolving around L.

To illustrate how this may be used for buffer dimensioning, suppose that each of the queues in the switch has its own non-shared buffer of size B, and we wish to choose B such that overflow in any of the queues is rare. Suppose the switch uses the maximum-weight scheduling algorithm, for which $L(\mathbf{Q}) = (\sum_n Q_n^2)^{1/2}$. It is straightforward to verify that $\max_n Q_n \leq L(\mathbf{Q})$, hence

$$\mathbb{P}(\max_n Q_n \geq B) \leq \mathbb{P}(L(\mathbf{Q}) \geq B) \approx e^{-BI}$$

and this may be used to choose a suitable B. See also [13] for a discussion of how to design a scheduling algorithm such that this inequality is nearly tight, and see [10] for an equivalent analysis when it is $\sum_n Q_n$ rather than $\max_n Q_n$ that we wish to control, for example, if there is shared memory between all the queues or if we want to control average queueing delay.

Note that the quantity I, known as the rate function, depends on the traffic model. In Section 15.2.2 we argued that a Poisson model is an appropriate approximation when there are small buffers. Cruise [5] has shown that the Poisson approximation is still appropriate when buffers are somewhat larger, large enough for (15.10) to hold. More work is needed to understand how I is modified when traffic is bursty, as in the model from Section 15.2.4.

5. Nice traffic

In Section 15.2.2 we found a differential equation model for the aggregate of many TCP flows. The precise form of the equation was not important for the analysis; what matters is that the aggregate traffic should adapt smoothly over time, and it should respond gradually to small changes in congestion. Any replacement for TCP should have the same properties. A single TCP flow using a high-capacity link does not respond suitably: it cuts its window by half in response to a single drop, and this cannot be modeled by a differential equation.

We also recommend that any high-speed replacement for TCP should space its packets out, so that aggregate traffic is approximately Poisson over short timescales. If a flow were able to dump arbitrarily big bursts of packets onto the network, it would be difficult to design a suitable response function D as in point 3, or to calculate a robust rate function as in point 4.

References

[1] Guido Appenzeller, Isaac Keslassy, and Nick McKeown. Sizing router buffers. In *SIGCOMM*, 2004. Extended version available as Stanford HPNG Technical Report TR04-HPNG-060800.

[2] N. Beheshti, Y. Ganjali, and N. McKeown. Obtaining high throughput in networks with tiny buffers. In *Proceedings of IWQoS*, 2008.

[3] Neda Beheshti, Yashar Ganjali, Ramesh Rajaduray, Daniel Blumenthal, and Nick McKeown. Buffer sizing in all-optical packet switches. In *Proceedings of OFC/NFOEC*, 2006.

[4] Jin Cao and Kavita Ramanan. A Poisson limit for buffer overflow probabilities. In *IEEE Infocom*, 2002.

[5] R. J. R. Cruise. Poisson convergence, in large deviations, for the superposition of independent point processes. *Annals of Operations Research*, 2008.

[6] M. Enachescu, Y. Ganjali, A. Goel, N. McKeown, and T. Roughgarden. Routers with very small buffers. In *Proceedings of IEEE INFOCOM*, 2006.

[7] Yossi Kanizo, David Hay, and Isaac Keslassy. The crosspoint-queued switch. In *Proceedings of IEEE INFOCOM*, 2009.

[8] F. P. Kelly and R. J. Williams. Fluid model for a network operating under a fair bandwidth-sharing policy. *Annals of Applied Probability*, 2004.

[9] Gaurav Raina and Damon Wischik. Buffer sizes for large multiplexers: TCP queueing theory and instability analysis. In *EuroNGI*, 2005.

[10] Devavrat Shah and Damon Wischik. The teleology of scheduling algorithms for switched networks under light load, critical load, and overload. Submitted, 2009.

[11] A. L. Stolyar. MaxWeight scheduling in a generalized switch: state space collapse and workload minimization in heavy traffic. *Annals of Applied Probability*, 2004.

[12] Vijay G. Subramanian, Tara Javidi, and Somsak Kittipiyakul. Many-sources large deviations for max-weight scheduling. arXiv:0902.4569v1, 2009.

[13] V. J. Venkataramanan and Xiaojun Lin. Structural properties of LDP for queue-length based wireless scheduling algorithms. In *Proceedings of Allerton*, 2007.

[14] Curtis Villamizar and Cheng Song. High performance TCP in ANSNET. *ACM/SIGCOMM CCR*, 1994.

[15] Arun Vishwanath, Vijay Sivaraman, and Marina Thottan. Perspectives on router buffer sizing: recent results and open problems. *ACM/SIGCOMM CCR*, 2009.

[16] D. Wischik. Buffer sizing theory for bursty TCP flows. In *Proceedings of IZS*, 2006.

16 Stochastic network utility maximization and wireless scheduling

Yung Yi[†] and Mung Chiang[‡]

[†]Korea Advanced Institute of Science and Technology, South Korea [‡]Princeton University, USA

Layering As optimization Decomposition (LAD) has offered a first-principled, top-down approach to derive, rather than just describe, network architectures. Incorporating stochastic network dynamics is one of the main directions to refine the approach and extend its applicability. In this chapter, we survey the latest results on Stochastic Network Utility Maximization (SNUM) across session, packet, channel, and topology levels. Developing simple yet effective distributed algorithms is another challenge often faced in LAD, such as scheduling algorithms for SNUM in wireless networks. We provide a taxonomy of the results on wireless scheduling and highlight the recent progress on understanding and reducing communication complexity of scheduling algorithms.

16.1   Introduction

The papers [44, 45] by Kelly *et al.* presented an innovative idea on network resource allocation – Network Utility Maximization (NUM) – that has led to many research activities since. In the basic NUM approach, an optimization problem is formulated where the variables are the source rates constrained by link capacities and the objective function captures design goals:

$$\text{maximize } \sum_i U_i(x_i)$$
$$\text{subject to } \boldsymbol{R}\boldsymbol{x} \leq \boldsymbol{c}, \tag{16.1}$$

where the source rate vector $\boldsymbol{x}$ is the set of optimization variables, one for each of the sources indexed by i, the $\{0,1\}$ routing matrix $\boldsymbol{R}$ and link capacity vector $\boldsymbol{c}$ are constants, and $U_i(\cdot)$ is the utility function of source i. Decomposing the above problem into several sub-problems enables a distributed algorithm to be developed elegantly, where each of the links and sources controls its local variable, such as link price or source rate, based on local observables, such as link load or path price.

Next-Generation Internet Architectures and Protocols, ed. Byrav Ramamurthy, George Rouskas, and Krishna M. Sivalingam. Published by Cambridge University Press. © Cambridge University Press 2011.

Substantial work has followed in terms of theory, algorithms, applications, and even commercialization based on the NUM model of networks. Now, the NUM mentality has been extended and used to model a significantly large array of resource allocation problems and network protocols, where utility may depend on rate, latency, jitter, energy, distortion, etc., may be coupled across users, and may be any non-decreasing function, referred to as *generalized NUM* (although most papers assume smooth and concave utility functions). They can be constructed based on a user behavior model, operator cost model, or traffic elasticity model.

Network Utility Maximization serves as a modeling language and a starting point to understand *network layering* as optimization decomposition [18]: the cross-layer interactions may be characterized and layered protocol stacks may be designed by viewing the process of "layering," i.e., the modularization and distribution of network functionalities into layers or network elements, as decomposition of a given (generalized) NUM problem into many subproblems. Then, these subproblems are "glued together" by certain functions of the primal and dual variables. The mindset of LAD (Layering As optimization Decomposition) has influenced the active research area of joint control of congestion, routing, scheduling, random access, transmit power, code and modulation, etc., as well as serving as a language to explain the benefits of innovative mechanisms such as back-pressure algorithms and network coding. Alternatives of decomposing the same NUM formulation in different ways further lead to the opportunities of enumerating and comparing alternative protocol stacks. The theory of decomposition of NUM thus becomes a foundation to understand, in a conceptually simple way, the complexities of network architectures: "who does what" and "how to connect them."

Among the current challenges in LAD are stochastic dynamics and complexity reduction, the focus of this chapter.

In the basic NUM (16.1) and the associated solutions, it is often assumed that the user population remains static, with each user carrying an infinite backlog of packets that can be treated as fluid, injected into a network with static connectivity and time-invariant channels. Will the results of NUM theory remain valid and the conclusions maintain predictive power under these stochastic dynamics? Can new questions arising out of stochastic factors also be answered? Incorporation of stochastic network dynamics into the generalized NUM/LAD formulation, referred to as *SNUM (Stochastic NUM)*, often leads to challenging models for those working in either stochastic network theory or distributed optimization algorithms. The first half of this chapter surveys the results over the last 10 years on the questions above. We classify them based on the different levels of stochastic dynamics: session, packet, constraints, or even mixture of those.

The second half of this chapter zooms in on an important and challenging component in realizing LAD in wireless networks. A key message from the recent research efforts on joint congestion control, routing, and scheduling in the LAD research of wireless networks indicates that scheduling may be the hardest part. Scheduling may require heavy computational overhead and/or significant amount of message passing. It motivates researchers to study various forms of algorithms with performance guarantee and capability of operating in a distributed manner. This survey focuses on understanding and reducing complexity of scheduling algorithms, including the three-dimensional tradeoff among throughput, delay, and complexity, and utility-optimal random access in the form of adaptive CSMA (Carrier-Sensing Multiple Access).

We use standard $\mathbb{R}^N$ and $\mathbb{R}^N_+$ for the N-dimensional real and non-negative Euclidean spaces, respectively. We use the calligraphic font $\mathcal{S}$ to refer to a set, and the bold-face fonts $\boldsymbol{x}$ and $\boldsymbol{X}$ to refer to a vector and a matrix, respectively. We introduce more notation as needed in the chapter.

16.2   LAD (Layering As optimization Decomposition)

16.2.1   Background

Network architecture essentially determines functionality allocation, i.e., "who does what" and "how to connect them." The study of network architecture is often more influential, harder to change, but less understood than that of resource allocation. Functionality allocations can happen, for example, between the network management system and network elements, between end-users and intermediate routers, and between source control and in-network control such as routing and physical resource sharing. Deciding a network architecture involves exploration and comparison of alternatives in functionality allocation.

Layering adopts a modularized and often distributed approach to network coordination. Each module, called layer, controls a subset of the decision variables, and observes a subset of constant parameters and the variables from other layers. Each layer in the protocol stack hides the complex behaviors inside itself and provides a service and an interface to the upper layer above, enabling a scalable, evolvable, and implementable network design.

The framework of "Layering As optimization Decomposition" starts by a constrained optimization model, like the one below:

$$
\begin{aligned}
\text{maximize} \quad & \sum_s U_s(x_s, P_{e,s}) + \sum_j V_j(w_j) \\
\text{subject to} \quad & \boldsymbol{R}\boldsymbol{x} \le \boldsymbol{c}(\boldsymbol{w}, \boldsymbol{P}_e), \\
& \boldsymbol{x} \in \mathcal{C}_1(\boldsymbol{P}_e), \ \boldsymbol{x} \in \mathcal{C}_2(\boldsymbol{F}) \text{ or } \in \boldsymbol{\Pi}, \\
& \boldsymbol{R} \in \mathcal{R}, \ \boldsymbol{F} \in \mathcal{F}, \ \boldsymbol{w} \in \mathcal{W} \\
\text{variables} \quad & \boldsymbol{x}, \boldsymbol{w}, \boldsymbol{P}_e, \boldsymbol{R}, \boldsymbol{F},
\end{aligned}
\tag{16.2}
$$

where
- x_s denotes the rate for source s, and w_j denotes the physical layer resource at network element j,
- U_s and V_j are utility functions that may be any nonlinear, monotonic functions,

- R is the routing matrix, and $c(w, P_e)$ are the logical link capacities as functions of both physical layer resources w and the desired decoding error probabilities P_e, in which functional dependency, the issue of signal interference, and power control can be captured,
- $C_1(P_e)$ is the set of rates constrained by the interplay between channel decoding reliability and other hop-by-hop error control mechanisms like ARQ,
- $C_2(F)$ is the set of rates constrained by the medium access success probability, where F is the contention matrix, or the schedulability constraint set Π,
- the sets of possible physical layer resource allocation schemes, of possible scheduling or contention-based medium access schemes, and of single-path or multi-path routing schemes are represented by $\mathcal{W}, \mathcal{F}, \mathcal{R}$, respectively.

Holding some of the variables constant and specifying some of these functional dependencies and constraint sets will then lead to a special class of this NUM formulation. In general, utility functions and constraint sets can be even richer than those in the problem (16.2).

16.2.2 Key ideas and procedures

One possible perspective to rigorously understand layering is to integrate the various protocol layers into a single coherent theory, by regarding them as carrying out an asynchronous distributed computation over the network to implicitly solve a global optimization problem, e.g., (16.2). Different layers iterate on different subsets of the decision variables using local information to achieve individual optimality. Taken together, these local algorithms attempt to achieve a global objective, where global optimality may or may not be achieved. Such a design process of modularization can be quantitatively understood through the mathematical language of *decomposition theory* for constrained optimization [77]. This framework of "Layering As optimization Decomposition" exposes the interconnections between protocol layers as different ways to modularize and distribute a centralized computation.

Different *vertical decompositions* of an optimization problem, in the form of a generalized NUM, are mapped to different *layering schemes* in a communication network. Each decomposed subproblem in a given decomposition corresponds to a layer, and certain functions of primal or Lagrange dual *variables* (coordinating the subproblems) correspond to the *interfaces* among the layers. *Horizontal decompositions* can be further carried out within one functionality module into *distributed computation and control* over geographically disparate network elements. The LAD based on these generalized NUM problems puts the end-user utilities in the "driver's seat" for network design. For example, benefits of innovations in the physical layer, such as modulation and coding schemes, are now characterized by the enhancement to applications rather than just the drop in bit error rates, which the users do not directly observe. Implicit message passing (where the messages have physical meanings and may need to be measured anyway) or explicit message passing quantifies the information sharing and decision coupling required for a particular decomposition.

Existing protocols can be reverse-engineered as an optimizer that solves the generalized optimization problem in a distributed way, e.g., TCP [37, 45, 47, 48, 61, 62, 63, 69, 92, 108], BGP [30], and MAC [51, 100]. In forward engineering, since different decompositions lead to alternative layering architectures, we can also tackle the question "how and how not to layer" by investigating the pros and cons of decomposition methods. By comparing the objective function values under various forms of optimal decompositions and suboptimal decompositions, we can seek "separation theorems" among layers: conditions under which layering incurs no loss of optimality, or allows the suboptimality gap of the overall design to be no more than that of a particular layer.

Despite the conceptual, mathematical, and practical progress made on LAD over the last few years, there are substantial challenges in the field, including that of stochastic dynamics, non-convexity, high dimensionality, and communication complexity. This chapter now surveys some of the latest results tackling some of these challenges.

16.3 Stochastic NUM (Network Utility Maximization)

16.3.1 Session-level dynamics

16.3.1.1 System model

Consider a network where sessions (classified by, e.g., a pair of source and destination addresses) are randomly generated by users and cease upon completion. First assume a finite set $\mathcal{S}$ of S classes of sessions that arrive at the system according to a Poisson process of intensity λ_s, $s \in \mathcal{S}$, sessions/s, and the exponentially distributed file sizes with mean $1/\mu_s$, $s \in \mathcal{S}$, bits. Denote by $\rho_s = \lambda_s/\mu_s$ bit/s the *traffic intensity* of the class-s sessions. We also denote by $N_s(t)$ the number of active class-s sessions at time t. Then the network state is $\mathbf{N}(t) = (N_1(t), \ldots, N_S(t))$ that is a random process. The constraint set $\mathcal{R}$, often called *rate region* in the research literature, is the set of achievable resource vectors $\phi = (\phi_1, \ldots, \phi_S)$ (provided by the resource allocation algorithms considered) where ϕ_s is the total rate allocated to class-s sessions. The form of the rate regions is diverse and can be either fixed or time-varying, and either convex or non-convex, depending on the system model and resource allocation algorithms.

Resource allocation algorithms allocate network resources (which are typically bandwidths or rates) to different session classes according to the current network state $\mathbf{N}(t)$, the utility function, and the rate region. Our interest is the resource allocation based on NUM, whose rates are the solution of the following optimization problem: at time t,

$$\text{maximize} \sum_s N_s(t)U_s(\phi_s/N_s(t)),$$
$$\text{subject to} \quad \phi \in \mathcal{R}, \tag{16.3}$$

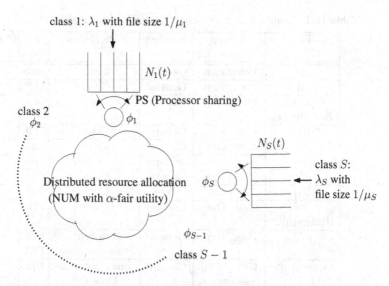

Figure 16.1 Distributed resource allocation by solving NUM under dynamic user populations with session arrivals. Session departure time is shaped by file size as well as service rate, i.e., the resource allocation vector $\{\phi_s\}$ obtained by solving NUM, where each session in class s receives the rate $\phi_s/N_s(t)$.

where the utility functions U_s are assumed to satisfy some technical conditions (e.g., twice differentiability and concavity). Our particular interest lies in the fairness of resource allocation by focusing on special classes of utility functions, referred to as α-fair utility functions that are parameterized by $\alpha \geq 0$: $U^\alpha(\cdot) = (\cdot)^{1-\alpha}/(1-\alpha)$ for $\alpha \geq 0$, and $\log(\cdot)$, for $\alpha = 1$ [69], and a feasible allocation $\boldsymbol{x}$ is called α-fair if, for any other feasible allocation $\boldsymbol{y}$, $\sum_s (y_s - x_s)/x_s^\alpha \leq 0$. The notion of α-fairness includes maxmin fairness, proportional fairness, and throughput maximization as special cases.

Figure 16.1 depicts the framework of session-level dynamics research. Sometimes we assume timescale separation in that resource allocation is much faster than session-level dynamics, i.e., whenever $\boldsymbol{N}(t)$ changes, the resource allocation is finished immediately as the solution of (16.3).

16.3.1.2 Performance metric: session-level stability

A main research focus in session-level dynamics is, for a given resource allocation, to compute the session-level stability region, which is the set of traffic intensities ρ with which the number of active sessions is finite over time. Then, the maximum stability region is the union of the stability regions achieved by all possible resource allocations: for any traffic intensity vector outside this set, no resource allocation algorithm can stabilize the network at session level.

Under the assumption of Poisson arrival and exponential file size, the system can be modeled by a Markov chain. Then, mathematically, stability means that the Markov process $\{\boldsymbol{N}(t)\}_{t=0}^\infty$ is positive-recurrent (under the technical

Table 16.1. Summary of main results on session-level stability.

Work	Arrival File size dist.	Topology	Rate Regions	U_i	U shape (i.e., α)
[22]	Pois., Exp.	General	Conv.	Same	$\alpha = 1, \alpha \to \infty$
[4]	Pois., Exp.	General	Conv.	Diff.	General
[53, 93] (Fast timesc.)	Pois., Exp.	General	Conv.	Same	$\alpha \geq 1$
[109]	Gen., Exp.	General	Conv.	Diff.	General
[11]	Gen., Gen.	General	Conv.	Same	$\alpha \to \infty$
[49]	Gen., Pha.	2×2 grid	Conv.	Same	$\alpha = 1$
[67]	Gen., Pha.	General	Conv.	Same	$\alpha = 1$
[31]	Gen., Gen.	Tree	Conv.	Same	General
[19] (Rate stab.)	Gen., Gen.	General	Conv.	Diff.	$\alpha \to 0^+$
[55]	Pois., Exp.	General	Non-conv. Time-var. conv.	Diff. Diff.	General General
Open prob.	Gen., Gen. Fast timesc.	General	Conv., non-conv. Time-var.	Diff.	General Non-concave

conditions of aperiodicity and irreducibility). In a general network topology, it is challenging to prove the session stability of NUM-based resource allocation. It is a multi-class queueing network with service rates dependent on the solution to NUM, which in turn relies on the number of active flows. A technique popularly used is called fluid-limit scaling [20], where a suitable scaling (by time and space) of the original system can make the limiting system deterministic rather than random, facilitating the stability proof.

16.3.1.3 State of the art on stability results

The existing research results on session-level stability are summarized in Table 16.1, with different network topologies, shapes of rate region, utility function, and arrival process and file size distributions. The case for time-varying rate regions will be discussed separately in Section 16.3.4.

Polytope and general convex rate region

The first analysis of session-level stability focuses on wired networks with fixed routing, supporting data traffic only [22, 4]. For such networks, the rate region is a polytope formed by the intersection of a finite number of linear capacity constraints, i.e., $\mathcal{R} = \{\rho | R\rho \leq c\}$. For this rate region, assuming timescale separation between resource allocation algorithms and session-level dynamics, it is shown that (i) all α-fair allocations with $\alpha > 0$ provide session-level stability if and only if the vector representing the traffic intensities of session classes lies in the rate region, and (ii) this stability region is also the maximum stability region. An immediate extension is to allow a general convex rate region. It is proved in [5] that the rate region is also the stability region for α-fair allocations, $\alpha > 0$. In some networks, session-level dynamics may operate fast on the same timescale as resource allocation, and hence the assumption on instantaneous convergence of the resource allocation algorithm, with respect to session arrivals, may not hold. Session-level stability without the timescale separation assumption is studied in [53, 93].

General non-convex rate region

There are also many practical scenarios in which the rate regions are non-convex, usually due to the limited capability of underlying resource allocation algorithms. For example, a simple random access may lead to a continuous but non-convex rate region, and quantized levels of parameter control lead to a discrete, thus non-convex, rate region. For non-convex rate regions, it is generally impossible to derive an explicit and exact stability condition due to dependency of the stability condition on session arrival and departure processes, where in particular the departure process is determined by the solutions of non-convex optimization problems.

The initial results in this case try to (i) compute bounds for specific topologies and allocation mechanisms, e.g., [3, 64], (ii) provide exact conditions in a recursive form for a particular class of networks, e.g., [9, 38, 99, 3, 64], or (iii) characterize the stability condition for two classes of sessions and discrete rate region, but not a general non-convex rate region [6]. Recently, the authors in [55] characterize sufficient and necessary conditions for session-level stability for $\alpha > 0$ in networks with an arbitrary number of classes over a discrete rate region. In this case, there exists a gap between the necessary and sufficient conditions for stability. However, these conditions coincide when the set of allocated rate vectors are continuous, leading to an explicit stability condition for a network with continuous non-convex rate regions, summarized as follows. For continuous $\mathcal{R}^\alpha$, where $\mathcal{R}^\alpha$ is the set of rate vectors that can be chosen by the α-fair allocation, the stability region of α-fair allocation is the smallest coordinate-convex set[1] containing $c(\mathcal{R}^\alpha)$, where $c(\mathcal{Y})$ denotes the smallest closed set containing $\mathcal{Y}$. Note that now the stability region varies for different values of α, in contrast to the stability result in the case of a convex rate region.

General arrival and general file size distribution

Another challenging extension is to remove the assumption of Poisson arrivals and/or exponential file size distribution, which is often unrealistic. In such cases, we generally lose the Markovian property, and keeping track of residual file size is not a scalable approach. New fluid limits and tight bounding of the drift in new Lyapunov functions need to be established for the proof of stability.

Again, using the fluid-limit technique in [20], [109] relaxes the assumption of Poisson arrivals, by studying a general stationary and a bursty network model. In [46], a fluid model is formulated for exponentially distributed workload to study

---

[1] A set $\mathcal{Y} \subset \mathbb{R}^n_+$ is said to be coordinate-convex when the following is true: if $b \in \mathcal{Y}$, then for all a: $0 \le a \le b$, $a \in \mathcal{Y}$.

the "invariant states" as an intermediate step for obtaining diffusion approxima-
tion for all $\alpha \in (0, \infty)$. In [31], the fluid model is established for α-fair rate allocation, $\alpha \in (0, \infty)$, under general distributional condition on arrival process and
service distribution. Using this fluid model, they have obtained characterization
of "invariant states," which led to stability of network under α-fair allocation,
$\alpha \in (0, \infty)$, when the network topology is a tree.

For general network topologies, three recent works have tackled this difficult
problem of stochastic stability under general file size distribution, for different special cases of utility functions: [11] establishes stability for maxmin fair
(corresponding to $\alpha \to \infty$) rate allocation, and [67] establishes stability for proportional fair (corresponding to $\alpha = 1$) rate allocation for Poisson arrival and
phase-type file size distribution. Using the fluid model in [31] but under a different scaling, [19] establishes the rate stability of α-fair allocation for general file
size distribution for a continuum of α: α sufficiently close to (but strictly larger
than) 0, and a partial stability result for any $\alpha > 0$ fair allocation policy. It is
also proved that α-fair allocation is rate stable over a convex rate region scaled
down by $1/(1 + \alpha)$, for all $\alpha > 0$, general topology, and possibly different utility
functions for different users. The general problem of session-level stability still
remains open.

16.3.2   Packet-level dynamics

Compared to session-level dynamics, packet-level dynamics takes a shorter timescale and appears in various forms. This chapter focuses on packet-level randomness due to dynamic traffic patterns (e.g., real-time flows), in-network random
queueing such as AQM (Active Queue Management), and noisy feedbacks.

NUM with packet-level randomness

To enable tractable analysis, much research on NUM uses deterministic fluid
approximations that are especially valid under the regime of a large number of
flows. In [88], the authors prove that stochastic delay-difference equations, which
model the system with a primal-based proportional fair congestion controller [45]
under stochastic noise, converge to a deterministic functional differential equation: the trajectory of the average rate (over flows) at the router converges to
that of the deterministic model with the noise part replaced just by its mean.
The result in [88] supports the popular ways of modeling methodology based
on deterministic approximation, at least for log utility functions over a single
link with homogeneous delay. The convergence to a deterministic fluid differential equation occurs over a finite time horizon, and convergence for the infinite
time horizon is also proved under additional technical conditions. The work in
[88] is extended to "TCP-like" controllers in [23]. Other works [1, 103] also justify a deterministic feedback system model. The authors in [103] proved that
the queue dynamics of the router with an AQM mechanism can be accurately
approximated by a sum of a deterministic process and a stochastic process, again

under a large-flow regime. Examples of AQM include Random Early Detection (RED) [29] that marks and/or drops the packet at intermediate routers to signal congestion to the sources. It is also shown in [1] that as both the number of sources and the link capacity increase, the queue process converges to a deterministic process, using a stochastic model for N TCP Reno sources sharing a single bottleneck link with capacity Nc implementing RED.

One of the sources of randomness in the Internet is the set of real-time flows that require a certain QoS, e.g., packet loss probability. The authors in [112] examine the effect of a congestion control mechanism on the QoS of real-time flows. In particular, they study the tradeoff between "aggressiveness" of marking functions at the intermediate router and the achieved QoS for real-time flows using LDP (Large Deviation Principle) technique. This result indicates that elastic users, who are typically not much interested in up and down of transmission rates, but in long-term throughput, helps real-time users with obtaining better QoS.

Application-layer burstiness

Another aspect of packet-level stochastic dynamics is to understand the effect of application-layer burstiness on congestion control at the transport layer. For example, in [13], the authors consider a single link accessed by HTTP flows that alternates between think and transfer times randomly. There are multiple classes of HTTP flows, where a class is identified by the mean of its think and transfer time. Only HTTP in transfer times are assigned a certain amount of network bandwidths by NUM allocations, where the length of transfer time depends on the active number of HTTP flows in transfer times. They proved that again under a large-flow regime, the average throughput, i.e., the throughput aggregated over active flows of each type normalized by the total number of flows of that type, turns out to solve a utility maximization problem with a modified utility function at the transport layer.

Stochastic noisy feedback

In the research on distributed implementation of the NUM problem, feedback is often assumed to be perfectly communicated among the network elements. However, in practice, perfect feedback is impossible, mainly due to probabilistic marking and dropping, contention-induced loss of packets, limited size of messages, etc. In [116], the authors study the impact of stochastic noisy feedback on distributed algorithms, where they considered a primal-dual algorithm in the following general form:

$$x_s(t+1) = \left[x_s(t) + \epsilon(t) \Big(L_{x_s}(\boldsymbol{x}(t), \boldsymbol{\lambda}(t)) \Big) \right]_{\mathcal{D}},$$

$$\lambda_l(t+1) = \left[\lambda_l(t) + \epsilon(t) \Big(L_{\lambda_l}(\boldsymbol{x}(t), \boldsymbol{\lambda}(t)) \Big) \right]_0^\infty,$$

where x_s is the source transmission rate of session s, λ_l is the shadow link price, L is gradient update, ϵ is step size, and $[\cdot]_\mathcal{D}$ is the projection onto the feasible set $\mathcal{D}$.

Noisy feedback adds noise to the gradients L_{x_s} and L_{λ_l}, which become stochastic. Two cases are considered for analysis: (i) *unbiased feedback noise* and (ii) *biased feedback noise* to the gradient estimator. For unbiased feedback noise, it is established, via a combination of the stochastic Lyapunov Stability Theorem and local analysis, that the iterates generated by distributed NUM algorithms converge to an optimum with probability one, whereas for biased feedback noise, in contrast, they converge to a contraction region around the optimal point. These results are extended to an algorithm with multiple timescales based on primal-decomposition. These results confirm those on feedback models with deterministic error [17, 68].

16.3.3   Constraint-level dynamics

In the seminal paper [102] by Tassiulas and Ephremides, a joint-control policy of routing and scheduling that is oblivious to the statistics of arrival process and achieves maximum stability region was proposed. Now consider the case where the arrivals are random but outside the stability region, and a certain amount of utility is defined for transmission of a unit volume of data. Combining the NUM framework with stability requirement is generally non-trivial, due to the fact that rate regions are difficult to characterize by distributed controllers, and can be time-varying due to, e.g., time-varying channels and mobility. This section discusses this research topic that is often referred to as "utility maximization subject to stability" under constraint-level dynamics.

The general problem formulation in this topic is given by the following optimization problem:

$$\begin{aligned} \text{maximize} \quad & \sum_s U_s(\bar{x}_s) \\ \text{subject to} \quad & \bar{x} \in \Lambda, \end{aligned} \tag{16.4}$$

where $\bar{x}$ is the long-term session rate vector, averaged over instantaneous rate $x(\tau)$ at time τ, i.e., $\bar{x}_s = \lim_{t \to \infty} \frac{1}{t} \int_{\tau=0}^{t} x(\tau) d\tau$, and the Λ is the *average* rate region. The Λ is the set of arrival rates for which there exists a control policy to stabilize the system, which is in turn the average of the instantaneous rate regions $\mathcal{R}(t)$.

There are two main formulations, and the associated distributed algorithms, for utility maximization subject to stability: (i) node-based [27, 74, 15, 28] and (ii) link-based ones [54], which we compare in Table 16.2. This section focuses only on the node-based algorithm.

We first introduce the notation: $\text{tx}(l)$ and $\text{rx}(l)$ are the transmitter and the receiver of a link l, respectively, and $Q_{s,v}(t)$ denotes the length of the session s queue at node v at slot t. We also denote by $Q_s(t)$ the queue length of the session s's source node. We now describe the algorithm in Figure 16.2. The main tool

Table 16.2. Comparison of node-based and link-based algorithms.

Algorithm	Λ	Routing	Congestion price	Delay	Queue
Node-based	larger	not fixed	price only at each src queue	large (back-pressure)	per-destination or per-session queues at a node
Link-based	smaller	fixed	aggregate link prices over the path	small	single queue at a link

At each time-slot t, the system runs the following per-layer algorithms:

Congestion control. The source of session s determines its transmission rate $x_s(t)$ by:

$$x_s(t) = U_s'^{-1}(Q_s(t)/V),\qquad (16.5)$$

for some parameter $V > 0$.

Routing. On each link l, decide the session $s_l^*(t)$ that has the MDB (Maximum Differential Backlog), i.e.,

$$s_l^*(t) = \arg\max_{s\in S}\Big(Q_{s,\mathrm{tx}(l)}(t) - Q_{s,\mathrm{rx}(l)}(t)\Big),\quad \forall l \in L.\qquad (16.6)$$

The link l will forward the packets of session $s_l^*(t)$ on this link at the rate of what the scheduling layer below decides. Let $Q_l(t) = \max_{s\in S}\Big(Q_{s,\mathrm{tx}(l)} - Q_{s,\mathrm{rx}(l)}\Big)$.

Scheduling. Allocate rates to links with the rate schedule $r^*(t)$ maximizing the aggregate "weight" (thus referred to as max-weight), i.e.,

$$r^*(t) = \max_{r\in\mathcal{R}(t)}\sum_{l\in L} Q_l(t)r_l.\qquad (16.7)$$

Figure 16.2 Optimal node-based back-pressure algorithm.

for developing this algorithm is LAD that decomposes the optimization problem (16.4) into three different subproblems, each of which is solved by algorithms in one layer in Figure 16.2. The layers interact through queue sizes.

We observe that it is easy to implement (16.5) and (16.6) using only local information. However, the problem in (16.7) is generally hard to implement, and may not be solvable in a distributed manner with reasonably low complexity, unless special structures of the rate region $\mathcal{R}(t)$ are specified and exploited. We will discuss the details and the recent advances of low complexity, distributed implementation of scheduling algorithms in the second half of this chapter.

With regard to the rate control at sources, the controller in (16.5) is referred to as *dual controller* since it is developed from the dual decomposition of (16.4) [27, 74, 15]. There are also other types of controllers mainly based on different decomposition methods and different timescale assumptions on source controller and price updates, one of which is the primal-dual controller [97, 28, 94]. These two types of congestion controllers are first studied in the wired Internet congestion control (see [61, 60, 92] for details). Briefly speaking, a dual controller consists of a gradient-type algorithm for shadow price updates and a static source rate algorithm, i.e., two different timescales for shadow price updates and source rate updates. In the primal-dual algorithm, the source rates and the shadow prices are updated at the same timescale.

The parameter V controls the tradeoff between delay and utility optimality. By choosing sufficiently large V (or sufficiently small step-size), the achieved utility, denoted by $\bar{U}$, can be made arbitrarily close to the optimal utility U^*. However, there may be some cost for large V, such as delay measured by the total average queue lengths [74, 73]. This tradeoff parameter V appears in different forms for different types of algorithms, e.g., K in [28] and the step-size in [15].

Another source of dynamics in constraints is due to topological changes in the network caused by mobility or energy depletion of wireless nodes. The authors in [78] extend the result of [102] by considering topological changes modeled by some stationary process (that has the same timescale as channel- and packet-level dynamics), and show that joint max-weight scheduling and back-pressure routing provably achieving throughput-optimality in the static topology in [102] also achieve throughput-optimality in the system with dynamic topologies. The work in [79, 114] also studied the optimal or near-optimal scheduling/routing schemes when time-varying channel and topology information are not instantly available and delayed.

16.3.4   Combinations of multiple dynamics

Combinations of more than one type of stochastic model also naturally arise. We focus on the question of session-level stability of NUM with a time-varying constraint set here. Existence of multiple stochastic dynamics requires careful treatment of different timescales under which each dynamics is operated. Three timescales are our major interest: (i) T_s: session arrivals, (ii) T_c: constraint set variations, and (iii) T_r: convergence of resource allocation algorithms. We use the notation $A = B$ ($A > B$) when A's timescale is similar to (slower than) B's timescale. We first make a reasonable assumption that $T_s \geq T_r$ and $T_s \geq T_c$, i.e., the session-level dynamics is much slower than other dynamics. Then, the remaining possible five cases, some of which still remain open, are discussed as follows.

T1. $T_c = T_r = T_s$. In [53], the authors study the session-level stability when the three timescales are similar, where it is proved that the maximum stability of α-fair allocation still holds for this case for $\alpha \geq 1$.

T2. $T_c = T_r < T_s$. When the timescales of rate region variations and the resource allocation algorithms are similar, the resource allocation algorithms can harness the rate region variations, by allocating resources *opportunistically*. A typical example of such systems is channel-aware scheduling in cellular networks [2, 59, 8], where fading variations of the channels are exploited to achieve more throughput. The session-level stability is analyzed in this regime in [8], again in the context of wireless cellular systems.

T3. $T_c < T_r < T_s$. This regime corresponds to fast variations of rate regions due to fast channel fluctuations. In this case, even the resource allocation algorithm does not track resource variations, but only sees the *average* of time-varying resource. The stochastic nature of resource variations are masked and invisible to resource allocation algorithms and session-level dynamics. The constant link capacity assumption in many papers on wireless networks implicitly presumes this regime.

T4. $T_r < T_c = T_s$. This regime has been studied recently in [55]. This regime clearly prevents the system from being opportunistic and the constraint-level variations can be exploited only at the expense of compromising delay perceived by the sources. The authors in [55] studied the session-level stability (of α-fair allocation) in this regime, and proved that, similar to the case for non-convex rate regions, the stability region depends on α. In particular, for a two-class network, the authors characterize the pattern of dependence of the stability region on α, by proving that there exists a tradeoff between fairness and stability, i.e., fairness can be enhanced at the expense of reduced network stability. This is in contrast to the case of fixed and convex rate regions, where fairness has no impact on stability.

T5. $T_c < T_r = T_s$. This regime again makes it possible to assume no channel variations with only average channel behavior observable to the resource allocation algorithm and the session-level dynamics. This regime can be regarded as a special case of **T1**.

16.4   Wireless scheduling

Distributed scheduling for MAC (Medium Access Control) – addressing the question of "who talks when in an interference environment" – is often the most challenging part of LAD-based design in wireless networks and assumes various forms of SNUM formulations.

Medium Access Control for wireless networks has been one of the most active research areas for 40 years. In the rest of this chapter, we present a taxonomy of those research results in the context of LAD, such as those since the seminal paper [102], and highlight the recent progress on quantifying and reducing complexity measured by message passing.

Given the vast landscape of scheduling algorithms, a "problem-tree" in Figure 16.3 serves as a representation of the taxonomy that we will follow in the rest of the chapter.

L1. In a local contention neighborhood, transmitters and receivers are located within one hop of each other, whereas in end-to-end connections, congestion control and routing over multiple hops need to be carried out jointly with

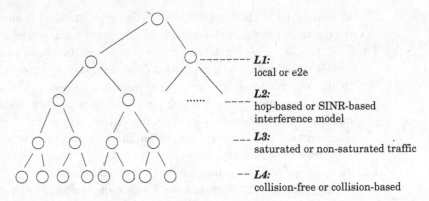

Figure 16.3 A tree of problem formulations in wireless scheduling.

scheduling. With the LAD methodology described in the last section, we will now only focus on the local contention neighborhood in this section.

L2. The K-hop interference model states that no two links whose transmitters or receivers are within K hops of each other can transmit simultaneously without colliding, whereas the SINR (Signal-to-Interference-Noise-Ratio)-based interference model states that collision happens only when the received SINR drops below a threshold. We focus only on the K-hop interference model here. The SINR-based model leads to scheduling problems that are often more challenging and less understood in the current literature.

L3. In unsaturated systems, there is arrival of traffic with finite workload to each node, and (queue, or rate) stability is a key metric, whereas in saturated systems, there is infinite backlog behind each node, and the utility function of the equilibrium rate is often the objective function to be maximized.

L4. Collision-free scheduling can be implemented either at a centralized scheduler or through message passing, whereas collision-based random access may or may not involve message passing.

We will structure this section primarily according to the type of algorithm, and secondarily according to the traffic arrival models, focusing on local contention neighborhood and the K-hop interference model.

In the rest of this section, a wireless network $G(V, L)$ is considered, where V and L are the sets (and also the numbers) of nodes and links, respectively. Time is discretized into slots, $t = 0, 1, \ldots$ For scheduling-based algorithms, we denote $\mathcal{S} \subset \{0,1\}^L$ to be all schedules. A *schedule,* $\boldsymbol{S} = (S_l \in \{0,1\} : l = 1, \ldots, L) \in \mathcal{S}$, is a binary vector representing the schedule, where $S_l = 1$ if the link l is scheduled, and 0 otherwise. We denote by $I(l)$ the set of interfering links of a link l. The maximum link-level rate region $\Lambda \subset \mathbb{R}_+^L$ (also referred to as *throughput region*) is the smallest coordinate-convex set of $\mathcal{S}$. For random access algorithms, we denote by $(p_l : l \in L)$ the access probabilities over links.

16.4.1   Collision-free algorithms

16.4.1.1 MW (max-weight)

As mentioned in the last section, scheduling in multi-hop networks with the objective of throughput-guarantee for unsaturated arrivals dates back to the seminal work by Tassiulas and Ephremides [102], where an algorithm, referred to as *max-weight*, stabilizing the system whenever possible (i.e., *throughput-optimal*), is proposed.

At each slot t, the links in $\boldsymbol{S}^\star(t)$ computed as follows activate transmissions:

$$\boldsymbol{S}^\star(t) = \arg\max_{\boldsymbol{S} \in \mathcal{S}} W(\boldsymbol{S}), \qquad W(\boldsymbol{S}) \triangleq \sum_{l \in L} Q_l(t) S_l. \tag{16.8}$$

$W(\boldsymbol{S})$ denotes the *weight* of a schedule $\boldsymbol{S}$, corresponding to the aggregate sum of queue lengths of links in $\boldsymbol{S}$. Max-weight chooses a schedule with the maximum weight (that may not be unique) at each slot. Solving MW can be reduced to the NP-hard WMIS (Weighted Maximum Independent Set),[2] thus is computationally intractable and typically requires centralized computation.

The power of MW lies in the generality of its application and unawareness of arrival statistics (i.e., making scheduling decisions based on the instantaneous queue lengths). For single-hop, unsaturated sessions, it is proved that any mean arrival vector $\boldsymbol{\lambda} \in \Lambda$ can be stabilized by MW, thus is throughput-optimal. It was extended to end-to-end sessions with both saturated [16] and unsaturated data [102], even under time-varying channels [15, 74].

Maximum weight scheduling is also complex: exponential computational complexity at a centralized scheduler. Randomization and approximation have been two main ideas used to reduce the computational complexity to polynomial, and to turn the centralized calculation into a distributed algorithm with message passing.

16.4.1.2 RPC (Randomized Pick-and-Compare)

The RPC algorithm [101] leads to the possibility of achieving throughput optimality with polynomial complexity. A generalized version of RPC, γ-RPC, $\gamma > 0$, is described first as follows:

---

[2] For a special structure of $\mathcal{S}$, a polynomial-time solution may be possible, e.g., $O(L^3)$ complexity over one-hop interference model, in which case MW is simply the WMW (Weight Maximum Matching) problem.

At each time slot t, the γ-RPC first generates a random schedule $S'(t)$ satisfying **C1**, and then schedules $S(t)$ defined in **C2**.

C1 (Pick). $\exists 0 < \delta \leq 1$, s.t. $\mathbb{P}[S'(t) = S|Q(t)] \geq \delta$, for some schedule S, where $W(S) \geq \gamma W^\star(t)$.

C2 (Compare). $S(t) = \arg\max_{S=\{S(t-1),S'(t)\}} W(S)$.

We explain the intuition of RPC when $\gamma = 1$. Instead of finding the max-weight schedule at each slot, it is enough to find a schedule with probabilistic guarantee of finding an MW schedule (**Pick**), and sustain the good quality of schedules by selecting a schedule that has the larger weight between the schedule randomly chosen at this slot and that at the previous slot (**Compare**). Complexity reduction without sacrificing throughput is possible due to *infrequent* computation of schedules. For $0 < \gamma < 1$, all operations are the same by replacing MW schedules by γ-optimal schedule in terms of weight in the pick operation, and turns out to stabilize any arrival vector in $\gamma\Lambda$ [83, 111].

The authors in [70] proposed a distributed, throughput-optimal scheme using "gossip" algorithms. The key idea is that even the steps of **Pick** and **Compare** do not need to be accurate in order for the stability region to be maximized. In [26], end-to-end sessions are considered, where for the pick operation, each node tosses a coin for medium access and resolves contention through a RTS-CTS type signaling, and for compare operation, it uses the "conflict graph" of the previous schedule and the current random schedule, and a spanning tree is constructed in a distributed manner for distributed comparison of the weights of two schedules. In [83] for unsaturated, single-hop traffic under a one-hop interference model, each node randomly decides to be a *seed*, and then the seeds find a random *augmentation* with a maximum length, say k (a system parameter) for distributed pick operation. An augmentation is basically an alternating sequence of the links in the previous schedule and the current random schedule. Then, the network configures the final schedule by choosing "old" or "new" links in each augmentation for distributed compare operation. The parameter k trades off between throughput and complexity. A more general interference model has been studied in [110]. The authors in [41] applied a *graph partitioning* technique such that the entire network is partitioned into multiple clusters among which no interference exists. Then, each cluster runs RPC in parallel, such that the complexity is polynomial due to the sufficient number of partitions. For networks with polynomial growth, the proposed algorithm can also trade off between throughput and complexity in an arbitrary manner, e.g., ϵ-gap towards optimal throughput leads to the increasing complexity with decreasing ϵ, $\epsilon > 0$.

16.4.1.3 Maximal/greedy algorithms

A natural approach to reduce the complexity of MW is to take a simpler, polynomial-time solvable algorithm that may "roughly" approximate MW in terms of weight at each slot. Three basic algorithms – maximal, greedy, locally-greedy – are described first in the following:

Step 1. Start with an empty schedule and a set $\mathcal{N} = \mathcal{L}$;

Step 2. Choose a link $l \in \mathcal{N}$ in the following manner, and remove from $\mathcal{N}$ the links interfering with link l,

  • **Maximal:** a random link l,
  • **Greedy:** the link l that has the largest queue length,
  • **Locally-greedy:** a random link l that has a locally-longest queue length;[3]

Step 3. Repeat Step 2 until $\mathcal{N}$ is empty.

The best complexities of maximal, greedy, and locally-greedy algorithms known to date are $O((\log L)^4)$, $O(L)$, and $O(L \log V)$, respectively, for the one-hop interference model [80].

Generally, maximum weight scheduling does not allow PTAS (Polynomial-Time Approximation Scheme), i.e., there does not exist polynomial-time algorithms that can approximate WMIS with arbitrarily small performance gap. Thus, for a general network topology, only a specific ratio of partial throughput guarantee is achieved (see Section 16.4.3.1 for more discussions). Initial studies were conducted for single-hop, unsaturated sessions, and one-hop interference model. The first paper that studies maximal scheduling is [21] in the switching system, and [14] for wireless scheduling, where maximal scheduling, which just selects a maximal schedule randomly, achieves 1/2 of the throughput region in the worst case. Greedy and locally-greedy scheduling guarantee 1/2 of the weight from MW at *each* slot, which differs from maximal scheduling. However, the three algorithms are equivalent in terms of worst-case throughput performance [54, 111]. It has been further shown that with end-to-end, saturated sessions, with greedy scheduling, the maximum link-level throughput region can be carried over to the throughput of end-to-end sessions in conjunction with certain routing and congestion control schemes. For a general K-hop interference model, it has been shown in [107, 90] that the worst-case performance for a general K-hop interference model is not that bad: the lower bound on throughput is obtained by $1/\theta$, where θ is the maximum number of non-interfering links in any interference set.

---

[3] A link is said to have a locally-longest queue in $\mathcal{N}$, if its queue length is greater than any of its interfering links also in $\mathcal{N}$.

There are two directions of extensions since the above research: (i) better approximation approach, and (ii) study of greedy algorithms for important special classes of network topologies.

In the direction of (i), in [85], sequential maximal scheduling is applied to achieve 2/3 of the throughput region for trees and $K = 1$. Max-min fair rates within the feasible region of maximal scheduling are attained and throughput loss characterized through "interference degree" in [84]. In [32], further relaxations of maximal scheduling are considered, including one that collects local neighborhood queue information with the complexity independent of network size or topology, and another whose complexity depends on maximum degree.

The direction of (ii) is motivated by [24], which shows that throughput optimality can be achieved by greedy scheduling if a notion of local pooling condition is met. The local pooling condition of a graph consisting of the set of links $\mathcal{L}$ means that every subset $\mathcal{L}' \subset \mathcal{L}$ should satisfy the following: for all $\boldsymbol{\mu}, \boldsymbol{\nu} \in$ the convex hull of the set of all maximal schedules of links in $\mathcal{L}'$, there must exist some link $k \in \mathcal{L}'$ such that $\mu_k < \nu_k$. Then, [12] applied it to wireless mesh networks and [117] extends the local pooling idea to general multi-hop networks. In [39], local pooling factors are calculated for various graphs and a local pooling condition is shown to hold for greedy maximal matching over trees and under the K-hop interference model.

16.4.2   Collision-based algorithms

16.4.2.1 Random access based on slotted aloha

We first consider slotted Aloha (S-Aloha). The algorithm at the MAC-level can be simply described as: each link accesses the channel with the probability p_l, if backlogged. The key question here is the way of adapting $p_l(t)$ over timeslots for good throughput.

Let $\mu_l(\boldsymbol{p})$ be the (average) rate for the access probability vector $\boldsymbol{p} = (p_l : l \in L)$. Then, the average rate over link l is given by when l accesses but its interfering links do not access, i.e.,

$$\mu_l(\boldsymbol{p}) = p_l \prod_{l \in I(l)} (1 - p_l). \tag{16.9}$$

Denote the rate vector $\boldsymbol{\mu}(\boldsymbol{p}) = (\mu_l(\boldsymbol{p}) : l \in L)$. The link-level throughput region of S-Aloha assuming that the queues are saturated (referred to as *maximum saturated region*) is given by the set of all achievable rates for all combinations of $\boldsymbol{p}$, i.e.,

$$\Lambda_{\text{sat}} = \{\boldsymbol{v} \in [0,1]^L | \exists \boldsymbol{p} \in \mathcal{P}, \text{s.t. } \boldsymbol{v} \leq \boldsymbol{\mu}(\boldsymbol{p})\}. \tag{16.10}$$

It is clear that $\Lambda_{sat} \subset \Lambda$, where recall that Λ is the maximum throughput region achieved by MW. See Figure 16.4 for the difference of Λ and Λ_{sat} of a simple network with two interfering links.

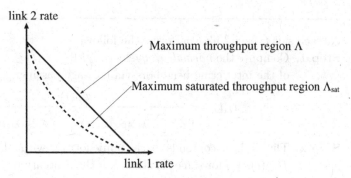

Figure 16.4 Λ and Λ_{sat}.

The authors in [42] considered single-hop, saturated sessions, and studied how to adapt $p_l(t)$ to achieve log-utility (i.e., proportional fair) optimal point, based on message passing to determine $p_l(t)$. It follows the standard NUM framework, but the parameters of utility functions are the average rates as in (16.9). Generalization to α-utility functions has been done in [51]. The current back-off based random access (which is essentially the same as the slotted-Alpha algorithm) has also been reverse-engineered with a non-cooperative game [100]. The aforementioned work under single-hop sessions has been extended to multi-hop sessions under the fixed routing for proportional fairness [105] and α-fairness [50], so that joint congestion control and random access is developed, where still message passing with the interfering links is required to determine the access probability. A recent work in [71, 72] showed that we can avoid step-size tuning and reduce message passing significantly, or even have zero complexity, for certain topologies, using distributed learning of the collision histories.

The research above uses an optimization-based framework, where the dual-based approach is applied for saturated users. For unsaturated and/or multi-hop sessions, queue length may be necessary to either determine the next-hop route of packets or stabilize the system. In fact, the price or Lagrange multiplier in the optimization-based framework essentially corresponds to the queue length of queue-based framework. The work in [34, 95] deals with single-hop sessions, which is extended to multi-hop sessions with variable routing in [56] using the back-pressure mechanism.

16.4.2.2 Random access based on constant time control phase

This line of work studies non-saturated arrivals, and a timeslot is now divided into two parts: control and data slot, where a control slot has M mini-slots (M is a system parameter). The M mini-slots are used to sense the neighborhood activities, and decide the schedule of data transmission. Let $A(v)$ be the incident links to node v. We assume a one-hop interference model, which can be readily extended to a K-hop model.

At each slot t, each link l performs the following:

Step 1. Compute the *normalized queue length* $0 \leq x_l(t) \leq 1$ using queue lengths of the interfering neighbors via message passing:

$$x_l(t) \triangleq \frac{Q_l(t)}{\max\left[\sum_{k \in A(tx(l))} Q_k(t), \sum_{k \in A(rx(l))} Q_k(t)\right]}. \quad (16.11)$$

Step 2. The link l contends each mini-slot m with the probability $p_l = f(x_l(t), M)$ for some function f, if the contention signals from its interfering links are not sensed prior to the mini-slot m.

The function $f(\cdot)$, which we call *access function*, controls the aggressiveness of media access and has to be appropriately chosen to strike a good balance between collision (due to aggressiveness) and under-utilization of media. The algorithms in the literature are categorized by the shape of access functions. Two types of contention functions have been considered so far, where $g(M)$ is an increasing function of M:

$$\text{Type I}: \quad f(x_l(t), M) = g(M)\frac{x_l(t)}{M},$$

$$\text{Type II}: \quad f(x_l(t), M) = 1 - \exp\left(-g(M)\frac{x_l(t)}{M}\right). \quad (16.12)$$

In Type I algorithms, $g(M) = 1$ is considered in [40], which showed that the worst-case throughput ratio is $1/3 - 1/M$.[4] The authors in [40] further showed that $g(M) = (\sqrt{M} - 1)/2$ results in the throughput ratio of $1/2 - 1/\sqrt{M}$. In Type II algorithms, the authors in [32] showed that with $g(M) = \log(2M)/2$, the achieved throughput ratio is at least $1/2 - \log(2M)/2M$ which improves the bound of Type I algorithms and is known to be the best so far.

All algorithms above essentially try to find a *maximal* schedule using the queue lengths of a neighborhood, based on which they set access probability appropriately. That is why the best worst-case throughput performance is roughly the same as that of maximal scheduling for sufficiently large M.

16.4.2.3 Adaptive CSMA (A-CSMA)

Despite the many nice properties of random access based algorithms discussed in earlier sections, they still have some limitations in terms of message passing (Section 16.4.2.2) or small throughput region (Section 16.4.2.1). We naturally ask the following question: *Can we achieve optimality using the schemes based on random access that do not require any message passing?* Optimality here

---

[4] The authors in [52] proposed a slightly different algorithm, which is equivalent to $g(M) = 1$ of Type I access function, in terms of the achieved throughput.

refers to utility-optimality in a saturated traffic case, or rate stability (a weaker notion than queue stability) in an unsaturated traffic case. The problem seems to be challenging, but the answer is positive, and the key idea is to adopt the CSMA (Collision Sense Multiple Access) scheme and adaptively control access probabilities and channel holding times of the links according to only local queue lengths. We present the algorithm for single-hop sessions with saturated traffic, and other cases will be discussed later.

Consider a simple CSMA mechanism, where each link $l \in L$ has two parameters λ_l and μ_l, denoted by $\text{CSMA}(\lambda_l, \mu_l)$: after a successful transmission, $tx(l)$ randomly picks a back-off counter according to some distribution of mean λ_l; it decrements the counter only when the channel is sensed idle; and it starts transmitting when the back-off reaches 0, and remains active for a duration μ_l.

The key part of A-CSMA is the way of controlling λ_l and μ_l based on the queue-length evolution over time. Time is slotted and transmitters update their parameters at the beginning of each slot, formally described by:

At each slot t, the $tx(l)$, $l \in L$ runs $\text{CSMA}(\lambda_l[t], \mu)$, where

$$q_l[t+1] = \left[q_l[t] + \frac{b[t]}{W'(q_l[t])} \left(U'^{-1}(\frac{W(q_l[t])}{V}) - S_l[t] \right) \right]_{q^{\min}}^{q^{\max}}, \quad (16.13)$$

$$\lambda_l[t+1] = \mu^{-1} \exp(W(q_l[t+1])), \quad (16.14)$$

for diminishing step size $b[t]$, some increasing function $W(\cdot)$, and positive parameter $V > 0$ chosen by the system designer, and the $S_l[t]$ corresponds to the total number of packets served at link l.

The equation (16.13) simply represents a virtual queue dynamics (lower- and upper-bounded by some constants). The adaptive choice of CSMA parameters is done as in Equation (16.14). Note that no message passing is necessary for A-CSMA, and only a sensing capability that is typically autonomous by measuring interference from neighbors, is required for implementation.

Adaptive CSMA above can be shown to be arbitrarily close to optimality, in terms of rate stability for unsaturated traffic or utility for saturated traffic. We first consider an ideal case in which the back-off timer and holding time are continuous so that there exist no collisions. The key idea of near-optimality of A-CSMA is as follows: for a fixed $(\lambda_l, \mu_l), l \in L$, the system evolves according to a reversible Markov process (e.g., see [25] and references therein) with its state being a schedule at time t, say, $m(t) \in S$. Then, the stationary distribution of a reversible Markov process $m(t)$ has a product-form with the intensity of links (i.e., $\rho_l = \lambda_l \times \mu$) as key parameters and is also insensitive to the holding time distribution, allowing us to use a constant holding time μ and only control λ_l. It turns out that choosing intensities according to (16.14) enables the system to concentrate on the max-weight with arbitrarily high probability. The key challenge for proving convergence without assuming timescale separation, however, is that the queue lengths still change dynamically while CSMA is running (thus the parameters of CSMA also change).

Historically, the case for the fixed access parameters $(\lambda_l, \mu_l : l \in L)$ can be considered generally in the theory of loss networks [43]. It has been demonstrated that even non-adaptive CSMA could also achieve strong throughput performance close to throughput-optimality [25, 81, 7].

Turning to random access with adaptive channel access rate, [35] first proposed a simulated-annealing based approach to general distributed scheduling. A similar idea has been applied to wireless scheduling in several recent papers: for saturated [36, 58, 57] and unsaturated arrivals [91] with [36] first developing a utility-optimal CSMA algorithm for wireless networks. For unsaturated sessions with the objective being throughput-optimality, the authors in [91] proved that when the weight function $W(\cdot)$ is sufficiently slow (e.g., $\log\log(\cdot)$ in [91]), the random access algorithm is asymptotically optimal. For saturated sessions, utility-optimality is shown in [36, 58, 57], where the weight function $W(\cdot)$ is relaxed and does not have to be very slow, since the decreasing step size $b[t]$ is responsible for making the dynamics of $(q_l[t], l \in \mathcal{L})_{t=0}^{\infty}$ sufficiently slow, such that the throughput over links is averaged out.

The proof is based on *stochastic approximation with continuous-time controlled Markov noise*, where $(\boldsymbol{q}[t])_{t=0}^{\infty}$ is slowly updated when the step size $b[t]$ is sufficiently small and the (non-homogeneous) Markov process with the kernel controlled by $\boldsymbol{q}[t]$. In this case, we can prove that the system converges to an ordinary differential equation with schedules at the stationary regime, whose converging point is in turn a solution of the slightly modified optimization problem from the original one. The difference gap between the modified and the original algorithm is a decreasing function of V, so as to be reduced arbitrarily by increasing V. In addition, the issues arising when the ideal continuous-time algorithms without collisions are applied to the practical discrete-time setting with collisions, such as their performance gap, avoiding collisions, trade off between short-term fairness and efficiency, have been studied in [58, 76, 57].

Related work also includes [66, 65] that study the throughput- and utility-optimality in the sense that the number of users is sufficiently large and the sensing period is sufficiently small, extending the fixed point analysis of CSMA in the single-hop wireless networks.

16.4.3   Performance–complexity tradeoff

The previous discussion focused primarily on throughput as represented through stability region or utility function. Other performance issues, especially delay, as well as various measures of complexity, such as time complexity,[5] are also important although relatively under-explored. We start this section on these topics with a 2D tradeoff discussion.

16.4.3.1 2D throughput–complexity tradeoff

A first goal here is to develop a family of parameterized algorithms that can arbitrarily trade off between throughput and complexity. The question is on the parameterized algorithms that can *arbitrarily trade off from 0 to 1 with polynomial complexity.*

We first explain why realizing this arbitrary tradeoff in parameterized algorithms needs a non-trivial study. A candidate approach is to take a computation-theoretic approach of finding approximating algorithms to the max-weight scheduling (the WMIS problem), i.e., finding "ϵ-optimal" algorithms that achieve $(1 - \epsilon)W^*$ with polynomial-time complexity being a decreasing function of ϵ (note that the algorithm guaranteeing $(1 - \epsilon)W^*$ at each slot achieves $(1 - \epsilon)$ of the throughput region [54]). The difficulty of such an approach comes from the fact that the WMIS problem does not allow PTAS (Polynomial Time Approximation Scheme), and only permits special approximation ratios such as $1/2$ of greedy scheduling under the one-hop interference model. We survey the recent advances on this subject by categorizing them into two approaches, briefly mentioned earlier in Sections 16.4.1.3 and 16.4.1.2.

The first approach is to consider only a special, yet wide class of network topologies that allow PTAS. The authors in [82] divide an entire graph into disjoint clusters of links also called "stripping," such that each cluster does not interfere with other clusters. With this "graph partitioning" technique, the schedule computations inside each cluster occur in a parallelized manner, leading to a reduction of complexity. A similar idea of graph partitioning has been used in [41] in conjunction with the approach explained next.

Another approach is to use γ-RPC in Section 16.4.1.2 for general networks that enables us to reduce complexity without loss of throughput. In other words, throughput ratios that cannot be given by the WMIS approximation can be "stuffed" by RPC with the corresponding γ. In [83], the authors propose a family of distributed scheduling algorithms parameterized by, say k, and show that the algorithm k achieves $k/(k + 2)$ of the maximum throughput region, under the one-hop interference model. The k-algorithm requires $4k + 2$ rounds as control overhead, where one round is measured by the time for sending a message and receiving an ACK to and from neighboring nodes. The work in [83] has been extended to the case of the general M-hop interference model in [110].

---

[5] The number of operations in centralized scheduling or the frequency of message passing in distributed scheduling.

16.4.3.2 3D throughput–delay–complexity tradeoff

How is it possible that time complexity can be reduced arbitrarily without affecting throughput? A simple answer is that throughput is defined with respect to the asymptotic concept of stability, and a price to pay for complexity reduction is delay. The authors in [111] quantify the 3D tradeoffs among throughput, complexity, and delay, and disclose the pros and cons of randomized techniques in a general setup, as summarized in Figure 16.5.

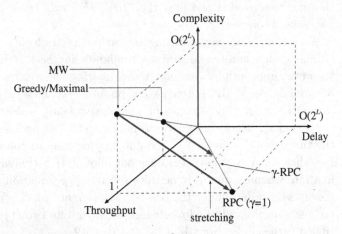

Figure 16.5 3D tradeoff among throughput, delay, and complexity.

The key intuition is as follows: consider the following simple scheduling, which we call *m-stretched MW*: on every $m < \infty$ slots, an optimal max-weight schedule is computed and updated, and between the slots of max-weight schedules, the schedule at the previous slot is used without any change. The m-stretched MW is probably throughput-optimal, since the throughput region, measured by stability region of arrivals, is an asymptotic concept only defined over very long time, so that infrequent computation of optimal schedules does not affect throughput. Note that since max-weight schedules are computed every m slots, the amortized complexity (per slot) can be reduced to $O(2^L/m)$. However, infrequent schedule update has an adverse impact on delay which linearly increases with m. Reduction of complexity from exponential to polynomial entails exponential increase of delay.

The RPC algorithm shares a similar approach as m-stretched MW. Randomly selecting a schedule that has positive probability δ of finding a max-weight schedule (**Pick**), it finds a max-weight schedule every $1/\delta$ slots on average. From **Compare**, a schedule as good as the schedule at the previous slot in terms of their weights is used, which is necessary for distributed, randomized implementation. Then, again to reduce complexity from exponential to polynomial, it is required that $\delta = O(1/2^L)$, which is verified by all algorithms based on RPC, leading to exponential delay increase.

16.4.3.3 Delay characterization and reduction

Analyzing performance or developing delay optimal scheduling is challenging due to intractable coupling of queues over links. Note that the queueing-theoretic interpretation of wireless scheduling is a type of constrained queueing system, where scheduling rules induce high coupling of queueing dynamics among links due to interferences, introducing technical difficulties especially for general network topologies. Thus, research in the literature mainly relies on *approximation* or (order-wise) *bound* techniques, which we divide into four categories explained next.

Lyapunov bounds

Bounds (on the total average queue lengths of links, i.e., $\sum_l \mathbb{E}[Q_l(t)]$) are tractable as a "by-product" of throughput proof by intelligently manipulating Lyapunov drifts used to prove stability. The bound is first studied for non-saturated traffic [75, 41, 111] for different system models and scheduling rules. For non-saturated (but not infinitely backlogged) traffic, the tradeoff between achieved utility and queue-length bound has been studied in [74, 73, 28, 15] (see Section 16.3.3), where especially in [73], it was shown that the order-optimal tradeoff is $O(V)$ queue length bound vs. $O(\log V)$ utility gap, where V is the tradeoff system parameter. The 3D tradeoff research in Section 16.4.3 is also based on this bound technique, which generalizes [83, 82]. Lyapunov bounds are often very loose, thus difficult to be applied to practical systems as a real delay performance.

Large deviations analysis

In the setting based on large deviations, one is interested in algorithms that maximize the asymptotic decay-rate of queue overflow probability, as measured by $\mathbb{P}[\max_i Q_i(0) > B]$, for some threshold B, and $Q_i(0)$ is the stationary queue length of link i [87, 115, 98, 104]. This research has been conducted only in a single-cell wireless network, since, again due to dimensionality and other technical difficulties, the proof techniques are not scalable for general multi-hop networks. In [98], it was shown that a concept called "exponential rule" achieves optimal decay-rate.

Heavy-traffic analysis

Unlike the previous two approaches, heavy traffic analysis focuses on the network model with bottleneck links (i.e., arrival volumes are almost at system capacities at these links). The key idea is to approximate the original system $Q(t) = \{Q_l(t)\}$ by considering a sequence of systems, appropriately scaled in time and space, and studying the limiting system. The limiting system is often an interesting one that allows mathematical tractability. The key technical challenge lies in an important intermediate step of proving *state space collapse* for reduction of dimensions (from L to one), permitting, i.e., multi-dimensional queue-length vectors to be expressed as a scaled version of a one-dimensional *workload* process. The research on heavy traffic analysis in scheduling has been conducted in [96, 86, 89] that essentially validates the delay-optimality of max-weight scheduling or its variants in the sense of a long-term average workload or path-wise. The major techniques are borrowed from the heavy traffic analysis in multi-class queueing networks, e.g., [10, 106]. A more practical system that explicitly considers signaling complexity has been studied in [113], by modeling time of message passing with vacations. Explicit consideration of signaling complexity leads to different trade-offs among delay, throughput, and complexity, which in turn depends in the various regimes on the ratio of the time duration of signaling to that of data transmission.

System simplification

The final approach is to modify the original queueing system into a simpler one which, however, still captures useful features for delay analysis. Examples include [33] that uses the idea of (K, X)-*bottleneck* – a set of links X such that no more than K of them can simultaneously transmit, from which an efficient technique is developed to reduce such bottlenecks to a single queue system fed by a refined arrival process from the original arrival process. They proved that delay analysis of the simplified system is tractable from the standard queueing theory, and provide a good lower-bound on delay.

16.4.4   Future research directions

While there has been substantial progress in LAD with stochastic dynamics on the problem of wireless scheduling, there are still many open problems and under-explored topics.

Q1. How should we consider the "effective performance" of scheduling algorithms by deducing the impact of complexity, for general graphs?

Q2. We only have "achievability" curves in most cases of the 3D tradeoff space. What about achievability surface or converse curves or surfaces? Can we also characterize the tradeoff between performance and complexity in terms of computation, communication, spatial complexity, and message size?

Q3. Delay analysis has been conducted mainly based on approximation or asymptotics. Can we characterize delay more sharply and minimize it?

Q4. Can we understand algorithms without message passing (e.g., A-CSMA) better in terms of cost to be paid as well as their transient performance? How can we transfer theory to practice in actual implementaion and deployment?

Q5. How can we jointly control scheduling and power control in a distributed way? As two mechanisms controlling interference, one over the time axis and the other over the power axis, scheduling and power control for the SIR-based interference model remains a challenging exercise for LAD and SNUM.

Acknowledgements

We thank the coauthors of recent papers on scheduling algorithms, including S. Chong, J. Lee, E. Knightly, B. Nardelli, J. Liu, H. V. Poor, A. Proutiere, A. Stolyar, and J. Zhang for helpful discussions and collaboration. This work is supported in part by US Presidential Early Career Award for Scientists and Engineers grant N00014-09-1-0449, AFOSR grant FA9550-09-1-0134, IT R&D program of MKE/IITA [2009-F-045-01].

References

[1] Baccelli F., McDonald D. R., Reynier J. (2002). A mean-field model for multiple TCP connections through a buffer implementing RED. *Performance Evaluation* **49**, 1–4, 77–97.

[2] Bender P., Black P., Grob M., *et al.* (2000). CDMA/HDR: a bandwidth-efficient high-speed wireless data service for nomadic users. *IEEE Communications Magazine* **38**, 4, 70–77.

[3] Bonald T., Borst S., Hegde N., Proutière A. (2004). Wireless data performance in multicell scenarios. In *Proceedings of ACM Sigmetrics*.

[4] Bonald T., Massoulie L. (2001). Impact of fairness on internet performance. In *Proceedings of ACM Sigmetrics*.

[5] Bonald T., Massoulie L., Proutière A., Virtamo J. (2006). A queueing analysis of max-min fairness, proportional fairness and balanced fairness. *Queueing Systems* **53**, 1–2, 65–84.

[6] Bonald T., Proutière A. (2006). Flow-level stability of utility-based allocations for non-convex rate regions. In *Proceedings of the 40th Conference on Information Sciences and Systems*.

[7] Bordenave C., McDonald D., Proutière A. (2008). Performance of random multi-access algorithms, an asymptotic approach. *Proceedings of ACM Sigmetrics*.

[8] Borst S. (2003). User-level performance of channel-aware scheduling algorithms in wireless data networks. In *Proceedings of IEEE Infocom*.

[9] Borst S., Leskela L., Jonckheere M. (2008). Stability of parallel queueing systems with coupled rates. *Discrete Event Dynamic Systems*. In press.

[10] Bramson M. (1998). State space collapse with application to heavy traffic limits for multiclass queueing networks. *Queueing Systems* **30**, 1–2, 89–148.

[11] Bramson M. (2005). Stability of networks for max-min fair routing. In *Presentation at INFORMS Applied Probability Conference*.

[12] Brzesinski A., Zussman G., Modiano E. (2006). Enabling distributed throughput maximization in wireless mesh networks: a partitioning approach. In *Proceedings of ACM Mobicom*.

[13] Chang C. S., Liu Z. (2004). A bandwidth sharing theory for a large number of http-like connections. *IEEE/ACM Transactions on Networking* **12**, 5, 952–962.

[14] Chaporkar P., Kar K., Sarkar S. (2005). Throughput guarantees through maximal scheduling in wireless networks. In *Proceedings of the 43rd Annual Allerton Conference on Communication, Control and Computing.*

[15] Chen L., Low S. H., Chiang M., Doyle J. C. (2006). Joint optimal congestion control, routing, and scheduling in wireless ad hoc networks. In *Proceedings of IEEE Infocom.*

[16] Chen L., Low S. H., Doyle J. C. (2005). Joint congestion control and medium access control design for wireless ad-hoc networks. In *Proceedings of IEEE Infocom.*

[17] Chiang M. (2005). Balancing transport and physical layers in wireless multihop networks: jointly optimal congestion control and power control. *IEEE Journal on Selected Areas in Communications* **23**, 1, 104–116.

[18] Chiang M., Low S. H., Calderbank A. R., Doyle J. C. (2007). Layering as optimization decomposition. *Proceedings of the IEEE* **95**, 1, 255–312.

[19] Chiang M., Shah D., Tang A. (2006). Stochastic stability of network utility maximization: general file size distribution. In *Proceedings of Allerton Conference.*

[20] Dai J. G. (1995). On positive Harris recurrence of multiclass queueing networks: a unified approach via fluid limit models. *Annals of Applied Probability* **5**, 49–77.

[21] Dai J. G., Prabhakar B. (2000). The throughput of data switches with and without speedup. In *INFOCOM.*

[22] de Veciana G., Lee T., Konstantopoulos T. (2001). Stability and performance analysis of networks supporting elastic services. *IEEE/ACM Transactions on Networking* **1**, 2–14.

[23] Deb S., Shakkottai S., Srikant R. (2005). Asymptotic behavior of internet congestion controllers in a many-flows regime. *Mathematics of Operation Research* **30**, 2, 420–440.

[24] Dimaki A., Walrand J. (2006). Sufficient conditions for stability of longest queue first scheduling: second order properties using fluid limits. *Advances in Applied Probability* **38**, 2, 505–521.

[25] Durvy M., Thiran P. (2006). Packing approach to compare slotted and non-slotted medium access control. In *Proceedings of IEEE Infocom.*

[26] Eryilmaz A., Ozdaglar A., Modiano E. (2007). Polynomial complexity algorithms for full utilization of multi-hop wireless networks. In *Proceedings of IEEE Infocom.*

[27] Eryilmaz A., Srikant R. (2005). Fair resource allocation in wireless networks using queue-length-based scheduling and congestion control. In *Proceedings of IEEE Infocom.*

[28] Eryilmaz A., Srikant R. (2006). Joint congestion control, routing, and MAC for stability and fairness in wireless networks. *IEEE Journal on Selected Areas in Communications* **24**, 8, 1514–1524.

[29] Floyd S., Jacobson V. (1993). Random early detection gateways for congestion avoidance. *IEEE/ACM Transactions on Networking* **1**, 4 (August), 397–413.

[30] Griffin T. G., Shepherd F. B., Wilfong G. (2002). The stable paths problem and interdomain routing. *IEEE/ACM Transactions on Networking* **10**, 2, 232–243.

[31] Gromoll H. C., Williams R. (2007). Fluid limit of a network with fair bandwidth sharing and general document size distribution. *Annals of Applied Probability*. In press.

[32] Gupta A., Lin X., Srikant R. (2007). Low-complexity distributed scheduling algorithms for wireless networks. In *Proceedings of IEEE Infocom*.

[33] Gupta G. R., Shroff N. (2009). Delay analysis of multi-hop wireless networks. In *Proceedings of IEEE Infocom*.

[34] Gupta P., Stolyar A. (2006). Optimal throughput in general random access networks. In *Proceedings of CISS*.

[35] Hajek B. (1988). Cooling schedules for optimal annealing. *Mathematics of Operations Research* **13**, 2, 311–329.

[36] Jiang L., Walrand J. (2008). A CSMA distributed algorithm for throughput and utility maximization in wireless networks. Technical report, UCB.

[37] Jin C., Wei D. X., Low S. H. (2004). Fast TCP: motivation, architecture, algorithms, and performance. In *Proceedings of IEEE Infocom*.

[38] Jonckheere M., Borst S. (2006). Stability of multi-class queueing systems with state-dependent service rates. In *Proceedings of IEEE Value Tools*.

[39] Joo C., Lin X., Shroff N. B. (2008). Understanding the capacity region of the greedy maximal scheduling algorithm in multihop wireless networks. In *Proceedings of IEEE Infocom*.

[40] Joo C., Shroff N. B. (2007). Performance of random access scheduling schemes in multi-hop wireless networks. In *Proceedings of IEEE Infocom*.

[41] Jung K., Shah D. (2007). Low delay scheduling in wireless network. In *Proceeding of ISIT*.

[42] Kar K., Sarkar S., Tassiulas L. (2004). Achieving proportional fairness using local information in Aloha networks. *IEEE Transactions on Automatic Control* **49**, 10, 1858–1862.

[43] Kelly F. (1979). *Reversibility and Stochastic Networks*. Wiley.

[44] Kelly F. P. (1997). Charging and rate control for elastic traffic. *European Transactions on Telecommunications* **8**, 33–37.

[45] Kelly F. P., Maulloo A., Tan D. (1998). Rate control in communication networks: shadow prices, proportional fairness and stability. *Journal of the Operational Research Society* **49**, 237–252.

[46] Kelly F. P., Williams R. J. (2004). Fluid model for a network operating under a fair bandwidth-sharing policy. *Annals of Applied Probability* **14**, 1055–1083.

[47] Kunniyur S., Srikant R. (2000). End-to-end congestion control: utility functions, random losses and ECN marks. In *Proceedings of IEEE Infocom.*

[48] La R. J., Anantharam V. (2002). Utility-based rate control in the internet for elastic traffic. *IEEE/ACM Transactions on Networking* **10**, 2, 272–286.

[49] Lakshmikantha A., Beck C. L., Srikant R. (2004). Connection level stability analysis of the internet using the sum of squares (SOS) techniques. In *Proceedings of the 38th Conference on Information Sciences and Systems.*

[50] Lee J. W., Chiang M., Calderbank A. R. (2006a). Jointly optimal congestion and contention control based on network utility maximization. *IEEE Communication Letters* **10**, 3, 216–218.

[51] Lee J. W., Chiang M., Calderbank R. A. (2006b). Utility-optimal medium access control: reverse and forward engineering. In *Proceedings of IEEE Infocom.*

[52] Lin X., Rasool S. (2006). Constant-time distributed scheduling policies for ad hoc wireless networks. In *Proceedings of IEEE CDC.*

[53] Lin X., Shroff N. B. (2004). On the stability region of congestion control. In *Proceedings of the 42nd Annual Allerton Conference on Communication, Control and Computing.*

[54] Lin X., Shroff N. B. (2005). The impact of imperfect scheduling on cross-layer rate control in wireless networks. In *Proceedings of IEEE Infocom.*

[55] Liu J., Proutière A., Yi Y., Chiang M., Poor V. H. (2007). Flow-level stability of data networks with non-convex and time-varying rate regions. In *Proceedings of ACM Sigmetrics.*

[56] Liu J., Stolyar A., Chiang M., Poor H. V. (2008). Queue backpressure random access in multihop wireless networks: optimality and stability. *IEEE Transactions on Information Theory.* In submission.

[57] Liu J., Yi Y., Proutière A., Chiang M., Poor H. V. (2009a). Adaptive CSMA: approaching optimality without message passing. *Wiley Journal of Wireless Communications and Mobile Computing, Special Issue on Recent Advances in Wireless Communications and Networking.*

[58] Liu J., Yi Y., Proutière A., Chiang M., Poor H. V. (2009b). Maximizing utility via random access without message passing. Technical report, Microsoft Research Labs, UK. September.

[59] Liu X., Chong E. K. P., Shroff N. B. (2001). Opportunistic transmission scheduling with resource-sharing constraints in wireless networks. *IEEE Journal on Selected Areas in Communications* **19**, 10, 2053–2064.

[60] Low S., Srikant R. (2003). A mathematical framework for designing a low-loss low-delay internet. *Network and Spatial Economics, Special Issue on Crossovers between Transportation Planning and Telecommunications* **4**, 75–101.

[61] Low S. H. (2003). A duality model of TCP and queue management algorithms. *IEEE/ACM Transactions on Networking* **11**, 4, 525–536.

[62] Low S. H., Lapsley D. E. (1999). Optimization flow control, I: Basic algorithm and convergence. *IEEE/ACM Transactions on Networking*, 861–875.

[63] Low S. H., Peterson L., Wang L. (2002). Understanding vegas: a duality model. *Journal of ACM* **49**, 2 (March), 207–235.

[64] Luo W., Ephremides A. (1999). Stability of n interacting queues in random-access systems. *IEEE Transactions on Information Theory* **45**, 5, 1579–1587.

[65] Marbach P., Eryilmaz A. (2008). A backlog-based CSMA-mechanism to achieve fairness and throughput-optimality in multihop wireless networks. In *Proceedings of the 46th Annual Allerton Conference on Communication, Control and Computing*.

[66] Marbach P., Eryilmaz A., Ozdaglar A. (2007). Achievable rate region of CSMA schedulers in wireless networks with primary interference constraints. In *Proceedings of IEEE CDC*.

[67] Massoulie L. (2007). Structural properties of proportional fairness: stability and insensitivity. *Annals of Applied Probability* **17**, 3, 809–839.

[68] Mehyar M., Spanos D., Low S. H. (2004). Optimization flow control with estimation error. In *Proceedings of IEEE Infocom*.

[69] Mo J., Walrand J. (2000). Fair end-to-end window-based congestion control. *IEEE/ACM Transactions on Networking* **8**, 5, 556–567.

[70] Modiano E., Shah D., Zussman G. (2006). Maximizing throughput in wireless networks via gossiping. In *Proceedings of ACM Sigmetrics*.

[71] Mohsenian-Rad A. H., Huang J., Chiang M., Wong V. W. S. (2009a). Utility-optimal random access: optimal performance without frequent explicit message passing. *IEEE Transactions on Wireless Communications*. To appear.

[72] Mohsenian-Rad A. H., Huang J., Chiang M., Wong V. W. S. (2009b). Utility-optimal random access: reduced complexity, fast convergence, and robust performance. *IEEE Transactions on Wireless Communications* **8**, 2, 898–911.

[73] Neely M. J. (2006). Super-fast delay tradeoffs for utility optimal fair scheduling in wireless networks. *IEEE Journal on Selected Areas in Communications* **24**, 8, 1489–1501.

[74] Neely M. J., Modiano E., Li C. (2005). Fairness and optimal stochastic control for heterogeneous networks. In *Proceedings of IEEE Infocom*.

[75] Neely M. J., Modiano E., Rohrs C. E. (2002). Tradeoffs in delay guarantees and computation complexity for $n \times n$ packet switches. In *Proceedings of CISS*.

[76] Ni J., Srikant R. (2009). Distributed CSMA/CA algorithms for achieving maximum throughput in wireless networks. In *Invited talk in Information Theory and Applications Workshop*.

[77] Palomar D., Chiang M. (2006). Alternative decompositions for distributed maximization of network utility: framework and applications. In *Proceedings of IEEE Infocom*.

[78] Pantelidou A., Ephremides A., Tits A. L. (2005). Maximum throughput scheduling in time-varying-topology wireless ad-hoc networks. In *Proceedings of Conference on Information Sciences and Systems*.

[79] Pantelidou A., Ephremides A., Tits A. L. (2009). A cross-layer approach for stable throughput maximization under channel state uncertainty. *ACM/Kluwer Journal of Wireless Networks*. To appear.

[80] Preis R. (1999). Linear time 1/2-approximation algorithm for maximum weighted matching in general graphs. In *Proceedings of STOC*.

[81] Proutière A., Yi Y., Chiang M. (2008). Throughput of random access without message passing. In *Proceedings of CISS*.

[82] Ray S., Sarkar S. (2007). Arbitrary throughput versus complexity tradeoffs in wireless networks using graph partitioning. In *Proceedings of Information Theory and Applications Second Workshop*.

[83] Sanghavi S., Bui L., Srikant R. (2007). Distributed link scheduling with constant overhead. In *Proceedings of ACM Sigmetrics*.

[84] Sarkar S., Chaporkar P., Kar K. (2006). Fairness and throughput guarantee with maximal scheduling in multihop wireless networks. In *Proceedings of Wiopt*.

[85] Sarkar S., Kar K. (2006). Achieving 2/3 throughput approximation with sequential maximal scheduling under primary internference constraints. In *Proceedings of Allerton*.

[86] Shah D., Wischik D. J. (2006). Optimal scheduling algorithms for input-queued switches. In *Proceedings of IEEE Infocom*.

[87] Shakkottai S. (2008). Effective capacity and QoS for wireless scheduling. *IEEE Transactions on Automatic Control* **53**, 3, 749–761.

[88] Shakkottai S., Srikant R. (2004). Mean FDE models for Internet congestion control under a many-flows regime. *IEEE Transactions on Information Theory* **50**, 6 (June).

[89] Shakkottai S., Srikant R., Stolyar A. (2004). Pathwise optimality of the exponential scheduling rule for wireless channels. *Advances in Applied Probability* **36**, 4, 1021–1045.

[90] Sharma G., Mazumdar R. R., Shroff N. B. (2006). On the complexity of scheduling in wireless networks. In *Proceedings of ACM Mobicom*.

[91] Shin J., Shah D., Rajagopalan S. (2009). Network adiabatic theorem: an efficient randomized protocol for contention resolution. In *Proceedings of ACM Sigmetrics*.

[92] Srikant R. (2004). *The Mathematics of Internet Congestion Control*. Birkhauser.

[93] Srikant R. (2005). On the positive recurrence of a Markov chain describing file arrivals and departures in a congestion-controlled network. In *IEEE Computer Communications Workshop*.

[94] Stolyar A. (2006a). Greedy primal-dual algorithm for dynamic resource allocation in complex networks. *Queueing Systems* **54**, 203–220.

[95] Stolyar A. (2008). Dynamic distributed scheduling in random access network. *Journal of Applied Probability* **45**, 2.

[96] Stolyar A. L. (2004). Maxweight scheduling in a generalized switch: state space collapse and workload minimization in heavy traffic. *Annals in Applied Probability* **14**, 1, 1–53.

[97] Stolyar A. L. (2005). Maximizing queueing network utility subject to stability: greedy primal-dual algorithm. *Queueing Systems* **50**, 4, 401–457.

[98] Stolyar A. L. (2006b). Large deviations of queues under QoS scheduling algorithms. In *Proceedings of Allerton*.

[99] Szpankowski W. (1994). Stability conditions for some multi-queue distributed systems: buffered random access systems. *Annals of Applied Probability* **26**, 498–515.

[100] Tang A., Lee J. W., Chiang M., Calderbank A. R. (2006). Reverse engineering MAC. In *Proceedings of IEEE Wiopt*.

[101] Tassiulas L. (1998). Linear complexity algorithms for maximum throughput in radio networks and input queued switches. In *Proceedings of IEEE Infocom*.

[102] Tassiulas L., Ephremides A. (1992). Stability properties of constrained queueing systems and scheduling for maximum throughput in multihop radio networks. *IEEE Transactions on Automatic Control* **37**, 12, 1936–1949.

[103] Tinnakornsrisuphap P., La R. J. (2004). Characterization of queue fluctuations in probabilistic AQM mechanisms. In *Proceedings of ACM Sigmetrics*.

[104] Venkataramanan V. J., Lin X. (2007). Structural properties of LDP for queue-length based wireless scheduling algorithms. In *Proceedings of Allerton*.

[105] Wang X., Kar K. (2005). Cross-layer rate optimization for proportional fairness in multihop wireless networks with random access. In *Proceedings of ACM Mobihoc*.

[106] Williams R. J. (1998). An invariance principle for semimartingale reflecting brownian motions in an orthant. *Queueing Systems and Theory Applications* **30**, 1–2, 5–25.

[107] Wu X., Srikant R. (2006). Bounds on the capacity region of multi-hop wireless networks under distributed greedy scheduling. In *Proceedings of IEEE Infocom*.

[108] Yäiche H., Mazumdar R. R., Rosenberg C. (2000). A game theoretic framework for bandwidth allocation and pricing of elastic connections in broadband networks: theory and algorithms. *IEEE/ACM Transactions on Networking* **8**, 5, 667–678.

[109] Ye H., Ou J., Yuan X. (2005). Stability of data networks: stationary and bursty models. *Operations Research* **53**, 107–125.

[110] Yi Y., Chiang M. (2008). Wireless scheduling with $O(1)$ complexity for m-hop interference model. In *Proceedings of ICC*.

[111] Yi Y., Proutière A., Chiang M. (2008). Complexity in wireless scheduling: Impact and tradeoffs. In *Proceedings of ACM Mobihoc*.

[112] Yi Y., Shakkottai S. (2008). On the elasticity of marking functions in an integrated network. *IEEE Transactions on Automatic Control*. In press.

[113] Yi Y., Zhang J., Chiang M. (2009). Delay and effective throughput of wireless scheduling in heavy traffic regimes: vacation model for complexity. In *Proceedings of ACM Mobihoc*.

[114] Ying L., Shakkottai S. (2009). Scheduling in mobile wireless networks with topology and channel-state uncertainty. In *Proceedings of IEEE Infocom*.

[115] Ying L., Srikant R., Eryilmaz A., Dullerud G. E. (2006). A large deviations analysis of scheduling in wireless networks. *IEEE Transactions on Information Theory*, 5088–5098.

[116] Zhang J., Zheng D., Chiang M. (2008). The impact of stochastic noisy feedback on distributed network utility maximization. *IEEE Transactions on Information Theory* **54**, 2, 645–665.

[117] Zussman G., Brzesinski A., Modiano E. (2008). Multihop local pooling for distributed throughput maximization in wireless networks. In *Proceedings of IEEE Infocom*.

[118] Lee J., Lee J., Yi Y., Chong S., Proutière A., Chiang M. (2009). Implementing utility-optimal CSMA. In *Proceedings of Allerton Conference*.

[119] Nardelli B., Lee J., Lee K., Yi Y., Chong S., Knightly E. W., Chiang M. (2010). Technical report, Rice University.

17 Network coding in bi-directed and peer-to-peer networks

Zongpeng Li[†], Hong Xu[‡], and Baochun Li[‡]

[†]University of Calgary, Canada [‡]University of Toronto, Canada

Network coding has been shown to help achieve optimal throughput in directed networks with known link capacities. However, as real-world networks in the Internet are bi-directed in nature, it is important to investigate theoretical and practical advantages of network coding in more realistic bi-directed and peer-to-peer (P2P) network settings. In this chapter, we begin with a discussion of the fundamental limitations of network coding in improving routing throughput and cost in the classic undirected network model. A finite bound of 2 is proved for a single communication session. We then extend the discussions to bi-directed Internet-like networks and to the case of multiple communication sessions. Finally, we investigate advantages of network coding in a practical peer-to-peer network setting, and present both theoretical and experimental results on the use of network coding in P2P content distribution and media streaming.

17.1 Network coding background

Network coding is a fairly recent paradigm of research in information theory and data networking. It allows essentially every node in a network to perform information coding, besides normal forwarding and replication operations. Information flows can therefore be "mixed" during the course of routing. In contrast to source coding, the encoding and decoding operations are not restricted to the terminal nodes (sources and destinations) only, and may happen at all nodes across the network. In contrast to channel coding, network coding works beyond a single communication channel, it contains an integrated coding scheme that dictates the transmission at every link towards a common network-wise goal. The power of network coding can be appreciated with two classic examples in the literature, one for the wireline setting and one for the wireless setting, as shown in Figure 17.1.

Next-Generation Internet Architectures and Protocols, ed. Byrav Ramamurthy, George Rouskas, and Krishna M. Sivalingam. Published by Cambridge University Press. © Cambridge University Press 2011.

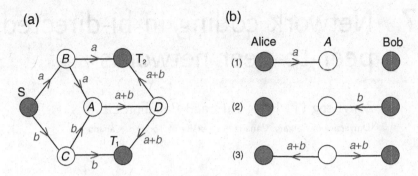

Figure 17.1 The power of network coding. (a) A wireline network where network coding help achieve a multicast throughput of 2 bps. (b) A wireless network where network coding help improve the data exchange rate between Alice and Bob.

Figure 17.1(a) shows a wireline network with a single multicast transmission, from a sender S to two receivers T_1 and T_2 simultaneously. These three terminal nodes are in black; the four white nodes are relay nodes. Each of the nine links in the network has the same unit capacity of 1 bps. With the assumption that link delays can be ignored, a multicast transmission scheme with network coding is depicted. Here a and b are two information flows each of rate 1 bps. Replication happens at nodes B, C and D. Encoding happens at relay node A, which takes a bit-wise exclusive-or upon the two incoming flows a and b, and generates $a + b$. The multicast receiver T_1 receives two information flows b and $a + b$, and can recover a as $a = b + (a + b)$. Similarly, T_2 receives a and $a + b$ and can recover b. Both receivers are therefore receiving information at 2 bps, leading to a multicast throughput of 2 bps. The reader is invited to verify that, without network coding, the throughput 2 bps cannot be achieved.

Figure 17.1(b) shows a wireless network, in which Alice and Bob each operates a laptop computer and communicate to each other through the help of a relay A (a third laptop computer or a base station). Each of the three nodes is equipped with an omnidirectional antenna; Alice and Bob are too far away from each other for direct communication, but can both reach the relay A in the middle. Assume Alice and Bob wish to exchange a pair of files. In the depicted transmission scheme, the exchange of a pair of packets (a from Alice and b from Bob) is achieved within three rounds, without interference between concurrent transmissions. It is an easy exercise to verify that without coding at the relay node A, four rounds would have been necessary.

In both examples above, the coding operation is *bit-wise exclusive-or*, which can be viewed as coding over the finite field $GF(2)$. A larger field $GF(2^k)$ can be used in general network coding, with typical values for k being 8 or 16. Since the seminal work of Ahlswede *et al.* [2] published in 2000, the benefit of network coding has been identified in a rather diverse set of applications, including for example: improving network capacity and transmission rates, efficient multicast algorithm design, robust network transmissions, network security, and P2P file dissemination and media streaming. Our focus in this chapter is on the potential for network coding to improve transmission throughput in various network models.

17.2   Network coding in bi-directed networks

Early research on network coding usually focused on directed network models, where each link in the network had a prefixed direction of transmission. A fundamental result for network coding in directed networks generalizes the celebrated max-flow min-cut theorem from one-to-one unicast flows to one-to-many multicast flows:

Theorem 17.1. [2] *For a given multicast session in a directed network with network coding support, if a unicast rate x is feasible from the sender to each receiver independently, then it is feasible as a multicast rate to all the receivers simultaneously.*

This result changed the underlying structure of multicast algorithm design, from a tree packing perspective (without coding) to a network flow perspective (with coding), and consequently reduced the computational complexity of optimal multicast from NP-hard to polynomial-time solvable. Both changes apply in undirected as well as in directed networks. It has been shown that the *coding advantage*, the ratio of achievable throughput with coding versus without coding, can be arbitrarily high in directed networks. In this chapter, we reveal a different picture in undirected networks and bi-directed networks , which are closer to the reality of Internet topologies.

17.2.1   Single multicast in undirected networks

A single communication session can be in the form of a one-to-one unicast, one-to-many multicast or one-to-all broadcast. Among these, multicast is the most general. Unicast and broadcast can be viewed as special cases of multicast, where the number of receivers equals one and the network size, respectively. Hence, for the case of a single communication session, we focus on multicast. We use a simple graph $G = (V, E)$ to represent the topology of a network, and use a function $C : E \rightarrow Z^+$ to denote link capacities. The multicast group is $M = \{S, T_1, \ldots, T_k\} \subseteq V$, with S being the multicast sender. In our graphical illustrations, terminal nodes in the multicast group are black, and relay nodes are white.

We use $\chi(N)$ to denote the maximum throughput of a multicast network N. A linear programming formulation for $\chi(N)$, based on Theorem 1.1, is given below [14]. Here the objective function is χ, the multicast throughput; $N(u)$ is the set of neighbor nodes of u, f_i is a network flow from the multicast sender S to receiver T_i, c is a variable vector storing link capacities for an orientation of the undirected network, and $\vec{T_i S}$ is a conceptual feedback link introduced for compact LP formulation.

$\chi(N) :=$ Maximize χ

Subject to:

$$\begin{cases} \chi \le f_i(\vec{T_i S}) & \forall i \\ f_i(\vec{uv}) \le c(\vec{uv}) & \forall i, \forall \ \vec{uv} \neq \vec{T_i S} \\ \sum_{v \in N(u)} (f_i(\vec{uv}) - f_i(\vec{vu})) = 0 & \forall i, \forall u \\ c(\vec{uv}) + c(\vec{vu}) \le C(uv) & \forall uv \neq T_i S \\ c(\vec{uv}), f_i(\vec{uv}), \chi \ge 0 & \forall i, \forall \ \vec{uv} \end{cases}$$

We compare $\chi(N)$ with two other parameters defined for a multicast network, the *packing number* and *edge connectivity*, and derive a bound on the coding advantage from the comparison results.

Packing refers to the procedure of finding pairwise edge-disjoint sub-trees of G, in each of which the multicast group remains connected. The packing number of a multicast network N is denoted as $\pi(N)$, and is equal to the maximum throughput without coding. The reason is that each tree can be used to transmit one unit information flow from the sender to all receivers, therefore the packing number gives the maximum number of unit information flows that can be transmitted. When relaxing flow rates on trees to be fractional, the packing number can be defined using the following linear program. Here $\mathcal{T}$ is the set of all multicast tree, $f(t)$ is a variable representing the amount of information flow one ships along tree t.

$\pi(N) :=$ Maximize $\sum_{t \in \mathcal{T}} f(t)$

Subject to:

$$\begin{cases} \sum_{uv \in t} f(t) \le C(uv) & \forall uv \in E \\ f(t) \ge 0 & \forall t \in \mathcal{T} \end{cases}$$

Connectivity refers to the minimum edge connectivity between a pair of nodes in the multicast group, and is denoted as $\lambda(N)$. It is also the minimum size of a cut that separates the communication group. Figure 17.2 illustrates the concept of these parameters using an example network. We next prove a main theorem that categorizes the relation among them.

Theorem 17.2. *For a multicast transmission in an undirected network,* $N = \{G(V, E), C : E \to Z^+, M = \{S, T_1, \ldots, T_k\} \subseteq V\},$

$$\frac{1}{2}\lambda(N) \le \pi(N) \le \chi(N) \le \lambda(N).$$

Proof: First, furnishing nodes with extra coding capabilities does not decrease the achievable throughput, hence $\pi(N) \le \chi(N)$. Furthermore, for a certain multicast throughput to be feasible, the edge connectivity from the sender to any receiver (unicast throughput) has to achieve at least the same value, therefore $\chi(N) \le$

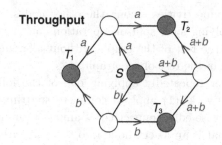

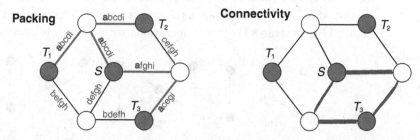

Figure 17.2 The three network parameters. In this particular network with unit capacity at each undirected link: $\pi(N) = 1.8$, nine trees (each labeled with a letter between 'a' and 'i') each of rate 0.2 can be packed; $\chi(N) = 2$, two unit information flows a and b can be delivered to all receivers simultaneously; $\lambda(N) = 2$, each pair of terminal nodes is 2-edge-connected.

$\lambda(N)$. We now have $\pi(N) \leq \chi(N) \leq \lambda(N)$, and will focus on the validity of $\frac{1}{2}\lambda(N) \leq \pi(N)$ in the rest of the proof. We first transform the multicast network into a broadcast one without hurting the validity of $\frac{1}{2}\lambda(N) \leq \pi(N)$, and then prove $\frac{1}{2}\lambda(N) \leq \pi(N)$ is true in the resulting broadcast network.

The transformation relies on Mader's Undirected Splitting Theorem [3]: *Let $G(V + z, E)$ be an undirected graph so that (V, E) is connected and the degree $d(z)$ is even. Then there exists a complete splitting at z preserving the edge-connectivity between all pairs of nodes in V.*

A split-off operation at node z refers to the replacement of a 2-hop path u-z-v by a direct edge between u and v, as illustrated in Figure 17.3. A complete splitting at z is the procedure of repeatedly applying split-off operations at z until z is isolated.

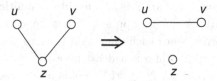

Figure 17.3 A split-off at node z.

The Undirected Splitting Theorem states that, if a graph has an even-degree non-cut node, then there exists a split-off operation at that node, after which the pairwise connectivities among the other nodes remain unchanged; and by repeatedly applying such split-off operations at this node, one can eventually isolate it from the rest of the graph, without affecting the edge-connectivity of any node pairs in the rest of the graph.

Now, consider repeatedly applying one of the following two operations on a multicast network: (1) apply a complete splitting at a non-cut relay node, preserving pairwise edge connectivities among terminal nodes in M; or (2) add a relay node that is an M-cut node into the multicast group M, i.e., change its role from a relay node to a receiver. Here an M-cut node is one whose removal separates the multicast group into more than one disconnected components. Figure 17.4 illustrates these two operations with a concrete example.

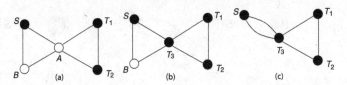

Figure 17.4 Transforming a multicast network into a broadcast network, where the validity of $\frac{1}{2}\lambda(N) \leq \pi(N)$ can be traced back. (a) The original multicast network, with unit capacity on each link. (b) The network after applying operation (2), moving the M-cut node A into the multicast group. Node A becomes receiver T_3. (c) The network after applying operation (1). A split-off was done at relay node B. A broadcast network is obtained.

In order to meet the even node degree requirement in the Undirected Splitting Theorem, we first double each link capacity in the input network, then scale the solution down by a factor of $1/2$ at the end. Each node has an even degree after doubling link capacities, and a split-off operation does not affect the parity of any node degree in the network. Therefore the Undirected Splitting Theorem guarantees that as long as there are relay nodes that are not cut nodes, operation (1) is possible. Furthermore, operation (1) does not increase $\pi(N)$. Therefore, if $\frac{1}{2}\lambda(N) \leq \pi(N)$ holds after applying operation (1), it holds before applying operation (1) as well. Operation (2), applied to M-cut nodes, does not affect either $\pi(N)$ or $\lambda(N)$. So, again we can claim that for operation (2), if $\frac{1}{2}\lambda(N) \leq \pi(N)$ holds after applying the operation, it holds before applying the operation as well.

As long as there are relay nodes in the multicast network, at least one of the two operations can be applied. If both operations are possible, operation (1) takes priority. Since each operation reduces the number of relay nodes by one, eventually we obtain a broadcast network with terminal nodes only.

The fact that $\frac{1}{2}\lambda(N) \leq \pi(N)$ holds in a broadcast network can be proved by applying Nash-Williams' Weak Graph Orientation Theorem [3]: *a graph G has an α-edge connected orientation if and only if it is 2α-edge connected.* By the Weak Graph Orientation Theorem, we can orient an undirected multicast with connectivity $\lambda(N)$ into a directed one, so that the network flow rate from any node to any other node (including in particular from the multicast sender to any multi-cast receiver) is at least $\frac{1}{2}\lambda(N)$. Then by Theorem 1.1, a multicast rate of $\frac{1}{2}\lambda(N)$ is feasible with network coding. Therefore we obtain $\chi(N) \geq \frac{1}{2}\lambda(N)$. Further-more, Tutte–Nash-Williams' Theorem [20] on spanning tree packing implies that $\pi(N) = \chi(N)$ in any broadcast network, and hence we also have $\pi(N) \geq \frac{1}{2}\lambda(N)$.

Finally, note that we obtained an integral transmission strategy after doubling each link capacity. Therefore, after we scale the solution back by a factor of $1/2$, the transmission strategy is half-integral.

Corollary 17.1. *For a multicast transmission in an undirected network, the coding advantage is upper-bounded by a constant factor of 2, with either fractional routing or half-integer routing.*

Proof: By Theorem 3, $\frac{1}{2}\lambda(N) \leq \pi(N)$ and $\chi(N) \leq \lambda(N)$ as long as half inte-ger routing is allowed. Therefore we conclude $\frac{1}{2}\chi(N) \leq \pi(N)$, i.e., the coding advantage $\chi(N)/\pi(N) \leq 2$. $\qquad\square$

17.2.2   The linear programming perspective

We have just derived a bound of 2 for the coding advantage through a graph-theoretic approach. A linear programming perspective for studying the coding advantage turns out to be also interesting, and can lead to the same proven bound as well as other insights.

Table 17.1. min Steiner tree IP and min-cost multicast LP.

Minimize $\sum_e w(e)f(e)$ Subject to:	Minimize $\sum_e w(e)f(e)$ Subject to:
$\begin{cases} \sum_{e \in \Gamma} f(e) \geq 1 \ \forall \text{ cut } \Gamma \\ f(e) \in \{0,1\} \quad \forall e \end{cases}$	$\begin{cases} \sum_{e \in \Gamma} f(e) \geq 1 \ \forall \text{ cut } \Gamma \\ f(e) \geq 0 \quad\quad \forall e \end{cases}$

Table 17.1 shows the linear integer program for the min Steiner tree problem on the left, where f is the variable vector and w is the constant link cost vector; Γ is a *cut*, or a set of links whose removal separates at least one receiver from the sender. The flow $f(e)$ can be assumed to be in the direction of originating from the sender component. On the right of the table is a linear program for min-cost multicast with network coding, with target throughput 1. The validity of this LP is based on Theorem 1.1. Note that different LP formulations for optimal multicast with network coding are possible and known, including link-based, path-based, and cut-based ones. The first has a polynomial size and is practical to solve, while the later two are often convenient for theoretical analysis.

It is interesting that the min-cost multicast LP is exactly the LP relaxation of the min Steiner tree IP. Therefore the coding advantage for cost is equivalent to the integrality gap of the Steiner tree IP. Agarwal and Charikar [1] further applied LP duality to prove that the maximum coding advantage for throughput is equivalent to the maximum integrality gap of the min Steiner tree IP. This reveals an underlying equivalence between the coding advantage in reducing cost and that in improving throughput, and also provides an interpretation of the power of network coding from the LP perspective: the flexibility of allowing arbitrary fractional flow rates. Furthermore, since it has been proven that the maximum integrality gap for the Steiner tree IP is 2 [1], one obtains an alternative proof for the bound 2 of the coding advantage; although it is not immediate whether this proof also works for half-integral flows.

17.2.3   Single multicast in Internet-like bi-directed networks

Given the fact that the coding advantage is finitely bounded in undirected networks but not so in directed ones, it is natural to ask which model is closer to real-world networks, and whether the coding advantage is bounded in such networks. A real-world computer network, such as the current-generation Internet, is usually bi-directional but not undirected. If u and v are two neighbor routers in the Internet, the amount of bandwidth available from u to v and that from v to u are fixed and independent. At a certain moment, if the $u \rightarrow v$ link is congested and the $v \rightarrow u$ link is idling, it is not feasible to "borrow" bandwidth from the $v \rightarrow u$ direction to the $u \rightarrow v$ direction, due to the lack of a dynamic bandwidth allocation module. Therefore, the Internet resembles an undirected network in that communication is bi-directional, and resembles a directed network in that each link is directed with a fixed amount of bandwidth.

A better model for the Internet is a *balanced directed* network. In a balanced directed network, each link has a fixed direction. However, a pair of neighboring nodes u and v are always mutually reachable through a direct link, and the ratio between $c(\vec{uv})$ and $c(\vec{vu})$ is upper-bounded by a constant ratio $\alpha \geq 1$. In the case $\alpha = 1$, we have an absolutely balanced directed network. This is rather close to the reality in the Internet backbone, although last-hop connections to the Internet exhibit a higher degree of asymmetry in upstream/downstream capacities. Based on the constant bound developed in the previous section, we can show that the coding advantage in such an α-balanced network is also finitely bounded.

Theorem 17.3. *For a multicast session in an α-balanced bi-directional network, the coding advantage is upper-bounded by $2(\alpha + 1)$.*

Proof: We first define a few notations. Let $N_{1:\alpha}$ be the α-balanced network; let N_1 be an undirected network with the same topology as $N_{1:\alpha}$, where $c(uv)$ in N_1 is equal to the smaller one of $c(\vec{uv})$ and $c(\vec{vu})$ in $N_{1:\alpha}$; let $N_{\alpha+1}$ be the undirected network obtained by multiplying every link capacity in N_1 with $\alpha + 1$. Then we have:

$$\pi(N_{1:\alpha}) \geq \pi(N_1) \geq \frac{1}{\alpha + 1} \pi(N_{\alpha+1}) \geq \frac{1}{\alpha + 1} \frac{1}{2} \chi(N_{\alpha+1})$$

$$\geq \frac{1}{2(\alpha + 1)} \chi(N_{1:\alpha}).$$

In the derivations above, the third inequality is an application of Theorem 1.1; the other inequalities are based on definitions. □

From Theorem 1.3, we can see that the more "balanced" a directed network is, a smaller bound on the coding advantage can be claimed. In the case of an absolutely balanced network, the bound is 4. In arbitrary directed networks, α may approach ∞, and correspondingly a finite bound on the coding advantage does not exist.

17.2.4   Towards tighter bounds

The constant bound of 2 for the coding advantage in undirected networks is not tight. So far, the largest coding advantage value observed is 9/8 for relatively small networks [14], and approaches 8/7 in a network pattern that grows to infinite size [1]. Closing the gap between 2 and 8/7 is an important open research direction. The significance here is twofold. First, it may provide a better understanding and more in-depth insights to network coding. Second, it may lead to advances in designing Steiner tree algorithms, which have important applications in operations research, VLSI design, and communication networks. Both the minimum Steiner tree problem and the Steiner tree packing problem are NP-hard, and it is known that for any constant $\alpha > 1$, a polynomial-time α-approximation algorithm exists for one of them if and only if it exists for the other. Note that, one may approximate the Steiner packing value $\pi(N)$ by computing the multicast throughput $\chi(N)$ instead. Such an approach yields a polynomial-time approximation algorithm, and the approximation ratio is precisely the tight upper-bound for the coding advantage. It is probable that the tight bound is closer to 8/7 than to 2. In that case, we might have obtained a better approximation algorithm for Steiner trees than the current best [18], which has an approximation ratio of 1.55. Initial progress has been made towards proving a tighter bound. In particular, it was proven that for a special set of combinatorial networks containing infinitely many network instances, the coding advantage is always bounded by 8/7 [19]. It is worth noting that so far, most known undirected network examples with >1 coding advantage are closely related to the three-layer combinatorial networks.

17.2.5   Multiple communication sessions

Drastic changes occur when we switch the context from single to multiple communication sessions, where both intra-session and inter-session network coding are possible and need to be jointly considered. The complexity of deciding the optimal network coding scheme or the optimal throughput becomes NP-hard, and linear coding is not always sufficient. A fundamental tradeoff accompanies network coding across sessions: the ability to exploit the diversity of information flows brought by network coding, versus the onus of eliminating "noise" introduced by network coding on the receivers which no longer share the exact same interest in information reception. An understanding towards such a tradeoff is only preliminary so far. In terms of demonstrating a higher coding advantage, no existing evidence shows that multiple sessions represent a better paradigm to be considered than single session.

For the special case where each session is a unicast, it is known that the coding advantage can be larger than 1 if either (a) the network is directed, or (b) integral routing is required. Due to the arbitrary asymmetry of connectivity between node pairs in opposite directions in (a), the gap can be as high as linear to the network size [10] and is therefore unbounded. In sharp contrast is the fact that so far, no example has been discovered for the case of an undirected network with fractional routing, where the coding advantage is larger than 1. It was conjectured in 2004 that network coding does not make any difference in this case [13, 9]. Arguments based on duality theory show that if the conjecture is false, then that implies a fundamental barrier on throughput-distance product in data communication can be broken by means of coding. The conjecture remains open today, with settlement obtained in special cases. It is worth noting that such settlement, even for the case of a 5-node fixed topology network, leverages tools from not only graph theory but also information theory, such as entropy calculus and information inequalities [12].

The case of multiple multicast sessions is most general and is even less understood. Furthermore, most existing results there pertain to directed networks only, and will not be discussed in this chapter. Due to space limitations, we have also chosen not to cover related studies of the coding advantage in other models, such as the case of average throughput and the case of wireless networks.

17.2.6   The source independence property of multicast

The source independence property refers to the fact that once the set of terminal nodes (including the sender and the receivers in the multicast group) is fixed in a given network, then the maximum achievable throughput is fully determined, regardless of which terminal node takes the sender role. Such a property is not true in directed networks, where bandwidth between neighbor nodes can be arbitrarily unbalanced. It holds in undirected networks for multicast without network coding, i.e., for tree packing. The definition of the packing number $\pi(N)$ does not specify which terminal node is the "sender" or the "root of the tree." The fact that source independence also holds in undirected networks for multicast with network coding is less obvious, but has been proven [14]. The proof is based on the observation that a valid multicast flow originating from one terminal node can be manipulated to construct a valid multicast flow of unchanged throughput that originates from another terminal node. More specifically, one just needs to reverse the network flow between the old sender and the new sender, as illustrated in Figure 17.5; the information content of the flow at each link does not need to be changed. It is interesting to observe that the definition of $\chi(N)$ relies on the selection of a special terminal node as the multicast sender, which should not be necessary by the source independence property. An equivalent, symmetrical definition of $\chi(N)$ that does not isolate a single terminal node with a special role is open.

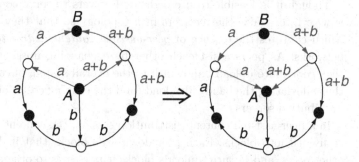

Figure 17.5 The source independence property of multicast with network coding: switching the source from a terminal A node to another terminal node B. The information flow between A and B are simply reversed.

We note that both the finite bound on the coding advantage and the source independence property hold in undirected networks but not in directed ones. We also showed that the finite bound holds in balanced directed networks. Now it is interesting to ask whether source independence also holds in an absolutely balanced directed network – we leave that as an exercise for the reader.

17.3   Network coding in peer-to-peer networks

Beyond theoretical studies, it is natural to assume a more pragmatic role and explore the feasibility of applying network coding to data communication over the Internet. It is intuitive that peer-to-peer (P2P) networks represent one of the most promising platforms to apply network coding, since end-hosts on the Internet (called "peers") have the computational power to perform network coding, and are not constrained by existing Internet standards that govern most network switches at the core of the Internet. We now turn our attention to the advantages (and possible pitfalls) of using network coding in peer-to-peer networks, with a focus on two applications: *content distribution* and *media streaming*.

17.3.1   Peer-assisted content distribution with network coding

If the Internet is modeled as a balanced directed network, we have shown that the coding advantage is upper-bounded in theory. In reality, however, more pragmatic factors come into play when we consider the fundamental problem of multicast sessions: link capacities are not known a priori, and the optimal transmission strategy – including the linear code to be applied – needs to be computed. In the Internet, a multicast session corresponds naturally to a session of *content distribution*, where information (such as a file) needs to be disseminated to a group of receivers.

Though it is feasible to use dedicated servers to serve content exclusively, it is wise to organize the receivers in a topology so that they serve one another, which captures the essence of *peer-assisted content distribution* in peer-to-peer networks. As peers assist each other by exchanging missing pieces of the file, they contribute upload bandwidth to the overall system of content distribution, thus alleviating the bandwidth load (and the ensuing costs) of dedicated content distribution servers.

In a peer-assisted content distribution session, the content to be disseminated is divided into blocks. Each peer downloads blocks that it does not have from other peers, and in turn, uploads blocks it possesses to others at the same time. Which block should a peer download, and from whom? Referred to as the *block scheduling* problem, this question needs to be addressed by designing protocols in a decentralized fashion, with local knowledge only. A poorly designed protocol may lead to the problem of *rare* blocks that are not readily available in the peer-to-peer network: those who have these rare blocks do not have the upload bandwidth to satisfy the demand for them.

To take advantage of network coding in peer-assisted content distribution, the first roadblock is the need to assign linear codes to network nodes. Randomized network coding, first proposed in [11], advocates the use of random linear codes, by assigning randomly generated coding coefficients to input symbols. With randomized network coding, receivers are able to decode with high probability, and no a-priori knowledge of the network topology is assumed.

Gkantsidis *et al.* [8] have proposed to apply the principles of randomized network coding to peer-assisted content distribution systems. The basic concept may be best illustrated in the example of Figure 17.6. The file to be disseminated is divided into n blocks $b_1, b_2, \ldots, b_n$. The source first generates random coefficients $c_1, c_2, \ldots, c_n$, and then performs random linear coding on the original blocks b_i using these coefficients. All other peers follow suit, by generating coded blocks with random linear coding, using random coefficients on existing coded blocks it has received so far. All operations are to take place in a Galois field of a reasonable size, e.g., 2^{16}, to ensure the linear independence of encoded blocks [11].

It has been shown in [8] that the use of network coding introduces a substantial performance gain. We intuitively show such potential using the example in Figure 17.7. Assume that peer A has received blocks 1 and 2 from the source. If network coding is not used, node B can download block 1 or 2 from A with the same probability. At the same time, assume C independently downloads block 1. If B decides to retrieve block 1 from A, then both B and C will have the same block, and the link between them cannot be utilized. With network coding, A blindly transmits a linear combination of the two blocks to B, which is always useful to C because the combination contains block 2 as well.

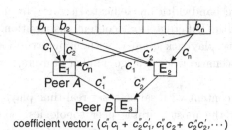

coefficient vector: $(c_1'' c_1 + c_2'' c_1', c_1'' c_2 + c_2'' c_2', \cdots)$

Figure 17.6 Peer-assisted content distribution with network coding.

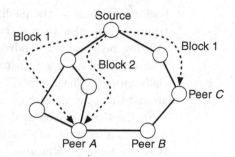

Figure 17.7 Advantages of network coding.

It is easy to see that random network coding offers three unique advantages to improve the performance of peer-assisted content distribution:

- First, it substantially increases the information contained in every block, as it transmits a random linear equation which may contain information about every block of the file. This has a high probability to be linearly independent to all existing blocks in the receivers, leading to a more efficient content dissemination process.
- Second, coding also greatly simplifies the problem of selecting the most appropriate (perhaps globally rarest) block to download, by blindly disseminating linear combinations of all existing blocks to downstream peers.
- Finally, it has been theoretically shown that network coding improves the resilience to highly dynamic scenarios of peer arrivals and departures [17]. This can be intuitively explained, since network coding eliminates the need to find rare blocks, the risk of "losing" these rare blocks when peers leave the system is no longer a concern.

To recover all n original blocks using n linearly independent coded blocks with high probability, a receiver needs to compute the inverse of the coefficient matrix, with a complexity of $O(n^3)$. As the number of blocks scales up with larger files, Chou et al. [4] proposed to divide the file into *generations*, and to perform network coding within the same generation. The performance of such generation-based network coding has been empirically evaluated in [7] and theoretically analyzed in [16], where network coding is shown to be highly practical for peer-assisted content distribution. Further, it has been shown [17] that generation-based network coding is still able to offer resilience to peer dynamics, even with just a small number of blocks in each generation.

17.3.2   Peer-assisted media streaming with network coding

Compared to content distribution, peer-assisted media streaming adds an additional requirement that the media content to be distributed needs to be played back in real time as the media stream is being received. Similar to content distribution, we wish to conserve bandwidth on dedicated servers by maximally utilizing peer upload bandwidth. Different from content distribution, we also wish to maintain a satisfactory playback quality, without interruptions, especially during a "flash crowd" scenario when a large number of users wish to join around the same time.

To effectively stream media content with satisfactory real-time playback quality, it is important to deploy the most suitable peer topologies. Some argue that *tree-based push* protocols, which organize peers into topologies consisting of one or multiple trees, are the best for minimizing the delay from the source to receivers (e.g., [21]). However, trees may be difficult to construct and maintain when peers join and leave frequently. Most real-world peer-assisted streaming protocols, in contrast, use *mesh-based pull* protocols, which organize peers into mesh topologies, with each peer having an arbitrary subset of other peers as its neighbors (e.g., [24]). Such simplicity affords much better flexibility: there is no need to maintain the topology, as long as a sufficient number of neighbors is always available.

In particular, mesh-based pull protocols work as follows. For each streaming session, a finite buffer at a peer is maintained, with segments ordered according to playback sequence. Outdated segments after playback are deleted from the buffer immediately. After a new peer joins the system, it waits to accumulate a certain number of segments to start playback, the delay of which is referred to as the *initial buffering delay*. During playback, a peer concurrently sends requests for missing segments in the buffer, and downloads (or "pulls") these segments from those who have them. To update the knowledge of which neighbor has the missing segments, a peer would need to exchange availability bitmaps of its buffer with neighbors, referred to as *buffer map exchanges*.

Would network coding be instrumental in peer-assisted media streaming? Wang and Li [22] first evaluated the feasibility and effectiveness of applying network coding in live peer-assisted streaming sessions, with strict timing and bandwidth requirements. Generation-based random network coding has been applied as a "plug-in" component into a traditional pull-based streaming protocol without any changes. Gauss–Jordan elimination has been used to make it possible for peers to decode generations on the fly as blocks are being received, which fits naturally into streaming systems. It has been discovered that network coding provides some marginal benefits when peers are volatile with their arrivals and departures, and when the overall bandwidth supply barely exceeds the demand.

With such mildly negative results against the use of network coding, one would argue that the advantages of network coding may not be fully explored with a traditional pull-based protocol. In [23], Wang and Li proposed R^2, which uses random push with random network coding, and is designed from scratch to take full advantage of the benefits of network coding.

In R^2, the media stream is again divided into generations, and each generation is further divided into blocks. In traditional mesh-based pull protocols, missing segments are requested by the downstream peer explicitly. Due to the periodic nature of buffer map exchanges, such requests can only be sent to those neighbors who have the missing segments in the previous round of exchanges. With network coding in R^2, however, since much larger generations are used, buffer maps that represent the state of generations – as opposed to blocks – can be much smaller. In addition, with larger generations than blocks, each generation takes a longer period of time to be received. As such, without additional overhead, buffer maps can be pushed to all neighbors as soon as a missing segment has been completely received, and there is no need to perform the exchange in a periodic fashion.

Since all buffer maps are up-to-date in R^2, a sender can simply select one of the generations – at random – still missing on the receiver. It then produces a coded block in this generation, and blindly *pushes* it to the receiver. An important advantage of network coding is "perfect collaboration": since all coded blocks within a generation are equally useful, multiple senders are able to serve coded blocks in the same missing generation to the same receiver, without the need for explicit coordination and reconciliation. An illustrative comparison between R^2 and traditional pull-based protocols is shown in Figure 17.8.

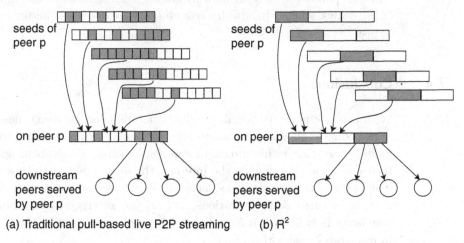

(a) Traditional pull-based live P2P streaming   (b) R^2

Figure 17.8 An illustrative comparison between a traditional pull-based streaming protocol and R^2.

Naturally, R^2 represents a simple design philosophy, rather than a strict protocol design. It allows for a flexible design space for more fine-tuned protocols. For example, the random push strategy can be tailored so that generations closer to playback deadlines are given higher priority to be served, in order to ensure timely playback. One possibility is to define and impose a *priority region*, which includes urgent segments immediately after the point of playback. Other prioritization and randomization strategies can also be incorporated. In this sense, R^2 is complementary to the usual algorithm design spaces, such as block selection and neighbor selection strategies.

With the use of generation-based network coding, R^2 enjoys two distinct advantages:

- First, it induces much less messaging overhead in buffer map exchange, leading to better performance in terms of playback quality and resilience. As buffer maps are pushed in real time when they change, neighboring peers receive timely feedback, and can proceed to serve missing segments with minimal delay. Indeed, Feng *et al.* [6] have theoretically corroborated the effectiveness of R^2, and pointed out that the lack of timely exchange of buffer maps may be a major factor that separates the actual performance of pull-based protocols from optimality.

- Second, equipped with random push and random network coding, R^2 offers shorter initial buffering delays and reduced bandwidth costs on dedicated streaming servers. This is due to the fact that multiple senders can serve the same receiver without the messaging overhead for explicit coordination purposes. With a stochastic analytical framework, Feng and Li [5] have analyzed both flash crowd and highly dynamic peer scenarios, and have shown that the design philosophy of using network coding in R^2 leads to shorter initial buffering delays, smaller bandwidth costs on servers, as well as better resilience to peer departures.

17.4   Conclusions

With network nodes performing coding operations on incoming messages, network coding is shown in theory to have the ability of improving throughput in multicast sessions within directed networks. When we think about applying the theory of network coding to the Internet, the most likely scenario is within the scope of peer-to-peer networks, since end-hosts are computationally capable of performing such coding operations, and are not governed by stringent Internet standards. How likely is it for network coding to achieve similar theoretical gains in peer-to-peer networks?

In this chapter, we have started with a theoretical perspective: by extending from directed to undirected and bi-directed networks, we note that the coding advantage – the throughput gain compared to not using network coding – is finitely bounded in undirected and bi-directed networks. Instead of improving throughput, network coding makes it computationally feasible to compute optimal strategies for achieving optimal throughput: it changes the underlying structure of multicast algorithm design, from a tree packing perspective (without coding) to a network flow perspective (with coding), and consequently reduces the computational complexity of optimal multicast from NP-hard to polynomial-time solvable.

Assuming a more pragmatic role, we have presented known results in two applications in peer-to-peer networks: peer-assisted content distribution and media streaming. In peer-assisted content distribution, we have shown that network coding substantially simplifies the problem of selecting the most appropriate block to download (referred to as "the block selection problem"), and improves the resilience to highly dynamic scenarios of peer arrivals and departures. In peer-assisted media streaming, we have shown that protocols need to be designed from scratch to take full advantage of network coding; R^2, a collection of protocol design guidelines to incorporate network coding, has led to less messaging overhead in terms of exchanging buffer availability information, as well as shorter initial buffering delays.

It appears that the overall message is very optimistic. Despite the computational complexity of network coding (even with the use of generations), one would envision that network coding can be applied in the near-term future to peer-to-peer networks, which have consumed a substantial portion of the bandwidth available in the Internet today. Substantial savings in bandwidth costs on servers alone may be able to justify the additional computational complexity on end-hosts. With Moore's Law that predicts increasingly abundant computational power, bandwidth, as a resource, will naturally be more scarce and needs to be carefully utilized. Network coding may very well be a useful tool to achieve more efficient bandwidth utilization in the Internet.

References

[1] Agarwal, A. and Charikar, M. (2004). On the Advantage of Network Coding for Improving Network Throughput. *Proc. IEEE Inform. Theory Workshop*, 2004.

[2] Ahlswede, R., Cai, N., Li, S. Y. R., and Yeung, R. W. (2000). Network Information Flow. *IEEE Trans. Inform. Theory*, 46(4):1204–1216, July 2000.

[3] Bang-Jensen, J. and Gutin G. (2009). Digraphs: Theory, Algorithms and Applications, 2nd edn., Springer.

[4] Chou, P. A., Wu, Y. and Jain, K. (2003). Practical Network Coding. *Proc. 42nd Annual Allerton Conference on Communication, Control and Computing*, 2003.

[5] Feng, C. and Li, B. (2008). On Large Scale Peer-to-Peer Streaming Systems with Network Coding. *Proc. ACM Multimedia*, 2008.

[6] Feng, C., Li, B., and Li, B. (2009). Understanding the Performance Gap between Pull-based Mesh Streaming Protocols and Fundamental Limits. *Proc. IEEE INFOCOM*, 2009.

[7] Gkantsidis, C., Miller, J., and Rodriguez, P. (2006). Comprehensive View of a Live Network Coding P2P System. *Proc. Internet Measurement Conference (IMC)*, 2006.

[8] Gkantsidis, C. and Rodriguez, P. (2005). Network Coding for Large Scale Content Distribution. *Proc. IEEE INFOCOM*, 2005.

[9] Harvey, N. J. A., Kleinberg, R. and Lehman, A. R. (2004). Comparing Network Coding with Multicommodity Flow for the k-Pairs Communication Problem. Technical Report, MIT CSAIL, November 2004.

[10] Harvey, N. J. A., Kleinberg, R., and Lehman, A. R. (2006). On the Capacity of Information Networks. *IEEE Trans. Inform. Theory*, 52(6):2345–2364, June 2006.

[11] Ho, T., Medard, M., Shi, J., Effros, M., and Karger, D. (2003). On Randomized Network Coding. *Proc. 41st Allerton Conference on Communication, Control, and Computing*, 2003.

[12] Jain, K., Vazirani, V. V., and Yuval, G. (2006) On The Capacity of Multiple Unicast Sessions in Undirected Graphs. *IEEE Trans. Inform. Theory*, 52(6):2805–2809, 2006.

[13] Li, Z. and Li, B. (2004). Network Coding: The Case of Multiple Unicast Sessions. *Proc. 42nd Annual Allerton Conference on Communication, Control, and Computing*, 2004.

[14] Li, Z., Li, B., and Lau, L. C. (2006). On Achieving Maximum Multicast Throughput in Undirected Networks. *IEEE/ACM Trans. Networking*, 14(SI):2467–2485, June 2006.

[15] Li, Z., Li, B., and Lau, L. C. (2009). A Constant Bound on Throughput Improvement of Multicast Network Coding in Undirected Networks. *IEEE Trans. Inform. Theory*, 55(3):997–1015, March 2009.

[16] Maymounkov, P., Harvey, N. J. A., and Lun D. S. (2006). Methods for Efficient Network Coding. *Proc. 44th Annual Allerton Conference on Communication, Control and Computing*, 2006.

[17] Niu, D. and Li, B. (2007). On the Resilience-Complexity Tradeoff of Network Coding in Dynamic P2P Networks. *Proc. International Workshop on Quality of Service (IWQoS)*, 2007.

[18] Robins, G. and Zelikovsky A. (2000). Improved Steiner Tree Approximation in Graphs. *Proc. 11th ACM-SIAM Symposium on Discrete Algorithms*, 2000.

[19] Smith, A., Evans, B., Li, Z., and Li, B. (2008). The Cost Advantage of Network Coding in Uniform Combinatorial Networks. *Proc. 1st IEEE Workshop on Wireless Network Coding*, 2008.

[20] Tutte, W. T. (1961). On the Problem of Decomposing a Graph into n Connected Factors. *Journal of London Math. Soc.*, 36:221–230, 1961.

[21] Venkataraman, V., Yoshida, K., and Francis, P. (2006). Chunkyspread: Heterogeneous Unstructured Tree-based Peer-to-Peer Muticast. *Proc. IEEE International Conference on Network Protocols (ICNP)*, 2006.

[22] Wang, M. and Li, B. (2007). Lava: A Reality Check of Network Coding in Peer-to-Peer Live Streaming. *Proc. IEEE INFOCOM*, 2007.

[23] Wang, M. and Li, B. (2007). R^2: Random Push with Random Network Coding in Live Peer-to-Peer Streaming. *IEEE J. Sel. Areas Commun.*, 25(9): 1655–1667, December 2007.

[24] Zhang, X., Liu, J., Li, B., and Yum, T.-S. P. (2005). CoolStreaming/ DONet: A Data-Driven Overlay Network for Efficient Live Media Streaming. *Proc. IEEE INFOCOM*, 2005.

18 Network economics: neutrality, competition, and service differentiation

John Musacchio[†], Galina Schwartz[‡], and Jean Walrand[‡]

[†]University of California, Santa Cruz, USA   [‡]University of California, Berkeley, USA

In 2007 Comcast, a cable TV and Internet service provider in the United States, began to selectively rate limit or "shape" the traffic from users of the peer-to-peer application Bit Torrent. The access technology that Comcast uses is asymmetric in its capacity – the "uplink" from users is much slower than the "downlink." For client-server applications like Web, this asymmetry is fine, but for peer-to-peer, where home users are serving up huge files for others to download, the uplink quickly becomes congested. Comcast felt that it had to protect the rest of its users from a relatively small number of heavy peer-to-peer users that were using a disproportionate fraction of the system's capacity. In other words, peer-to-peer users were imposing a negative externality by creating congestion that harmed other users.

This negative externality reduces the welfare of the system because users act selfishly. The peer-to-peer user is going to continue to exchange movies even though this action is disrupting his neighbor's critical, work-related video conference. Comcast thought that by singling out users of peer-to-peer applications, it could limit the ill effects of this externality and keep the rest of its users (who mostly don't use peer-to-peer) happy. Instead Comcast's decision placed them in the center of the ongoing network neutrality debate. Supporters of the network neutrality concept feel that the Internet access provider ought not to be allowed to "discriminate" between traffic of different users or different applications. Said another way, the provider of a network layer service should not act on information about the application layer that generated the traffic. Some supporters of net-neutrality argue that this principle of non-discrimination is not just a good architectural principle, but it is also vital for a number of societal reasons such as maintaining free-speech, and the ability for new, innovative content providers to enter the market and have the same potential reach as entrenched incumbents with more resources.

Related to the question of whether ISPs ought to be allowed to treat traffic from different applications differently, there are the ideas of service differentiation

Next-Generation Internet Architectures and Protocols, ed. Byrav Ramamurthy, George Rouskas, and Krishna M. Sivalingam. Published by Cambridge University Press. © Cambridge University Press 2011.

and quality of service. Service differentiation is not a new concept in communications networks. For instance, over a decade ago the designers of the Asynchronous Transfer Mode (ATM) networking technology understood that traffic from interactive applications like telephony is much more sensitive to delay than traffic from file transfers which is also more bursty. Therefore mixing the traffic of both types together has the potential to greatly reduce the utility of the entire system by hurting the users of interactive applications. Many have argued that these problems would disappear as the capacity of the network's links grew exponentially. The argument was that the network would be so fast that traffic of all types would see negligible delays, so the differences in traffic requirements would not matter. However as the network's capacity grew, so did the consumption of that capacity by new applications, like peer-to-peer and streaming video. When Comcast recently deployed a new voice over IP (VoIP) telephony service in their network, they indeed chose to keep that traffic protected from the congestion of other Internet traffic, and not coincidentally from traffic from other providers of VoIP services like Skype. The question of whether service differentiation ought to be included in the Internet might still be debated, but the reality is that it is being deployed today. The danger is that it will be deployed in an ad-hoc way by different ISPs, and the possible benefits of having a coherent architectural approach across ISPs will be lost. For instance the ability to independently select application service provider and Internet service provider might be lost. For example the Skype/Comcast "stack" for VoIP could be at a severe disadvantage compared to a Comcast/Comcast stack.

Another directly related set of questions is how revenue ought to be shared between the players of this "stack." What is the right way to share revenue between content providers and ISPs? In particular, should ISPs like Comcast be allowed to charge Skype a fee for offering services to Comcast subscribers? This is one of the questions we will examine in this chapter.

The potential of a better Internet is enormous. Today, the Internet delivers insufficient or inconsistent service quality for many applications. For instance, companies pay thousands of dollars for access to private networks for business-grade video conferencing because the quality on the public Internet in many cases is too unreliable. As the costs of energy rise, it is not hard to envision a day when most of the workforce will not be able to commute to their jobs every day. If that were to happen, the continued productivity of our economy will depend on the workforce having access to high quality, reliable, interactive applications. They would even be willing to pay for them – a few dollars to have a reliable HDTV video conference with colleagues would be worth it if it saves having to buy a $100 tank of gasoline for instance. Unfortunately, today's Internet does not give one the option of paying a little extra for quality.

In addition to the likely growth in importance of interactive applications, another important trend is the move toward cloud computing and the related trend of service-oriented architecture. With these approaches an organization can push parts of their computing and Information Technology (IT) infrastructure out of their own facilities and instead "outsource" them to specialized providers. This capability has the potential to greatly lower administrative costs, and also make an organization's IT system much more adaptable to changing needs. However, this approach requires a reliable, fast, and low latency network in order that these distributed services be as responsive for users as a more traditional approach with mostly on-site infrastructure.

As we have implied there are a large number of economic questions facing the Internet, and how they get resolved will arguably be the driving force behind the future evolution of the Internet. We cannot address all of the questions in depth in this chapter, but instead we will limit our discussion to three. First we will look at the question of revenue sharing between content providers and ISPs; an issue that we argue is core to the net-neutrality debate. Next we will discuss some fundamental modeling work that looks at the economic efficiency that results when multiple interconnected ISPs compete on price, and users seek lower prices and lower delays. Lastly, we examine some of the basic issues behind service differentiation.

Economic modeling is also central to the study of architectures that employ explicit congestion notification and charging in order to achieve fairness and utility maximization across a network of users with rate-dependent utility functions. These ideas are extremely useful both for understanding the behavior of existing protocols like TCP, and for the design of new networking technologies. In this volume, Section 4.3 by Kelly and Raina discusses the issues behind explicit congestion notification, and Section 3.2 by Yi and Chiang addresses the network utility maximization approach for wireless networks.

18.1   Neutrality

Today, an Internet service provider (ISP) charges both end-users who subscribe to that ISP for their last-mile Internet access as well as content providers that are directly connected to the ISP. However, an ISP generally does not charge content providers that are not directly attached to it for delivering content to end-users. One of the focal questions in the network neutrality policy debate is whether these current charging practices should continue and be mandated by law, or if ISPs ought to be allowed to charge all content providers that deliver content to the ISP's end-users. Indeed the current network neutrality debate began when the CEO of AT&T suggested that such charges be allowed [1].

To address this question, we develop a two-sided market model of the interaction of ISPs, end-users, and content providers. A more complete description of this work is available in [2]. The model is closely related to the existing two-sided markets literature as we detail later in this section. We model a "neutral" network as a regime in which ISPs are allowed to charge only content providers that buy their Internet access from them. We argue that with such a charging structure, the ISPs compete on price to attract content providers to buy their access from them, driving these prices to the ISP's costs. For simplicity we normalize the price content providers pay ISPs to be net of ISP connection cost, so in a "neutral" network only the end-users (and not the content providers) pay a positive price to the ISPs. In a "non-neutral" network, all ISPs are allowed to charge all content providers, and thus ISPs extract revenues from both content providers and end-users.

The question we address in this work is part of the larger debate on network neutrality, which includes diverse issues such as whether service differentiation should be allowed, or whether charges for content constitute an impingement of freedom of speech (see [3] and [4]). In 2006 there was a considerable divergence of opinions on the subject of net-neutrality. Indeed the issue was intensely debated by law and policy makers, and the imposition of restrictive network regulations on ISPs in order to achieve network neutrality seemed likely. In 2007 the situation began to change. In June of that year, the Federal Trade Commission (FTC) issued a report, forcefully stating the lack of FTC support for network neutrality regulatory restraints, and warning of "potentially adverse and unintended effects of regulation" [5]. Similarly, on September 7, 2007, the Department of Justice issued comments "cautioning against premature regulation of the Internet" [6]. However, US President Barack Obama, elected in 2008, voiced support for network neutrality, thus the debate is far from over. We do not attempt to address all of the questions in the network neutrality debate. We only study the issue of whether ISPs ought to be allowed to charge content providers for accessing the end-users.

Our model is based on the ideas of two-sided markets, and there is a large literature on the subject. For a survey of two-sided markets see for example works by Rochet and Tirole [7] and Armstrong [8]. Other work has used the ideas of two-sided markets to study network neutrality. For instance, Hermalin and Katz [9] model network neutrality as a restriction on the product space, and consider whether ISPs should be allowed to offer more than one grade of service. While Hogendorn [10] studies two-sided markets where intermediaries sit between "conduits" and content providers. In his context, net-neutrality means content has open access to conduits where an "open access" regime affords open access to the intermediaries. Weiser [11] discusses policy issues related to two-sided markets. The work we describe here is unique from other studies of network neutrality in that we develop a game-theoretic model to study investment incentives of the providers under each network regime.

18.1.1   Model

Figure 18.1 illustrates our setting. In the model, there are M content providers and N ISPs. Each ISP T_n is attached to end-users U_n ($n = 1, 2, \ldots, N$) and charges them p_n per click. The ISP T_n has a monopoly over its end-user base

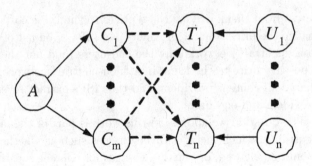

Figure 18.1 The direction of payments in the model. Each C_m is a content provider; each T_n is an ISP; and each U_n is a pool of users subscribing to ISP T_n. The A block denotes the set of advertisers that pay the content providers. The dotted lines indicate payments made only with two-sided pricing ("non-neutral").

U_n. Thus, the end-users are divided between the ISPs, with each ISP having $1/N$ of the entire market. This assumption reflects the market power of local ISPs. Each ISP T_n charges each content provider C_m an amount equal to q_n per click. Content provider C_m invests c_m and ISP T_n invests t_n.

Recall from the beginning of this section that we measure q net of content providers' access payment (for network attachment), which is set at marginal cost due to the competition amongst ISPs. Accordingly, we measure the content provider per-user charges to advertisers (which we denote as a) net of the content provider's access payment.

We characterize usage of end-users U_n by the number of "clicks" B_n they make. Since Internet advertising is most often priced per click, clicks are a natural metric for expressing advertising revenue. It is a less natural metric for expressing ISP revenue from end-users, because ISPs do not charge users per click but rather base their charges on bits. However, it is convenient to use only one metric and argue that one could approximate one metric from knowledge of the other using an appropriate scaling factor. The rate B_n of clicks of end-users U_n depends on the price p_n but also on the quality of the network, which in turn is determined by provider investments. The rate of clicks B_n, which characterizes end-user demand, depends on the end-user access price p_n and investments as

$$B_n = (1/N^{1-w})(c_1^v + \cdots + c_M^v)$$
$$\times \left[(1 - \rho)t_n^w + (\rho/N)(t_1^w + \cdots + t_N^w) \right] e^{-p_n/\theta}, \qquad (18.1)$$

where $\rho \in (0, 1)$, $\theta > 0$, and $v, w \geq 0$ with $v + w < 1$. For a given network quality (the expression in the curly brackets) the rate of clicks exponentially decreases with price p_n.

The term $c_1^v + \cdots + c_M^v$ is the value of the content providers' investments as seen by a typical end-user. This expression is concave in the investments of the individual providers, and the interpretation is that each content provider adds value to the network. Also note that the structure of the expression is such that end-users value a network in which content is produced by numerous content providers higher than a network in which the content is provided by a single provider with the same cumulative investment. Our end-users' preference for content variety is similar to that of the classical monopolistic competition model by Dixit and Stiglitz [12]. In the expression (18.1), the term in square brackets reflects the value of the ISP's investments for end-users. Clearly users U_n value the investment made by their ISP, but they may also value the investments made by other ISPs. For instance, a user of one ISP might derive more value by having a better connection with users of another ISP. In our model the parameter ρ captures this spill-over effect. When $\rho = 1$, end-users U_n value investments of all ISPs equally while when $\rho = 0$, they value only the investment of their ISP. When $\rho \in (0, 1)$ end-users U_n value investment of their ISP T_n more than investments of other ISPs $T_k \neq T_n$. The term ρ reflects the direct network effect of ISPs on end-users (not between them and content providers). This effect captures a typical network externality (see [13] for a discussion of investment spill-over effects). The factor $1/N^{1-w}$ is a convenient normalization. It reflects the division of the end-user pool among N providers and it is justified as follows. Suppose there were no spill-over and each ISP were to invest t/N. The total rate of clicks should be independent of N. In our model, the total click rate is proportional to $(1/N^{1-w})(N(t/N)^w)$, which is indeed independent of N.

The rate R_{mn} of clicks from end-users U_n to C_m is given by

$$R_{mn} = \frac{c_m^v}{c_1^v + \cdots + c_M^v} B_n. \tag{18.2}$$

Thus, the total rate of clicks for content provider C_m is given by

$$D_m = \sum_n R_{mn}.$$

We assume that content providers charge a fixed amount a per click to the advertisers. Each content provider's objective is to maximize its profit which is equal to revenues from end-user clicks net of investment costs. Thus

$$\Pi_{Cm} = \sum_{n=1}^N (a - q_n) R_{mn} - \beta c_m$$

where the term $\beta > 1$ is the outside option (alternative use of funds c_m).

Internet service provider T_n's profit is

$$\Pi_{T_n} = (p_n + q_n) B_n - \alpha t_n,$$

where $\alpha > 1$ is the outside option of the ISP. We assume providers of each type are identical and we will focus on finding symmetric equilibria for both one- and two-sided pricing.

18.1.2   The analysis of one- and two-sided pricing

To compare one-sided and two-sided pricing (neutral and non-neutral networks), we make the following assumptions.

(a) One-sided pricing (neutral network): in stage 1 each T_n simultaneously chooses (t_n, p_n). The price q_n charged to content providers is constrained to be 0. (Recall the discussion in Section 18.1.) In stage 2 each C_m chooses c_m.

(b) Two-sided pricing (non-neutral network): in stage 1 each T_n simultaneously chooses (t_n, p_n, q_n). In stage 2 each C_m chooses c_m.

In both cases, we assume that content providers observe ISP investments, and can subsequently adjust their investments based on the ISPs' choices. We justify this assumption by the difference in time and scale of the required initial investments. The investments of ISPs tend to be longer-term investments in infrastructure, such as deploying networks of fibre-optic cable. Conversely, the investments of content-providers tend to be shorter term and more ongoing in nature, such as development of content, making ongoing improvements to a search algorithm, or adding/replacing servers in a server farm.

18.1.2.1   Two-sided pricing

In a network with two-sided pricing (non-neutral network), each ISP chooses (t_n, p_n, q_n) and each content provider chooses c_m. To analyze the game we use the principle of backwards induction – analyzing the last stage of the game first and then working back in time supposing that the players in earlier stages will anticipate the play in the latter stages. A content provider C_m in the last stage of the game should choose the optimal c_m after having observed actions (t_n, p_n, q_n) from the preceding stage of the game. Because of a cancelation of terms, it turns out that content provider C_m's profit Π_{C_m} is independent of other content provider investments c_j, $j \neq m$. Therefore, each content provider's optimization is unaffected by the simultaneously made (but correctly anticipated in equilibrium) investment decisions of the other content providers. We therefore simply can find the optimal c_m as a function of (t_n, p_n, q_n), and since the ISPs should anticipate that the content providers will play their best response, we can substitute the function $c_m(t_n, p_n, q_n)$ into the expression for each ISP T_n's profit Π_{T_n}. Each ISP T_n gets to choose the investment and prices (t_n, p_n, q_n) that maximize his profit Π_{T_n}. Since the simultaneous decisions of each of the ISPs affect each other, in order to find a Nash equilibrium we need to identify a point where the best response functions intersect. By carrying out this analysis, we can find closed-form expressions for the Nash equilibrium actions and profits of the content and transit providers (see our working paper, [14] for the detailed derivations):

$$p_n = p = \theta - a; \tag{18.3}$$

$$q_n = q = a - \theta \frac{v}{N(1-v) + v};$$

$$t_n = t \text{ with } (Nt)^{1-v-w} = x^{1-v}y^v e^{-(\theta-a)/\theta};$$

$$c_m = c \text{ with } c^{1-v-w} = x^w y^{1-w} e^{-(\theta-a)/\theta};$$

$$\Pi_{C_m} = \Pi_C := \left(\frac{\theta v(1-v)}{N(1-v)+v}\right) \left[x^w y^v e^{-(\theta-a)/\theta}\right]^{\frac{1}{1-v-w}}; \tag{18.4}$$

$$\Pi_{T_n} = \Pi_T := \left(\frac{M\theta(N(1-v)(1-w\phi) - wv)}{N(N(1-v)+v)}\right) \left[x^w y^v e^{-(\theta-a)/\theta}\right]^{\frac{1}{1-v-w}};$$

where

$$x := \frac{M\theta w}{\alpha} \frac{N\phi(1-v)+v}{N(1-v)+v}; \quad y := \frac{\theta}{\beta} \frac{v^2}{N(1-v)+v}. \tag{18.5}$$

18.1.2.2 One-sided pricing

The analysis of the one-sided case is similar to that of the two-sided pricing (non-neutral), except that $q_n = 0$ as we argued in Section 18.1.1 for $n = 1, \ldots, N$. We use the same backwards induction approach that we described for analyzing the two-sided case. Doing that leads to the following solution for the content providers' and ISPs' actions and profits in equilibrium (see [14] for the derivations):

$$p_n = p_0 := \frac{\theta N(1-v)}{N(1-v)+v};$$

$$q_m = 0;$$

$$t_n = t_0, \text{ where } (Nt_0)^{1-v-w} = x^{1-v}y_0^v e^{-p_0/\theta};$$

$$c_m = c_0, \text{ where } c_0^{1-v-w} = x^w y_0^{1-w} e^{-p_0/\theta};$$

$$\Pi_{C_m} = \Pi_{C0} := a(1-v) \left[x^w y_0^v e^{-p_0/\theta}\right]^{\frac{1}{1-v-w}}; \tag{18.6}$$

$$\Pi_{T_n} = \Pi_{T0} := \left(\frac{M\theta(N(1-v)(1-w\phi) - wv)}{N(N(1-v)+v)}\right) \left[x^w y_0^v e^{-p_0/\theta}\right]^{\frac{1}{1-v-w}};$$

where x is given in (18.5), and $y_0 := av/\beta$.

In Section 18.1.4 we compare the profits and social welfare of the two regimes for a range of parameters.

18.1.3   User welfare and social welfare

In order to compare the one-sided and two-sided regimes, we would like to find expressions for the social welfare of each regime. Social welfare is simply the sum of the net payoffs of all the players involved – content providers, ISPs, and users. The welfare of each content provider and ISP is simply their profit, but for users we need to define their welfare. To do this, we will use the concept of consumer surplus, which is the total amount of the difference between what consumers would have been willing to pay for the service they consumed and what they actually had to pay. In order to make this calculation, we need a demand function that relates end-user price to the quantity consumed, and for our model that is the expression that relates total click rate to price. We compute the consumer surplus by taking the integral of the demand function from the equilibrium price to infinity. This integral is taken with the investment levels of content providers and ISPs fixed. We find

$$W_U(\text{two-sided}) = M\theta x^{w/(1-v-w)} y^{v/(1-v-w)} e^{-\frac{\theta-a}{\theta(1-v-w)}}.$$

The expression for the one-sided case is the same, but with y exchanged for y_0 and $\theta - a$ in the exponent exchanged with p_0. The ratio of the social welfare with one- vs. two-sided pricing has the form

$$\frac{W_U(\text{one-sided}) + N\Pi_T(\text{one-sided}) + M\Pi_C(\text{one-sided})}{W_U(\text{two-sided}) + N\Pi_T(\text{two-sided}) + M\Pi_C(\text{two-sided})}.$$

18.1.4   Comparison

To compare the revenue in the one- and two-sided cases, we define the ratio

$$r(\Pi_C) := \left(\frac{\Pi_C(\text{one-sided})}{\Pi_C(\text{two-sided})}\right)^{1-v-w},$$

where $\Pi_C(\text{two-sided})$ is the profit per content provider in the two-sided case as expressed in (18.4) and $\Pi_C(\text{one-sided})$ is the profit per content provider with one-sided pricing (18.6). We define $r(\Pi_T)$ similarly. We find

$$r(\Pi_T) = \left(\frac{\delta}{\pi}\right)^v e^{\pi-\delta}, \quad r(\Pi_C) = \left(\frac{\delta}{\pi}\right)^{1-w} e^{\pi-\delta}, \tag{18.7}$$

where

$$\pi := \frac{v}{N(1-v)+v} \quad \text{and} \quad \delta := \frac{a}{\theta}.$$

Figure 18.2 shows the ratios of revenues with one- vs. two-sided pricing for both content providers and ISPs. The figures show that for small or large values of a/θ, the ratio of advertising revenue per click to the constant characterizing price sensitivity of end-users, two-sided pricing is preferable to both content providers and ISPs. (Here we say "preferable" in that the revenues are larger,

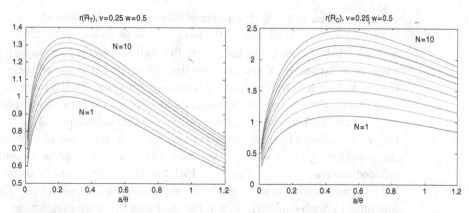

Figure 18.2 The ratios of profits ($v = 0.25, w = 0.5$) for different values of N.

though we have seen that the rate of return on investments is the same.) For mid-range values of a/θ, one-sided pricing is preferable to both, though the values of a/θ where the transition between one-sided being preferable to two-sided are not exactly the same for content providers and ISPs. Furthermore, as N, the number of ISPs increases, the range of a/θ values for which one-sided pricing is superior increases, while also the degree by which it is superior (in terms of revenues to content providers and ISPs) increases.

These results can be explained by the following reasoning. When a/θ is large, the content providers' revenues from advertising are relatively high, and the ISPs' revenue from end-users are relatively low. Because of this, the ISPs' incentives to invest are suboptimal (too low relative to the socially optimal ones), unless they can extract some of the content providers' advertising revenue by charging the content providers. Thus in the one-sided pricing case, the ISPs underinvest, making the rewards for them as well as content providers less than it could have been with two-sided pricing.

It is important to note that when a/θ is larger than 1, the price p charged end-users becomes negative in the two-sided case as can be seen from (18.3). If end-users were actually paid to click, intuition suggests that they would click an unbounded amount and therefore our exponential model of demand (18.1) would not be valid in this region of a/θ. However, one could interpret price p to be net any variable costs v to end-users – similar to how we define q. With this interpretation, the price p could be negative while the actual prices users see is positive, so long as $|p| < v$. We therefore show numerical results in our plots for a/θ as large as 1.2.

When a/θ is very small, the content providers' advertising revenue is relatively low, and the ISP's end-user revenue is relatively high. In order to get the content providers to invest adequately, the ISPs need to pay the content providers. That is why for small enough a/θ the price q actually becomes negative, representing a per click payment from the ISPs to the content providers.

Finally, when a/θ is in the intermediate range, in between the two extremes, both content providers and ISPs have adequate incentive to invest. However another effect comes into play – ISP free riding becomes an important factor when N is large. As N increases in the two-sided pricing case there are more ISPs that levy a charge against each content provider. As the price ISPs charge content providers increases, it becomes less attractive for content providers to invest. Thus an ISP choosing the price to charge content providers is balancing the positive effect of earning more revenue per click from content providers versus the negative effect of having fewer clicks because the content providers have reduced their investment. But each ISP sees the entire gain of raising its price, but the loss is borne by all N ISPs. Consequently, the ISPs overcharge the content providers in Nash equilibrium, and the degree of this overcharging increases with N. This is analogous to the tragedy of the commons where people overexploit a public resource. Another perhaps more direct analogy is the "castles on the Rhine effect" where each castle owner is incentivized to increase transit tolls on passing traffic excessively by ignoring the fact that the resulting reduction in traffic harms not only him, but also other castle owners. When all castles do the same, the traffic on the Rhine decreases [15]. The extent of this negative externality, and hence the degree of overcharging, increases with N.

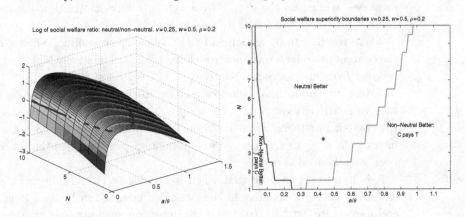

Figure 18.3 Left: the log of social welfare ratio (one-sided to two-sided). Right: regions of social welfare superiority (one-sided vs. two-sided).

Figure 18.3 shows a three-dimensional plot of the ratio of social welfare for one- vs. two-sided pricing. The plot shows how the ratio changes for different N and a/θ. The second panel of Figure 18.3 is a simplified version of the first panel. It depicts only the boundaries in the parameter space where one-sided pricing is preferable to two-sided and vice versa.

It is also worthwhile to note that the spill-over parameter ρ and the number of content providers M do not appear in the expression for the ratio of content provider revenue between the two regimes nor do they appear in the ratio of revenues for ISPs (18.7). This is in spite of the fact that ρ and M do appear in the expressions for both the one-sided and two-sided pricing equilibria. This suggests that the spill-over effect and number of content providers have little or no effect on the comparative welfare of the two regimes.

18.1.5 Conclusions

Our model shows how each pricing regime affects investment incentives of transit and content providers. In particular, the model reveals how parameters such as advertising rate, end-user price sensitivity, and the number of ISPs influence whether one- or two-sided pricing achieves a higher social welfare. From our results, when the ratio of advertising rates to the constant characterizing price sensitivity is an extreme value, either large or small, two-sided pricing is preferable. If the ratio of advertising rates to the constant characterizing price sensitivity is not extreme, then an effect like the "castles on the Rhine effect" becomes more important. Internet service providers in a two-sided pricing regime have the potential to overcharge content providers, and this effect becomes stronger as the number of ISPs increases.

Of course our model contains many simplifications. Among these simplifications is the assumption that ISPs have local monopolies over their users. We believe that if we had studied a model where each ISP is a duopolist, which better models the degree of choice most end-users have today, our results would have been qualitatively similar. However, there are a number of competing effects such a model might introduce. First, such a scenario would reduce the market power of ISPs over end-users, thus reducing the revenue the ISPs could extract from them. Since two-sided pricing provides ISPs with another source of revenue to justify their investments, this effect would tend to increase the parameter region for which two-sided pricing is social welfare superior. Second, a duopolist competing on product quality invests more than a monopolist, so this would tend to increase efficiency of one-sided pricing. Third, if the model were changed from having N to $2N$ ISPs, then the free riding or "castles on the Rhine" effect would grow, tending to reduce the welfare of the two-sided pricing case. The net effect of all these individual effects would of course depend on the detailed specifications of such a model.

Our two-sided pricing model implicitly assumes that in-bound traffic to a local ISP could be identified as originating at a particular content provider, in order for the ISP to levy the appropriate charge to the content provider. This assumption would not strictly hold if content providers had some way of reaching end-users of an ISP without paying this ISP for end-user traffic. For instance, if there were a second ISP that enjoyed settlement free peering with the first ISP, the content provider could route its traffic through the second ISP and thus avoid the first ISP charge for end-user access. This strategy might be facilitated by the fact that the end-users of both ISPs send traffic to each other, and perhaps the traffic from the content provider could be masked in some way to look like traffic originating from the second ISP's end-users. However, the communication protocols of the Internet require that packets be labeled with the origin (IP) address. It seems unlikely today that a large content provider could have the origin addresses of its traffic falsified in a way that would both prevent ISPs from being able to charge the content provider while still enabling end-users to send traffic in the reverse direction back to the content provider. However, it is certainly possible that technology would be developed to enable such a strategy in the future, especially if there were an economic incentive for developing it.

18.2   Competition

The subject of communication network economics is fundamentally complex largely because there are so many interactions. For instance, the Internet is built and operated by many different providers, most of which are making pricing, capacity investment, and routing decisions in order to further their own commercial interests. Users are also acting selfishly – choosing their ISPs, when and how much to use the network, and sometimes even their traffic's routing, in order to maximize their payoff – the difference between the utility they enjoy from the network minus what they have to pay. Studying how efficiently such a complex market system can operate is fundamentally important, because it is with this understanding that we will be able to evaluate the potential of new network architectures that may change the structure of the interactions between the agents of such a market.

There has been an enormous amount of research on the fundamentals of communications networks economics, yet there are still important gaps in our basic understanding. For instance, past work has looked at the effects of selfish routing by users, selfish inter-domain routing between ISPs, congestion effects, ISP price competition, and ISP capacity investment decisions. (There is a large amount of literature in each of these areas. For example, for selfish routing by users see [16, 17, 18]; for selfish inter-domain routing see [19]; for price competition see [20], [21], [22], and [23].) There are many properties that are of interest when one studies such models, but of particular importance is the efficiency loss of the system due to the selfish interactions of the agents as compared to a hypothetical, perfect, central agent making all the decisions. One way of quantifying the efficiency loss is the "price of anarchy," concept introduced by Koutsoupias and Papadimitriou [24]. The price of anarchy (PoA) is simply the ratio of the optimum total utility of the system divided by the worst utility achieved in Nash equilibrium.

A large number of price of anarchy results have been shown for general models that contain various combinations of a few of the above features we listed. Yet, there are few general results (price of anarchy bounds for a class of models) when one considers more than a few of these features – even when the simplest possible models are used for each feature. For instance, if one considers a model with selfish routing, elastic demand (users can send less traffic if the network is too expensive or slow), and price competition between ISPs, there are price of anarchy results for limited cases, but there is not yet a complete understanding of the problem. In the remainder of this section we look at this particular class of problems, and describe some recent results that push in the direction of getting a more general understanding.

The model we consider is based on a model first proposed and studied by Acemoglu and Ozdaglar [20] and later extended by Hayrapetyan, Tardos, and Wexler [22]. The model studies the pricing behavior of several providers competing to offer users connectivity between two nodes, and it has the features that (i) a provider's link becomes less attractive as it becomes congested; and (ii) that user demand is elastic – users will choose not to use any link if the sum of the price and latency of the available links is too high. In the first version of the model studied by Acemoglu and Ozdaglar, the user elasticity is modeled by assuming that all users have a single reservation utility and that if the best available price plus latency exceeds this level, users do not use any service. In this setting, the authors find that the price of anarchy – the worst case ratio of social welfare achieved by a social planner choosing prices to the social welfare arising when providers strategically choose their prices – is $(\sqrt{2}+1)/2$. (Or expressed the other way, as the ratio of welfare in Nash equilibrium to social optimum, the ratio is $2\sqrt{2}-2$.) In a later work, Ozdaglar and Acemoglu [25] extend the model to consider providers in a parallel-serial combination. Traffic chooses between several branches connected in parallel, and then for each branch, the traffic traverses the links of several providers connected serially.

Hayrapetyan *et al.* [22] consider a model in which user demand is elastic, which is the form of the demand model we will study here. The topology they consider is just the simple situation of competing links between two nodes. They derived the first loose bounds on the price of anarchy for this model. Later Ozdaglar showed that the bound is actually 1.5, and furthermore that this bound is tight [21]. Ozdaglar's derivation uses techniques of mathematical programming, and is similar to the techniques used in [20]. In [23], Musacchio and Wu provide an independent derivation of the same result using an analogy to an electrical circuit where each branch represents a provider link and the current represents flow.

The work we describe next generalizes the work of [23] by considering a more general topology of providers connected in parallel and serial combinations. A more detailed description of the work is available in [26]. As in the earlier work, we use a circuit analogy to derive bounds on the price of anarchy. Furthermore, our bounds depend on a metric that is a measure of how concentrated the market power is in the network. That metric is the reciprocal of the slope of the latency function of the network branch with the least slope divided by the harmonic mean of the slopes of all the branches. In terms of the circuit analogy, this is simply the ratio between the conductance of the most conductive branch to the conductance of the whole system.

18.2.1   Model

We consider a model in which users need to send traffic from a single source to a single destination, but there are several alternative paths that the users can choose from. The paths consist of parallel-serial combinations of Internet service providers as in the example illustrated by the left panel of Figure 18.4. The traffic incurs delay as it crosses each ISP $\mathbf{i}$ that is linear in the amount of traffic that is passing through that ISP. Thus if the flow through ISP $\mathbf{i}$ is $f_{\mathbf{i}}$ we suppose the delay is $a_{\mathbf{i}}f_{\mathbf{i}}$ where $a_{\mathbf{i}}$ is a constant. Furthermore, there is a fixed latency along each main branch, as illustrated by Figure 18.4. Note that we denote the index $\mathbf{i}$ in bold because it represents a vector of indices that determine the location of that ISP in the network.

Users control over which path(s) their traffic goes. A user chooses a path for her traffic selfishly and she cares both about the delay her traffic incurs as well as the price she pays. To account for both of these preferences, we suppose that users seek to use the path with the minimum price plus delay (just as in the model of [20] and others). We term the metric price plus delay, "disutility," just as in [22]. We make the further assumption that the traffic from each user represents a negligible fraction of the total traffic, thus when a single user switches routes, that one user's switch does not change the delays significantly. An equilibrium occurs when users cannot improve their disutility by changing paths, so therefore in such an equilibrium all used paths must have the same disutility and all unused paths must have a disutility that is larger (not smaller.) This kind of equilibrium in which users select routes under the assumption that their individual choice will have a negligible change on the congestion of the route they choose is called a Wardrop equilibrium [27].

As in the model of [22] we suppose that the demand is elastic so if the disutility is too high, some users do not connect to any network or reduce the amount of traffic they send. If d is the disutility on the used paths, then we suppose that the total amount of traffic that is willing to accept that disutility is $f = U^{-1}(d)$ where $U(\cdot)$ is a decreasing function. Furthermore, we suppose that $U(\cdot)$ is concave.

In summary, we suppose that given a set of prices from the ISPs (a price profile), the users selfishly select both routes and the amount of traffic to send and as a result, the network reaches a Wardrop equilibrium and furthermore the amount of traffic sent satisfies $f = U^{-1}(d)$. It turns out that for a given set of prices, there exists a unique Wardrop equilibrium that satisfies these properties. (We refer to $U(\cdot)$ as the "inverse demand" function and disutility curve interchangeably.)

Given this user behavior, we now turn to the ISPs. The ISPs play a game in which each ISP $\mathbf{i}$ chooses their price $p_{\mathbf{i}}$, and the resulting Wardrop equilibrium induced by price profile determines the amount of traffic each ISP carries. In turn, the profit each ISP earns is the product of their price times the traffic they carry. We are interested in comparing the social welfare in a Nash equilibrium of this game and comparing that to the maximum social welfare achievable if a

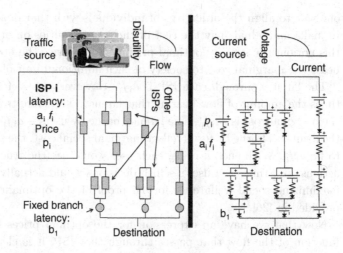

Figure 18.4 An analogous circuit for analyzing the network.

central agent were to choose the price profile. (Of course we are also interested in whether a Nash equilibrium even exists, and if so, if it is unique.) The social welfare of the game includes the sum of the ISP profits, as well as the welfare of users.

We use the standard notion of consumer surplus as our metric of user welfare. If the total rate offered Web traffic is f, the consumer surplus is found simply by taking $\int_0^f U(z)dz - fU(f)$. The integral has the following interpretation. From the disutility curve $U(\cdot)$, we see that the first ϵ units of traffic would be willing to pay a price as high as $k - \epsilon$ per unit, the next ϵ units of traffic would be willing to pay a price of $k - 2\epsilon$ per unit, and so on. Thus integrating the disutility curve from 0 to f captures the total amount the traffic is willing to pay, and then subtracting the amount it actually pays $fU(f)$, yields the surplus enjoyed by the traffic (users).

18.2.2   Circuit analogy

In order to analyze this game we are going to draw an analogy between this game and an electric circuit as suggested by Figure 18.4. It turns out that the relationships between prices and flows in our model are analogous to the Kirchoff voltage and current law relations between the voltages and currents in the circuit pictured in the figure.

To begin analyzing the system, we start by determining the prices that induce the socially optimum flow. The well-known theory of Pigovian taxes says that one way to align the objectives of individuals with that of society as a whole is to make individuals pay the cost of their externalities on others [28]. Following this notion, the socially optimal pricing should price the flow so that each user bears the marginal cost to society of each additional unit of new flow.

The latency through each ISP is $a_i f_i^*$. However, the social cost is the latency times the amount of flow bearing that latency. Therefore the cost is $a_i f_i^{*2}$. Thus the marginal cost is $2a_i f_i^*$. The latency borne by users is $a_i f_i^*$. Therefore to make the disutility borne by users reflect marginal social cost, the price they pay ought to be $a_i f_i^*$. With such a price, each user would see the true social cost of their actions, and thus each user's selfish decisions would actually be working toward maximizing social welfare. A formal proof of the optimality of these prices is provided in [26].

Now that we have an expression for the optimal prices for each ISP i as a function of the flow that passes through that ISP, it is clear that the optimal assignment of flows ought to satisfy the constraint that $p_i = a_i f_i^*$ for each ISP i in addition to the Wardrop equilibrium and the inverse demand relationship between flow and disutility. If we now turn to the circuit pictured in the right panel of Figure 18.4 and substitute a resistor of size a_i for each voltage source that represents price p_i, we have a circuit whose solution gives us the solution to the problem of finding the optimal assignment of flows (and thus prices) for all the ISPs. We call the circuit with these resistor substitutions the *optimal circuit*.

We now turn to understanding the Nash equilibrium of the game. First it is important to realize that providers connected serially can have very inefficient equilibria. Consider just two providers connected serially, each of which charges a price above the highest point on the disutility curve. The flow through these providers will be zero, and furthermore it is a Nash equilibrium for these providers to charge these prices. This is because one player cannot cause the flow to become positive by lowering his price, so each player is playing a best response to the other player's prices. In examples for which there is more than one provider connected serially on each branch, one can construct a Nash equilibrium in which no flow is carried, and therefore the price of anarchy across all Nash equilibria is infinite.

However, these infinite price of anarchy Nash equilibria seem unlikely to be played in a real situation. Intuitively, a provider would want to charge a low enough price so that it would be at least possible that he could carry some flow if the other providers on the branch also had a low enough price. With that in mind, we would like to identify a more restricted set of Nash equilibria that seem more "reasonable" and that have a price of anarchy that can be bounded. To that end, we define the notion of a *zero-flow zero-price* equilibrium. A zero-flow zero-price equilibrium is a Nash equilibrium for which players who are carrying zero-flow must charge a zero-price. The intuition motivating this definition is that a player who is not attracting any flow will at least try lowering his price as far as possible in order to attract some flow. It indeed turns out that for this restricted set of equilibria, the price of anarchy is bounded.

Just as we obtained the social optimum solution, we will seek to express the ratio of price to flow in Nash equilibrium. An ISP that is trying to maximize his profit must consider how much his flow is reduced when he increases his price. This could be very complicated to compute in a complex network, but fortunately we can use our circuit analogy to simplify the problem. A basic result of circuit theory is that a resistive electric circuit viewed from a port, such as the pair of circuit nodes that ISP i is connected to, can be reduced to an equivalent circuit containing a single resistor and voltage source [29]. Such an equivalent is known as the Thévenin equivalent. Thus we can abstract ISP i's "competitive environment" into a single resistor and voltage source. The Thévenin equivalent resistance is found by adding the resistances of resistors connected in series, and taking the reciprocal of the sum of reciprocals of those connected in parallel.

There are a few more details that we need to take into account, as Thévenin equivalents hold only for linear circuits, and we have a few nonlinearities – the diodes in each branch that keep an ISP's flow from being negative, and the inverse demand function. It turns out that the inverse demand function can be linearized at the Nash equilibrium point, and its slope s can be used as a "resistor" in the computation of the Thévenin equivalent. The diodes can be taken into account by only including resistors of branches that are "on" in Nash equilibrium.

With these considerations taken in mind, suppose that the Thévenin equivalent resistance player i sees is δ_i. If i unilaterally raises his price by $+\epsilon$ the flow he carries will be reduced by $\epsilon/(\delta_i + a_i) - o(\epsilon^2)$. (It turns out the term $-o(\epsilon^2)$ is negative because of the concavity of the inverse demand function.) Thus we have that

$$(p_i + \epsilon)\left(f_i - \epsilon/(\delta_i + a_i) - o(\epsilon^2)\right) = p_i f_i + [-p_i(\delta_i + a_i)^{-1} + f_i]\epsilon - o(\epsilon^2).$$

The above expression is less than the original profit $p_i f_i$ for all nonzero ϵ if and only if

$$\frac{p_i}{f_i} = \delta_i + a_i. \tag{18.8}$$

At this point we have shown a property of the Nash equilibrium flow and prices, but we have not shown that an equilibrium actually exists. It turns out that a Nash equilibrium does exist, and this is proved by construction in [26]. Another technical detail is that in some cases the Thévenin equivalent could be different for a small price increase than it is for a small price decrease. This is because a small price increase might induce flow in some branches that are otherwise off. For these cases, one can show that a best response must satisfy

$$\frac{p_i}{f_i} \in [\delta_i^+ + a_i, \delta_i^- + a_i],$$

where δ_i^+ and δ_i^- are the two different Thévenin equivalents.

From equation (18.8) we see that the price to flow ratio in Nash equilibrium is higher than it is for the social optimum prices. As we did when we constructed the "optimal circuit" we can construct something we call a *Nash circuit* by

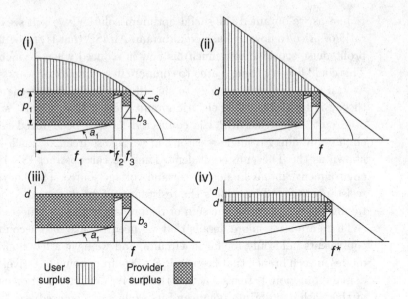

Figure 18.5 (i) The Nash equilibrium of game $\mathcal{G}$. (ii) The Nash equilibrium of game $\mathcal{G}_l$, where the disutility function of $\mathcal{G}$ has been linearized. (iii) The Nash equilibrium of the game $\mathcal{G}_t$, where the disutility function of $\mathcal{G}$ has been linearized and "truncated." (iv) The social optimum configuration of the game $\mathcal{G}_t$. Note that the flow $f^* > f$ and that link 2 is not used in the social optimum configuration of this example.

substituting resistors of size $\delta_i + a_i$ for the voltage sources representing price p_i for each ISP i. The solution to this circuit gives us the Nash equilibrium flows and prices.

Now that we have the social optimum and Nash equilibrium configurations of the system described by linear circuits, it is possible to derive closed-form expressions that lead to bounds on the price of anarchy. For the class of networks we call "simple parallel serial" which is a network of parallel branches each containing one or more serially connected providers, we can write Kirchoff's voltage laws in matrix form, and then from that derive quadratic forms for the total profit of the providers in both the social optimum and Nash cases. This is done in detail in [26].

It turns out that for any problem instance (consisting of a disutility function, a topology, and ISP latencies $\{a_i\}$) with Nash equilibrium flow f and disutility function slope $-s = U'(f)$, we can construct a new problem with a price of anarchy at least as high in the following way. We modify the disutility function to be flat for flows between 0 and f and then make it affine decreasing with slope $-s$ for higher flows. (This is an argument adapted from [22].) This is basically because the new problem instance has the same Nash equilibrium as the old problem, but now the user welfare is zero. The argument is illustrated by Figure 18.5. Because of this argument, we can restrict our attention to disutility functions of the shape shown in the lower part of the figure. After invoking this argument we can express the user welfare in social optimum for this shape of disutility function using a quadratic form.

It turns out that the algebra works out so that our bounds are found to be a function of a parameter y we term the *conductance ratio*. The conductance ratio is the conductance of the most conductive branch divided by the conductance of the network as a whole. The conductance ratio is therefore a measure of how concentrated the capabilities of the network are in a single branch. A conductance ratio near 1 means that most of the conductance of the system is concentrated in a single branch. The smaller the conductance ratio, the more that the overall conductance of the system is distributed across multiple branches. Thus, in a sense the conductance ratio reflects the market power or concentration of the system. As one would expect, the price of anarchy bounds that we find increase as the conductance ratio approaches 1. The following theorem comes from and is proven in [26].

Theorem 18.1. *Consider the game with a simple parallel-serial topology. Consider the following ratio*

$$y = \frac{\max_i 1/a_i}{\sum_i 1/a_i},$$

which is the conductance of the most conductive branch divided by the overall conductance. The price of anarchy for zero-flow zero-price Nash equilibria is no more than

$$\begin{cases} \dfrac{1}{4} \dfrac{m^2 + 2m(1+y) + (y-1)^2}{m} & y \leq 1 - m/3 \\ \dfrac{m^2(2-y) + m(4-y^2-y) + 2(y-1)^2}{8m - 6my} & y \geq 1 - m/3 \end{cases} \qquad (18.9)$$

where m is the maximum number of providers connected serially. Furthermore, the maximum of the above bound occurs when $y = 1$, and consequently the price of anarchy is no more than

$$1 + m/2. \qquad (18.10)$$

The bounds given in Theorem 18.1 are illustrated in Figure 18.6. Note how the price of anarchy falls as the conductance ratio falls, i.e., becomes less monopolistic. Also note the increase in the price of anarchy as the number of serially connected providers increases. This is an example of the well-known "double marginalization" effect in which serially connected providers tend to overcharge (as compared to social optimum) because they do not consider how their price increases will hurt the profits of the other serially connected providers.

The results of Theorem 18.1 are for simple parallel-serial topologies. For general parallel-serial topologies (with arbitrary groupings of ISPs connected in parallel and series), it turns out the same bounds hold, with an additional factor of 2. The argument is given in [26]. We do believe that the bound with a factor of 2 is not tight, but it remains an open problem to get a tighter bound.

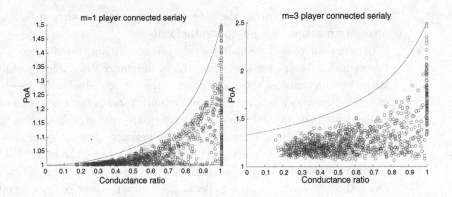

Figure 18.6 Simple parallel-serial PoA bound for the cases where there are either $m = 1$ or 3 providers connected serially. The bounds are plotted as a function of the conductance ratio – the ratio of the conductance of the most conductive branch to the conductance of the whole circuit. The points are the PoA of randomly generated example topologies.

18.3   Service differentiation

As we said in the introduction of this chapter, service differentiation is an important tool for improving the Internet by providing quality to those that need it most. Furthermore, for architectures that separate high-priority traffic from low-priority traffic, that separation can increase the utility of the system by preventing the mixing of traffic of different types that do not mix well (i.e., video conferencing and peer-to-peer file sharing). In addition to increasing user utility, service differentiation has the potential of increasing provider revenues. Walrand [30] demonstrates that with a simple model, and other works have shown this as well using different models.

Another interesting set of questions comes up when one considers the combination of service differentiation and ISP competition. Musacchio and Wu [31] consider such a combination and shows that architectures that support service differentiation by offering delay sensitive traffic with a priority queue lead to competition games between ISPs that have a lower price of anarchy than if the same ISPs were to use a shared queue, no differentiation architecture. That work supposes that the delay sensitive traffic comes from applications that are close to constant bit-rate (called "voice" for short), while the delay insensitive applications generate more bursty traffic (called "web" for short).

Musacchio and Wu [31] models the network with a simple queueing model and supposes a capacity constraint for each ISP. Using this approach we derive a space of feasible regions for the vector of web and voice traffic for both the shared and differentiated architecture cases. Basically with a shared architecture, as the fraction of web traffic is increased, the ISP needs to operate at a lower utilization to meet the delay constraint. However, if the ISP chooses not to provide good service to voice traffic, that ISP can operate at a much higher utilization. With a priority architecture, the ISP can operate at a high utilization regardless of the web vs. voice mix because the voice traffic is always protected. Using this approach we show that the price of anarchy can be as high as 2 for a shared architecture but only 1.5 for a priority architecture. The model in this study is still quite simplistic, so there are still many important open problems regarding the price of anarchy in differentiated services competition.

Despite the enormous potential advantages of the service differentiation there are a number of challenges. One set of challenges are in developing technical architectures to support service differentiation in a scalable way. A large amount of work has been done in this area, and many protocols and architectures have been proposed which we will not attempt to survey here. Another challenge is a coordination problem – if one ISP adopts a new architecture, the full benefit of that adoption might not be realized until other ISPs also adopt it. Several works have looked at the problem of getting new architectures adopted in the presence of an incumbent architecture (for example [32] and [33]).

Another challenge is that the move from a single service class to multiple service classes actually has the potential to cause some users to be worse off, even if in the aggregate the population of users benefits. This is because users who want mid-range quality and are happy with a single class of service might be forced to choose between a low-quality service or a high-priced, high-quality service in a differentiated system. Schwartz, Shetty, and Walrand study this issue, and they suggest [34] that the number of users who fall into this category could be reduced, while still getting most of the benefits of service differentiation.

Another challenge is that there can be a potential instability in systems that provide service differentiation. Consider a system with a priority queue and a best effort queue. Any "cycles" spent serving the priority queue come at the expense of the best effort queue, so if a lot of traffic uses the priority queue, the best effort traffic will see lower quality. Now consider the following dynamic. Suppose some users move from the best effort queue to the more expensive priority queue in order to get better service. This shift will degrade the quality of the best effort queue, causing more users to switch to the priority queue, which again in turn degrades the best effort queue. Depending on the precise utility functions of the users this process might not stop until all the users are using the priority queue. In other examples the mix might tip to everyone using best effort, or perhaps to some positive fraction using both. In summary, the situation is quite "tippy," and so a small change in modeling assumptions can lead to very different predicted outcomes.

This is a phenomenon that has been long recognized by other researchers. This phenomenon is one of the reasons that Odlyzko [35] proposes a Paris metro pricing (PMP) for Internet service classes. Like the Paris metro once did, Odlyzko's PMP scheme would offer a predetermined fraction of the resources to the different classes, but at different prices. The scheme is inherently more stable than a priority scheme, because the quality of each class depends only on the congestion of that class. However, one drawback is that one loses some statistical multiplexing gain. For instance, if there is instantaneous demand for the lower class but not the upper class, such a system would not offer its full capacity to the lower-class traffic.

Although the "tippiness" problem of schemes for which the congestion of one class affects the quality of the other classes makes these schemes difficult to analyze, the task is not impossible. A recent model of Schwartz, Shetty and Walrand (manuscript in submission as of 2009) includes these effects. Another work by Gibbens, Mason, and Steinberg [36] demonstrates that a different effect driven by competition between ISPs can lead to the market tipping to a single service class. Clearly, there is potential for future work in the area to better understand these phenomena.

Acknowledgement

The authors gratefully acknowledge the National Science Foundation for supporting this work under grants NeTS-FIND 0626397 and 0627161.

References

[1] E. Whitacre, At SBC, it's all about 'scale and scope.' *Business Week*, Interview by R. O. Crockett, (November 7, 2005).

[2] J. Musacchio, G. Schwartz, and J. Walrand, A two-sided market analysis of provider investment incentives with an application to the net neutrality issue. *Review of Network Economics*, **8**:1 (2009), 22–39.

[3] A. Odlyzko, Network neutrality, search neutrality, and the never-ending conflict between efficiency and fairness in markets. *Review of Network Economics*, **8** (2009), 40–60.

[4] R. B. Chong, The 31 flavors of the net neutrality debate: Beware the trojan horse. *Advanced Communications Law and Policy Institute, Scholarship Series*, New York Law School, (December 2007).

[5] Federal Trade Commission, Broadband connectivity competition policy. Report (June 2007).

[6] Department of Justice, Comments on network neutrality in federal communications commission proceeding. Press Release, (September 7, 2007).

[7] J.-C. Rochet and J. Tirole, Two-sided markets: A progress report. *RAND Journal of Economics*, **37**:3 (2006), 655–667.

[8] M. Armstrong, Competition in two sided markets. *RAND Journal of Economics*, **37**:3 (2006), 668–691.

[9] B. Hermalin and M. Katz, The economics of product-line restrictions with an application to the network neutrality controversy. *Information Economics and Policy*, **19** (2007), 215–248.

[10] C. Hogendorn, Broadband internet: Net neutrality versus open access. *International Economics and Economic Policy*, **4** (2007), 185–208.

[11] P. Weiser, Report from the Center for the New West putting network neutrality in perspective. Center for the New West discussion paper (January 2007).

[12] A. Dixit and J. Stiglitz, Monopolistic competition and optimum product diversity. *The American Economic Review*, **67**:3 (1977), 297–308.

[13] J. Thijssen, *Investment Under Uncertainty, Coalition Spillovers and Market Evolution in a Game Theoretic Perspective*, (Springer, 2004).

[14] J. Musacchio, G. Schwartz, and J. Walrand, *A Two-Sided Market Analysis of Provider Investment Incentives With an Application to the Net-Neutrality Issue*, School of Engineering, Universitiy of California, Santa Cruz, UCSC-SOE-09-08 (2009).

[15] J. A. Kay, Tax policy: A survey. *The Economic Journal*, **100**:399 (1990), 18–75.

[16] T. Roughgarden, Stackelberg scheduling strategies. In *Proceedings of the 33rd Annual Symposium ACM Symposium on Theory of Computing*, (2001), 104–13.

[17] T. Roughgarden, How bad is selfish routing? *Journal of ACM*, **49**:2 (2002), 236–259.

[18] T. Roughgarden, The price of anarchy is independent of the network topology. *Journal of Computer and System Sciences*, **67**:2 (2003), 341–364.

[19] J. Feigenbaum, R. Sami, and S. Shenker, Mechanism design for policy routing. *Distributed Computing*, **18** (2006), 293–305.

[20] D. Acemoglu and A. Ozdaglar, Competition and efficiency in congested markets. *Mathematics of Operations Research*, **32**:1 (2007), 1–31.

[21] A. Ozdaglar, Price competition with elastic traffic. *Networks*, **52**:3 (2008), 141–155.

[22] A. Hayrapetyan, E. Tardos, and T. Wexler, A network pricing game for selfish traffic. *Distributed Computing*, **19**:4 (2007), 255–266.

[23] J. Musacchio and S. Wu, The price of anarchy in a network pricing game. In *Proceedings of the 45th Annual Allerton Conference on Communication, Control, and Computing*, Monticello, IL, (2007), 882–891.

[24] E. Koutsoupias and C. H. Papadimitriou, Worst-case equilibria. In *Proceedings of the 16th Annual Symposium on Theoretical Aspects of Computer Science*, Trier, Germany, (1999), 404–413.

[25] D. Acemoglu and A. Ozdaglar, Competition in parallel-serial networks. *IEEE Journal on Selected Areas in Communications*, **25**:6 (2007), 1180–1192.

[26] J. Musacchio, *The Price of Anarchy in Parallel-Serial Competition with Elastic Demand*, School of Engineering, University of California, Santa Cruz, UCSC-SOE-09-20 (2009).

[27] J. Wardrop, Some theoretical aspects of road traffic research. *Proceedings of the Institute of Civil Engineers, Part II*, **1**:36 (1952), 352–362.

[28] A. C. Pigou, *The Economics of Welfare*, (Macmillan, 1920).

[29] L. Chua, C. Desoer, and E. Kuh, *Linear and Nonlinear Circuits*, (McGraw-Hill, 1987).

[30] J. Walrand, Economic models of communication networks. In *Performance Modeling and Engineering*, ed. Xia and Liu. (Springer, 2008), pp. 57–90.

[31] J. Musacchio and S. Wu, The price of anarchy in differentiated services networks. In *Proceedings of the 46th Annual Allerton Conference on Communication, Control, and Computing*, Monticello, IL, (2008), 615–622.

[32] J. Musacchio, J. Walrand, and S. Wu, A game theoretic model for network upgrade decisions. In *Proceedings of the 44th Annual Allerton Conference on Communication, Control, and Computing*, Monticello, IL, (2006), 191–200.

[33] Y. Jin, S. Sen, R. Guerin, K. Hosanagar, and Z.-L. Zhang, Dynamics of competition between incumbent and emerging network technologies. In *Proceedings of the ACM NetEcon08: The Workshop on the Economics of Networks, Systems, and Computation*, Seattle, WA, (2008), 49–54.

[34] G. Schwartz, N. Shetty, and J. Walrand, Network neutrality: Avoiding the extremes. In *Proceedings of the 46th Annual Allerton Conference on Communication, Control, and Computing*, Monticello, IL, (2008), 1086–1093.

[35] A. Odlyzko, Paris metro pricing for the internet. In *Proceedings of the 1st ACM conference on Electronic Commerce*, Denver, CO, (1999), 140–147.

[36] R. Gibbens, R. Mason, and R. Steinberg, Internet service classes under competition. *IEEE Journal on Selected Areas in Communications*, **18**:12 (2000), 2490–2498.

About the editors

Byrav Ramamurthy

Byrav Ramamurthy is an associate professor in the Department of Computer Science and Engineering at the University of Nebraska-Lincoln (UNL), USA, where he has been on the faculty since August 1998. He is author of the book *Design of Optical WDM Networks – LAN, MAN and WAN Architectures* and a co-author of the book *Secure Group Communications over Data Networks* published by Kluwer Academic Publishers/Springer in 2000 and 2004, respectively. He serves as an editor for *IEEE Communications Surveys and Tutorials* as well as the *IEEE/OSA Journal of Optical Communications and Networking*. He serves as the TPC Co-Chair for the premier networking conference, INFOCOM, to be held in Shanghai, China in 2011. Professor Ramamurthy is a recipient of the UNL College of Engineering and Technology Faculty Research Award for 2000 and the UNL CSE Department Student Choice Outstanding Teaching Award for Graduate-level Courses for 2002–2003 and 2006–2007. His research areas include optical networks, wireless/sensor networks, network security, and telecommunications.

George N. Rouskas

George Rouskas is a Professor of Computer Science at North Carolina State University, USA. He received the Ph.D. and M.S. degrees in Computer Science from the College of Computing, Georgia Institute of Technology, Atlanta, and a degree in Computer Engineering from the National Technical University of Athens (NTUA), Athens, Greece. Professor Rouskas has received several research and teaching awards, including an NSF CAREER Award, the NCSU Alumni Outstanding Research Award, and the ALCOA Foundation Research Achievement

Award, and he has been inducted in the NCSU Academy of Outstanding Teachers. He is the author of the book *Internet Tiered Services* (Springer, 2009). His research interests are in network architectures and protocols, optical networks, and performance evaluation.

Krishna M. Sivalingam

Krishna Sivalingam is a Professor in the Department of CSE, Indian Institute of Technology (IIT) Madras, Chennai, India. Previously, he was a Professor in the Department of CSEE at the University of Maryland, Baltimore County, Maryland, USA, with the School of EECS at Washington State University, Pullman, USA, and with the University of North Carolina Greensboro, USA. He has also conducted research at Lucent Technologies' Bell Labs in Murray Hill, NJ, and at AT&T Labs in Whippany, NJ. He received his Ph.D. and M.S. degrees in Computer Science from State University of New York at Buffalo in 1994 and 1990, respectively; and his B.E. degree in Computer Science and Engineering in 1988 from Anna University's College of Engineering Guindy, Chennai (Madras), India. His research interests include wireless networks, wireless sensor networks, optical networks, and performance evaluation. He holds three patents in wireless networks and has published many research articles, including more than 45 journal publications. He published an edited book on Wireless Sensor Networks in 2004 and edited books on optical WDM networks in 2000 and 2004. He is serving or has served as a member of the Editorial Board for several journals, including *ACM Wireless Networks*, *IEEE Transactions on Mobile Computing*, and *Elsevier Optical Switching and Networking*.

Index